Getting the Most out of ClassPad

by

Philip Todd,

Michael Siebold,

Brian Maguire

Saltire Software Inc.,

PO Box 1565, Beaverton OR, U.S.A.

ISBN 1-882564-01-4

Introduction

Have you ever watched the Vancouver hydrofoil pull out of Victoria's inner harbor?

It's sleeker than the MV Coho plugging in on its thrice daily run from Port Angeles, with racier lines and a smarter paint-job. But they are clearly two variants on the same basic ferry concept. Perhaps its wedge design gives it a knot or two advantage, at the loss of a few pounds' payload.

That's until it makes it out into the open waters of the Straight of San Juan de Fuca. The throttle opens up and the entire ferry lifts out of the water and goes onto its hydrofoils. You are reminded more of a float plane in the raucous moments before take off. This is as much aircraft as boat, and clearly no ordinary ferry.

At first glimpse, the ClassPad could appear to be a straightforward evolution on the basic concept of graphing calculator, which has been with us for over 20 years.

Sure, it has a pen and a larger screen - they had to get that, didn't they. How could you call it a "graphing" anything if it has the same number of pixels as the average cell phone or exercise bicycle. Just a sleeker design on the same basic concept.

Well, that's where you'd be wrong, ClassPad is not just a sleeker graphing calculator. It's quite a different creature. And the difference is just below the surface, waiting for you to open up the throttle.

That's what this book is for - to show you how to open up that throttle.

It's not a manual. You already get a manual free with the product. Why would you buy another one?

It's not classroom materials. Our goal is to introduce you to the cool stuff you can do with ClassPad, not how to learn a specific batch of knowledge with it. And to fully explore the strengths of the product, we draw from mathematical examples at levels all the way from elementary school to college.

But so does the manual, and you didn't throw that away.

Or did you?

For Example?

Suppose we wanted to solve a pair of simultaneous linear equations:

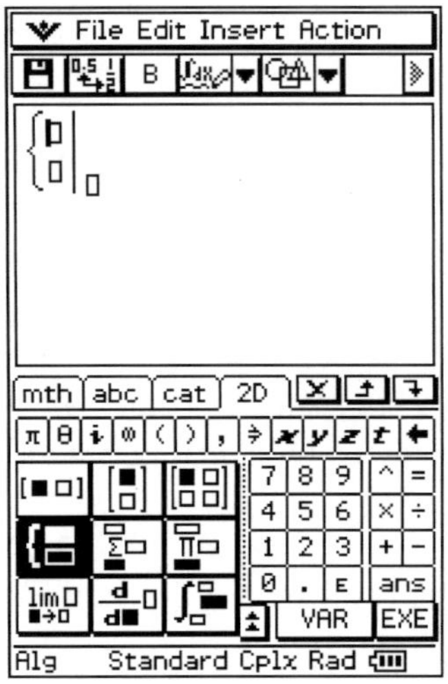

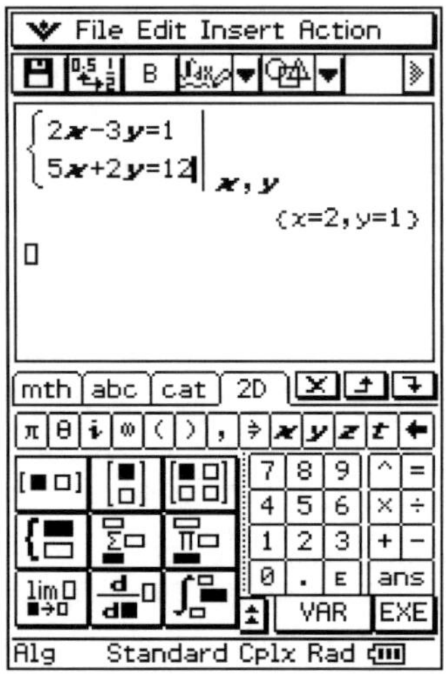

OK, very nice, the equations are very readable. And apparently I can go back and edit the original equations and have the calculation update. I agree this gives the device more horsepower, but I think the hydrofoil thing is a bit of a stretch.

Now select one of those equations and drag it into a geometry window:

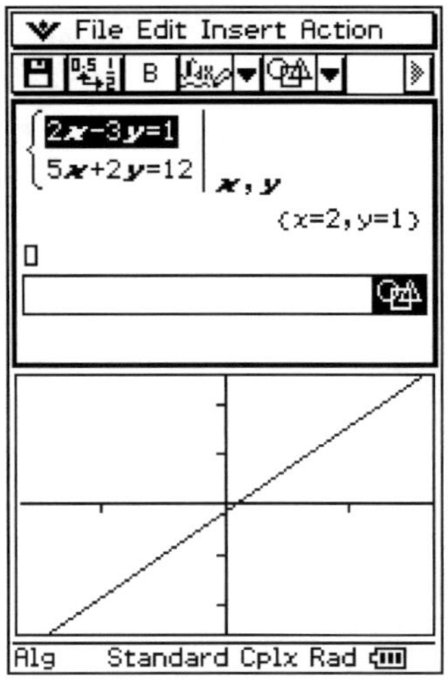

And the other?

Of course.

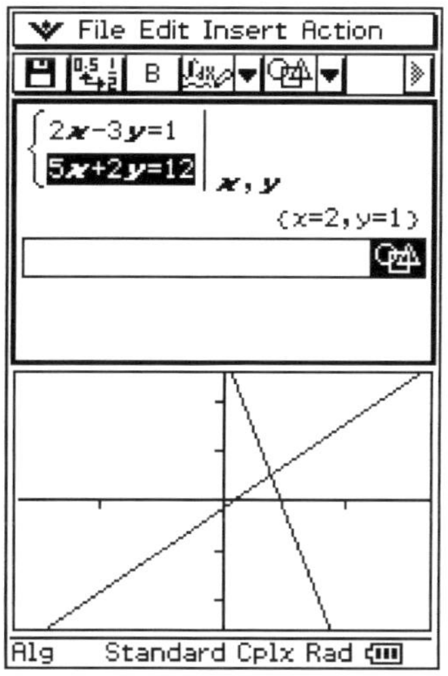

Can it create the intersection?

Second button from the right..

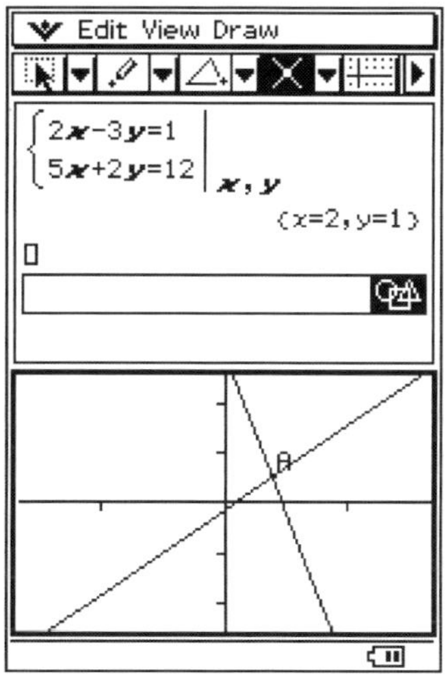

What next?

Drag the intersection point back to the math window:

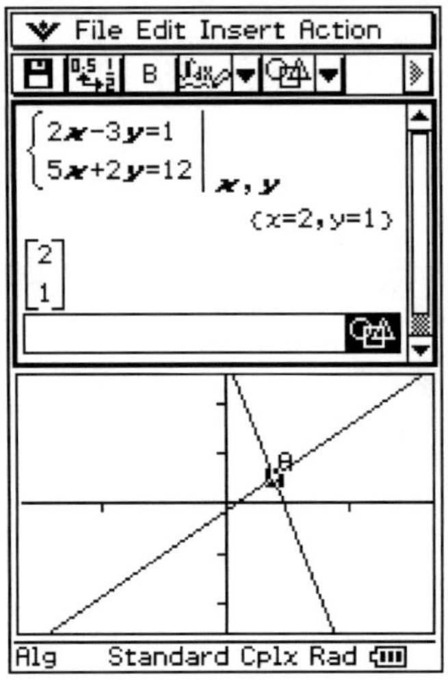

Cool! You're still wrong about the hydrofoil, though.

Oh?

This thing just took off..

ClassPad Overview

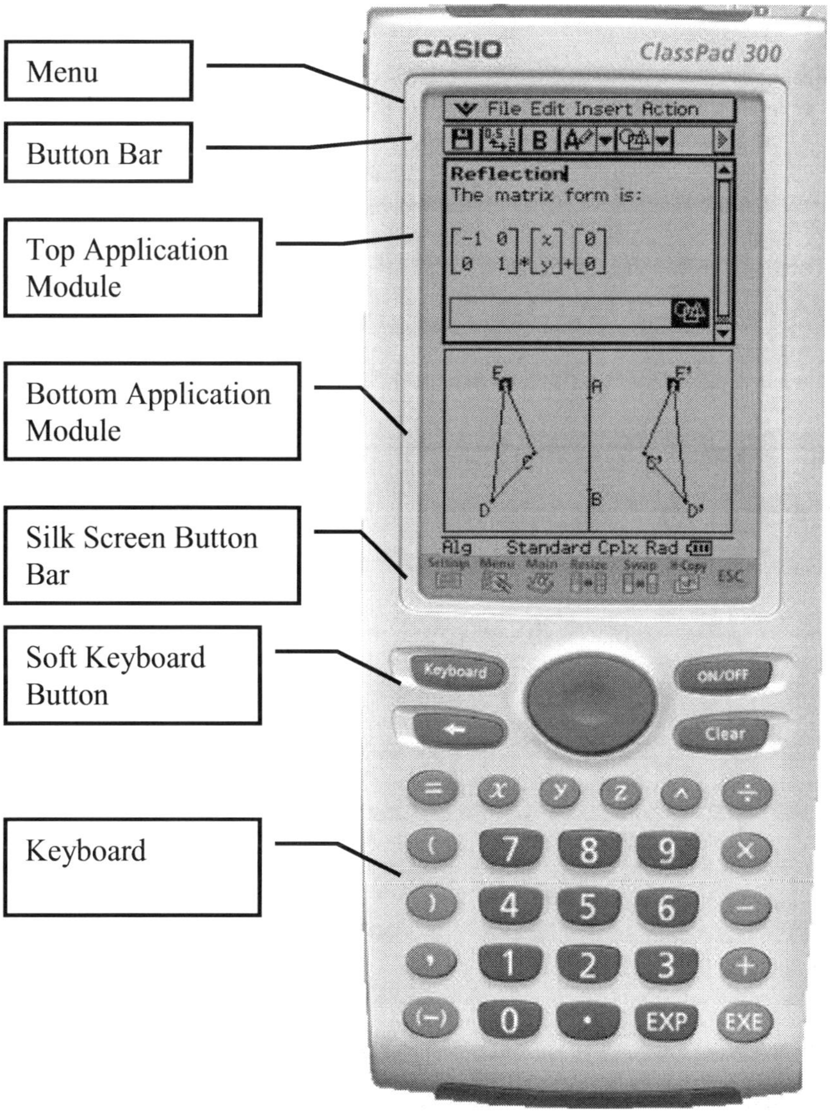

Anatomy of the ClassPad

ClassPad can have one or two windows open at any given time (full screen mode or split screen mode).

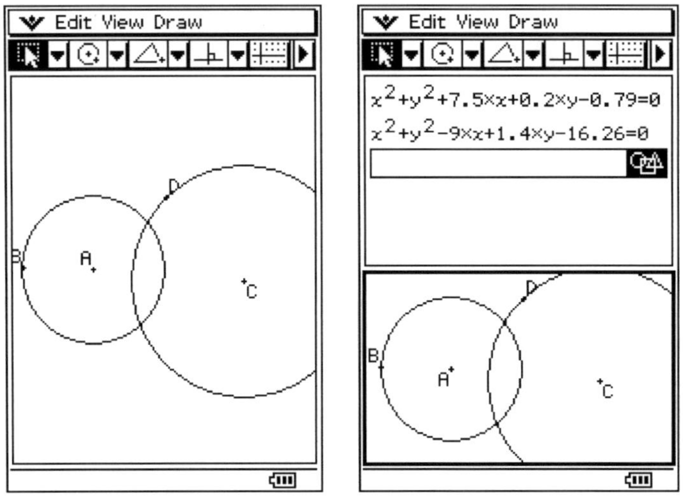

When ClassPad is in split screen mode, there is one active module, and one inactive module. You can tell which module is active by the fact that it has a slightly thicker border. The way to make an module active is by tapping on it.

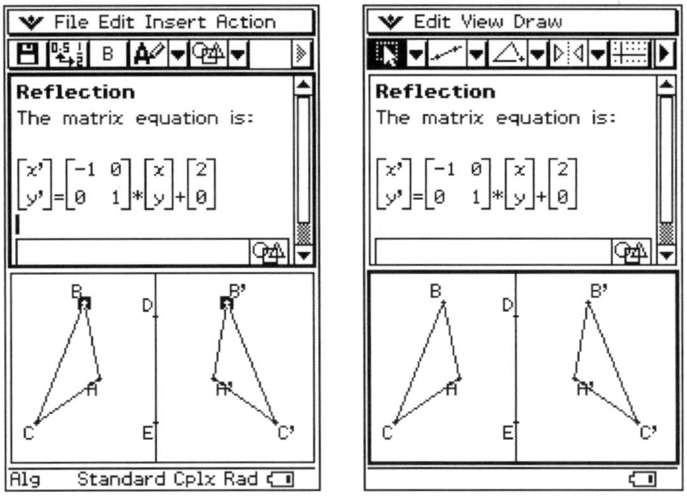

You will also notice that the menu and button bar change when you change the active module. The menu and button bar belong to the active module. If eActivity is active (the top window in the screenshot above), then the eActivity menu and button bar are displayed. If Geometry is active, then it is the Geometry menu and button bar which are displayed.

Zoom / Swap Windows

Two of the Silk Screen Buttons let you change the view between Full Screen and Half Screen, and to swap the top and bottom modules:

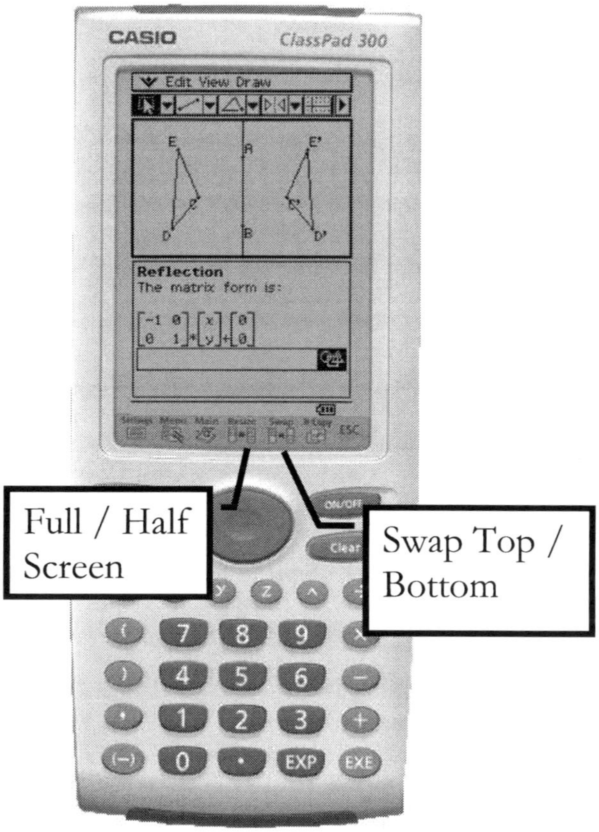

Soft Keypad

The Keypad Button toggles the Soft Keypad on and off:

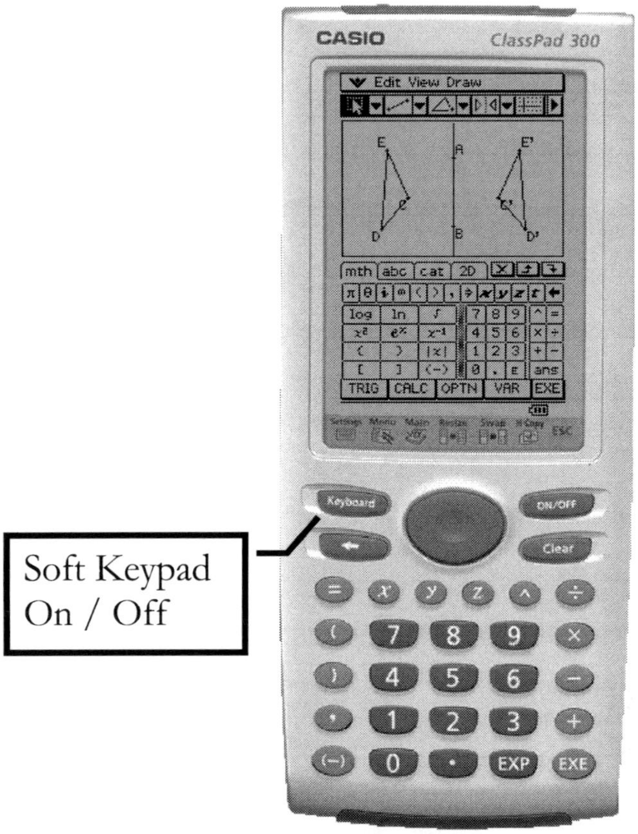

The Soft Keypad is too rich a topic to discuss in detail here, and is given the expansive treatment it deserves in a later section.

Cut Copy Paste

Cut Copy and Paste are system wide functionalities. Some Application Modules, such as eActivity and Geometry have these items on the Edit Menu.

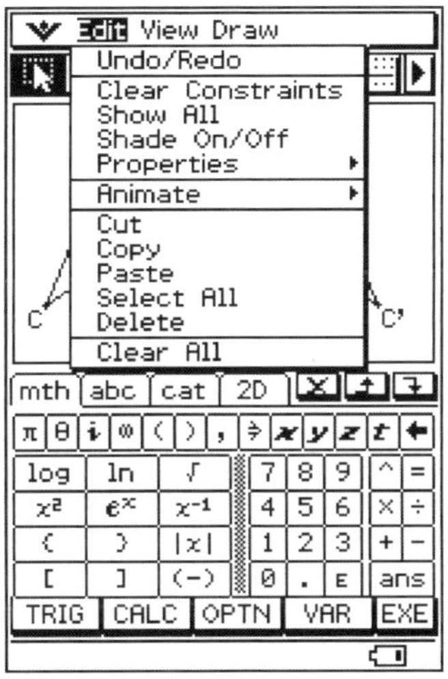

If this is not available, for example if you are in a dialog box with no menu, then you can use the Cut / Copy / Paste buttons on the Soft Keypad:

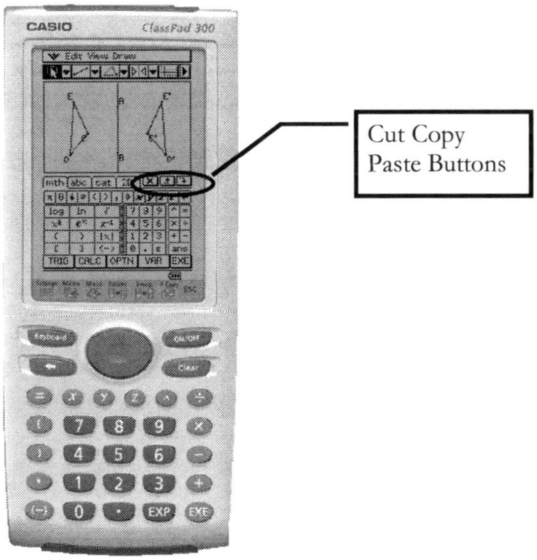

Applications / Application Modules

The overall ClassPad menu may be obtained by tapping the Menu Silk Screen Button:

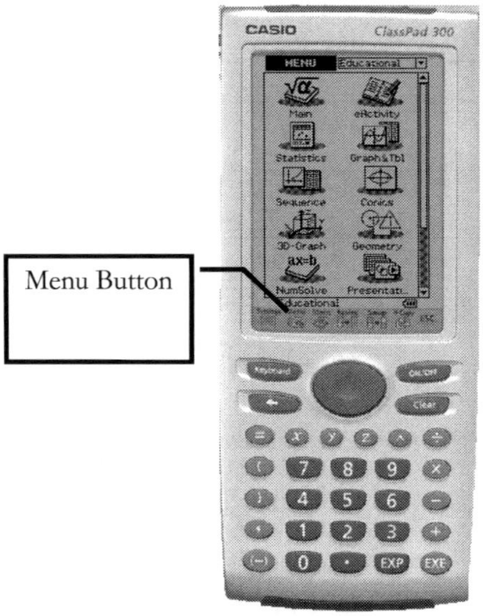

ClassPad applications are typically available in two forms: either as a lonely application called straight from the menu screen, or as an application module embedded in another application.

For example, Geometry can be called by clicking on the Geometry Icon in the menu screen:

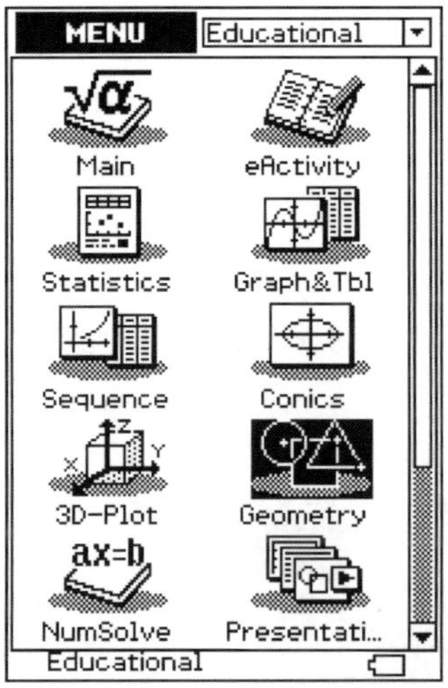

In which case you get a full screen Geometry Application, with no accompanying mathematics window.

Alternatively, you can first invoke the eActivity Application from the Menu screen, and open a Geometry Module from within eActivity:

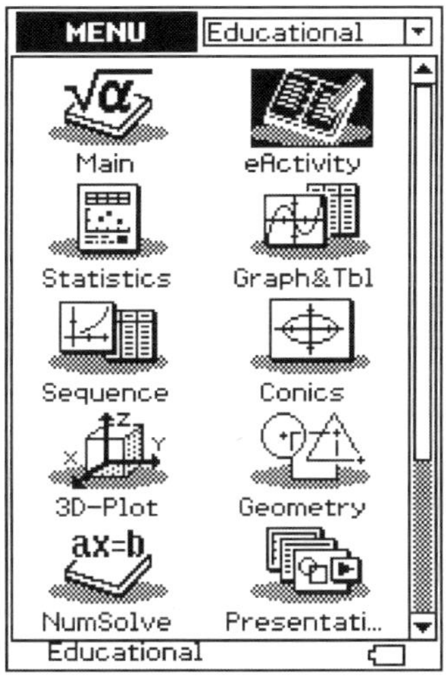

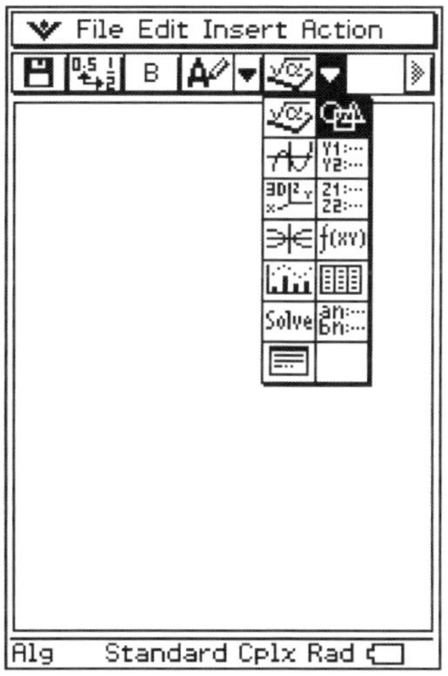

In which case, you get all the geometry functionality, but also the capability of accompanying it with mathematics, explanatory text, etc.

And one more click on the silk-screen resize button can make the geometry full screen if you want it.

Editing Basics

The editing features of a calculator are probably the least exciting topic we could discuss. Some are probably tempted to go galloping off into Graphing or other features and to skip this section entirely. But those will miss a few techniques which can help bypass the tedium of typing things into the calculator.

Start out by launching the eActivity application.

It's hard to ignore the blinking cursor. When editing, make sure that cursor is where you want the text to go. If it's not, just tap the spot with the pen. If that area accepts text, the cursor will appear.

Type in one billion – and just as we would expect, the numbers pop into place and the cursor marches forward:

If we change our mind and want one thousand, we can hit the backspace key a bunch of times. Or we can select six of the zeros and hit the backspace key. Try it out – drag the pen across six of those zeros. Notice that if you keep the pen pressed, moving the pen around moves the selection. Pressing the arrow pad also moves around the end of the selection. If you don't like where your selection started - tap in blank space to get the selection to go away, and swipe again. Hit the backspace to clear out the text that is selected.

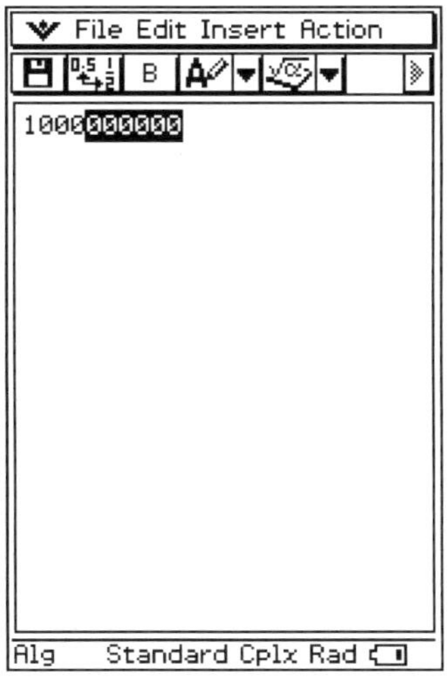

This is a good spot to try out the undo/redo feature. Tap the "Edit" menu, and click "Undo/Redo", and we'll get our original text back:

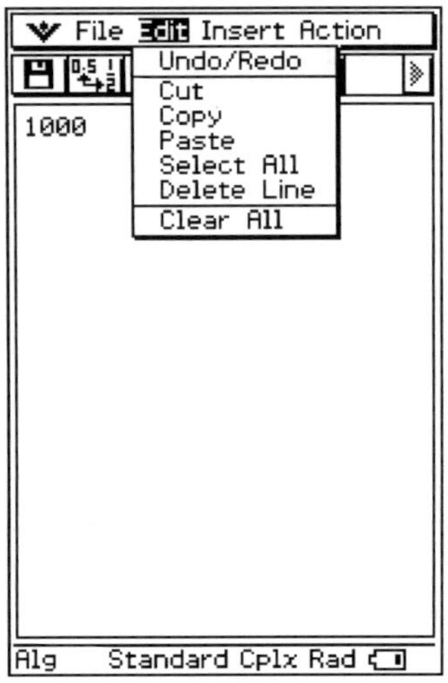

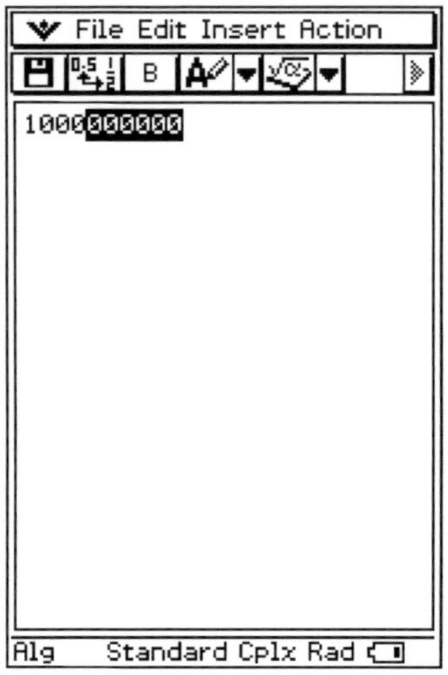

Drag and Drop

There are a few editing features which are so helpful, which open up so many possibilities that they deserve their own section – and Drag and Drop is one of these.

Within an expression, drag and drop operates like cut and paste. We can drag a fragment of an expression, and throw it somewhere else with a flick of our pen.

For example, if we were examining an equation, drag and drop makes rearranging it a snap. For example, if we want to collect terms of 2x+3-4y-x, we simply select the –x by dragging our pen over it:

Lift up the pen to end the drag. Press and hold the pen on the selection and drag it next to the other x term. Notice that the blinking cursor shows where the drop is going to occur.

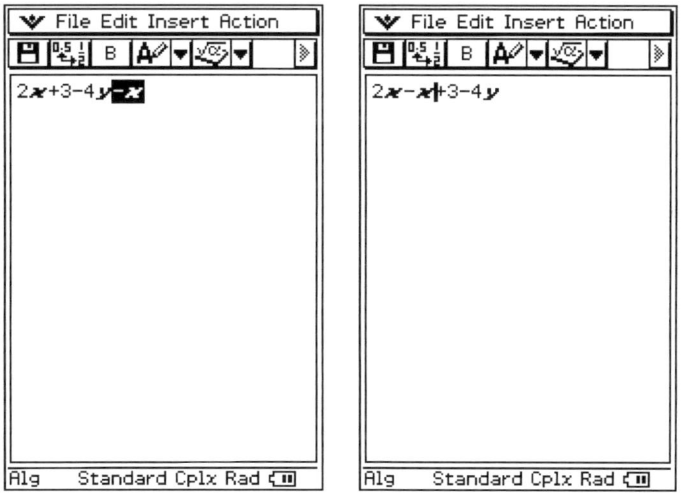

Between lines drag and drop acts like copy and paste. If we start from where we left off in the previous example, we can use drag and drop to quickly make a copy of the line. Press EXE to add a new line below the first. Drag across the entire line to select it, then drag and drop it down.

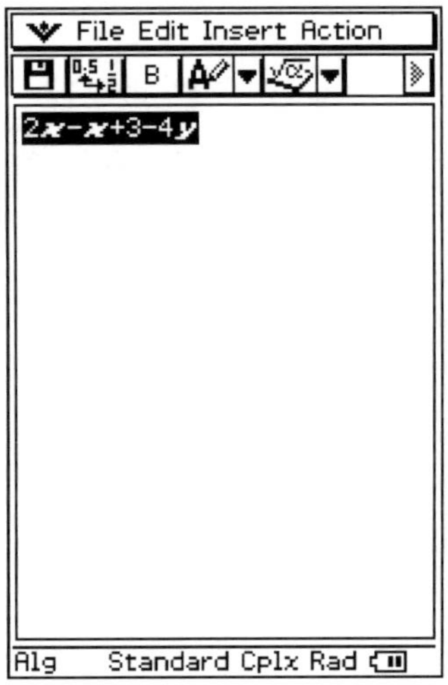

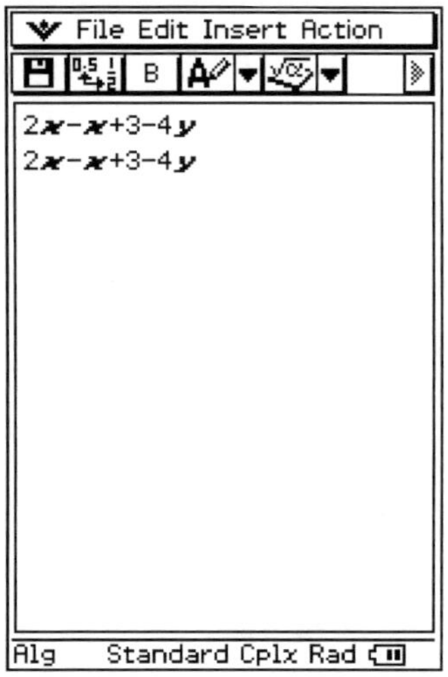

When first trying dragging and dropping, it helps to keep a couple things in mind. First, remember to lift up the pen after selecting – otherwise when you drag you'll just change the selection instead. Second, watch where the cursor appears as you drag. The cursor tells you that you are actually dragging, and it helps to see exactly where the dropped item will appear. Notice that if no cursor appears, you will not get the desired drag and drop effect. Keep hunting, you will eventually find a place where the cursor is prepared to appear.

Soft Keyboard

Many find the ClassPad puzzling at first glance. The generous screen, the elongated, sleek design make it stand out. But can it really be a scientific calculator? Where is the myriad of tiny keys? Where is the microscopic print and the impossibly abbreviated function names and symbols?

The ClassPad contains a tremendous library of functions – but they have been moved off the buttons and placed into the Soft Keyboard. A quick press of the Keyboard button (surprisingly easy to do with human-sized fingers) brings the keyboard into the lower half of the screen.

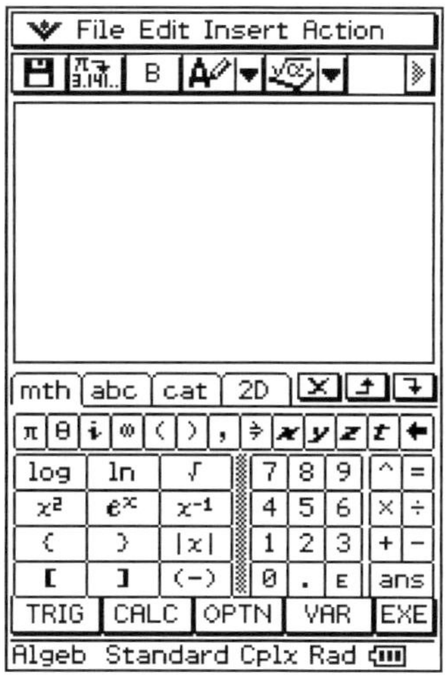

Instead of being overwhelmed by scores of functions screaming for attention, the keyboard is cunningly organized to present just what is needed. The tabs along the top divide the keyboard into four intuitive palettes – common math functions (MTH), general text entry (ABC), the function catalog (CAT), and the natural math (2D).

Quick cut, copy and paste keys are nestled just to the right of the tabs. Hit these keys to copy up a selection and paste it down.

Common Math (MTH)

Tapping on the first tab (MTH) causes the common math buttons to march into position. A good portion of these keys are duplicates of those found on the physical keypad. Often, it is faster to let the pen do all the work instead of switching between the physical keys and the pen.

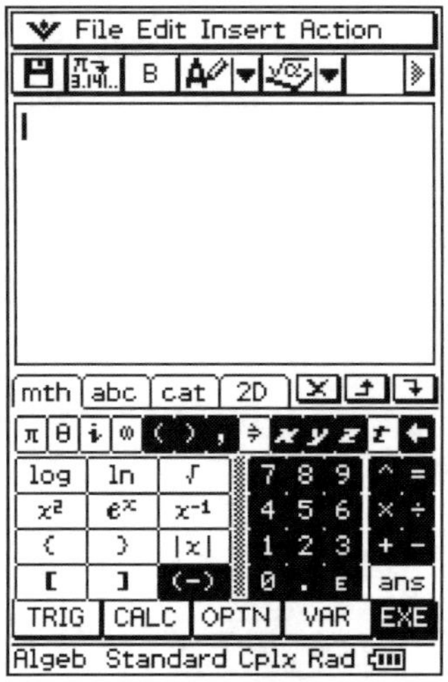

Single-Letter Variables

The bold and italicized x, y, z and t keys stand out from the rest. The ClassPad uses bold and italics on single-letter variables. This avoids a common pitfall that lurks in many standard calculators. Take a general formula for a line, for example. If we wanted to use

y = mx + b

in a standard calculator, we would be stepping into a mine field. Most calculators interpret 'mx' as a single variable called 'mx'. We would be forced to make the multiplication explicit.

y = m*x + b

The ClassPad's single-letter variables gives us the freedom to make the equation just like we would expect.

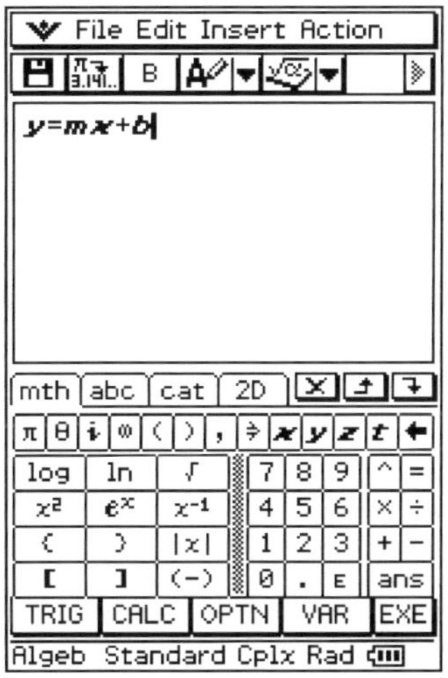

The beauty of this scheme is variables with names that have more than one letter are still available.

We used standard letters to spell out 'yintrcpt', making it clear that we do not mean y*i*n*t*r...

Layout Buttons

The diligent reader may have noticed something missing from the last example – where are the single-variables ***m*** and ***b***? Obviously, the entire alphabet of single-letter variables would crowd out most of the keys in the common math. Instead, the corps of variable buttons wait just behind the VAR layout button. Pressing VAR allows the variables to invade, temporarily displacing some of the common buttons.

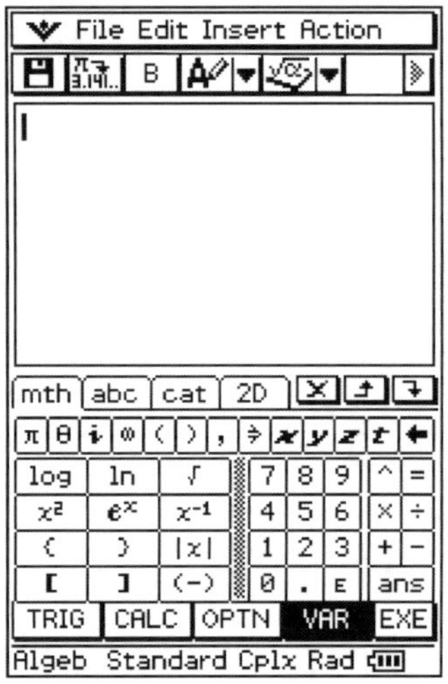

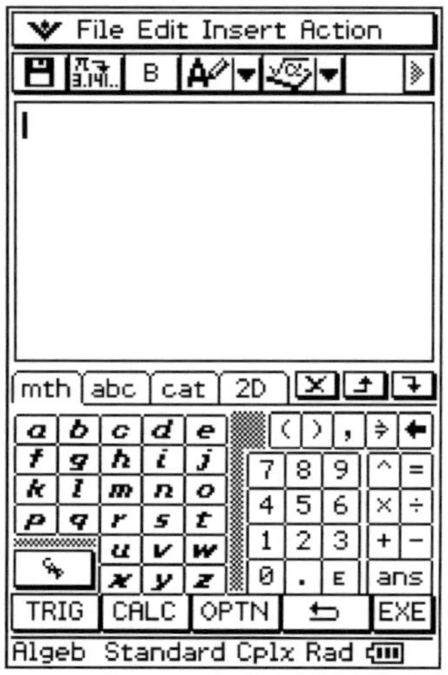

A caps lock button joins the fray – eager to provide upper case versions of single-letter variables. Another look at the VAR button reveals it has been replaced by something like a u-turn symbol. Pressing this button puts the variables into full retreat.

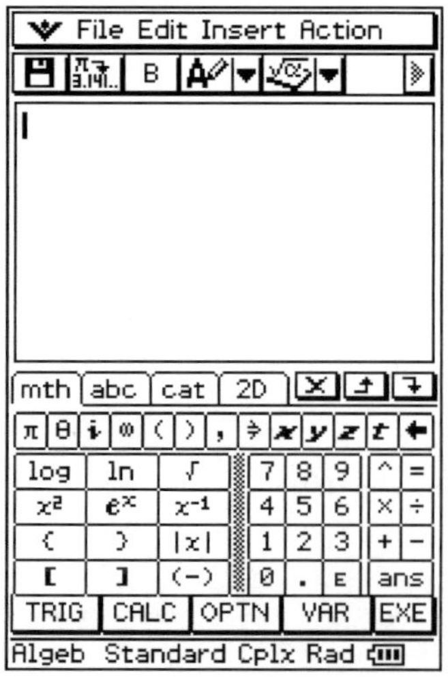

The variable set is not the only layout lurking behind the standard math. TRIG conceals the common trigonometric functions, CALC has sum, product and some probability, and finally inequality and series buttons wait beneath OPTN.

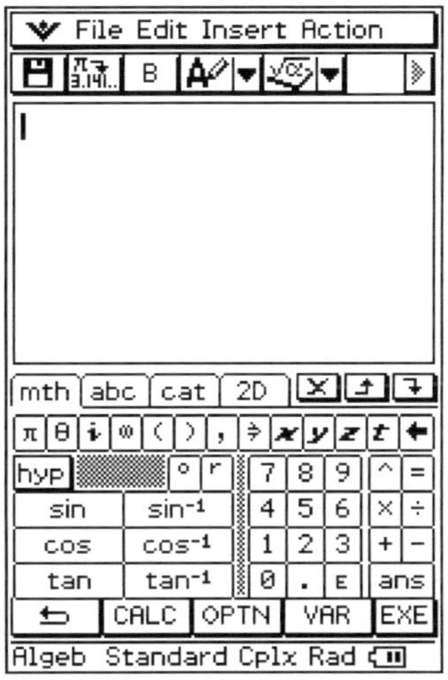

TRIG layout

CALC layout

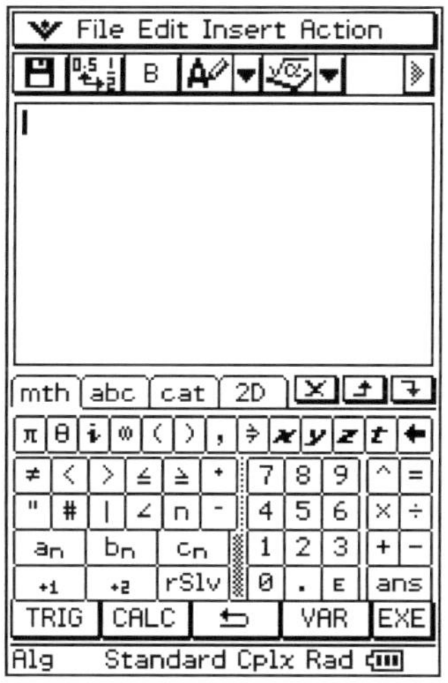

OPTN layout

General Text (ABC)

Considering the bewildering array of convoluted methods for text entry, it is not surprising that many shudder at the idea of composing even a few words with their calculator. The ClassPad allows this problem to be avoided altogether with its companion set of PC tools – the text can be composed at the computer and then downloaded to the calculator. Otherwise the ABC tab presents a refreshingly simple method for direct text input.

The caps lock and shift buttons on the lower left operate just as expected, offering upper case versions and the standard set of symbols. Of course, an extra pen is not required to hold the shift while pressing another key. The shift stays active until a key is selected or the shift is pressed again.

The standard alphabet is often inadequate – it is distracting when Greek symbols, set operators and other specialized characters have to be spelled out. The ClassPad layout buttons are the solution – αβγ for Greek and language-specific characters, **MATH** for items including operators and subscript characters, and **SMBL** for currency, temperature and other symbols.

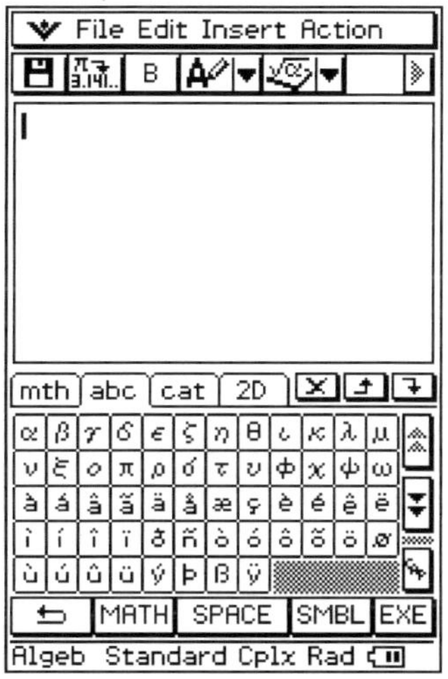

αβγ layout

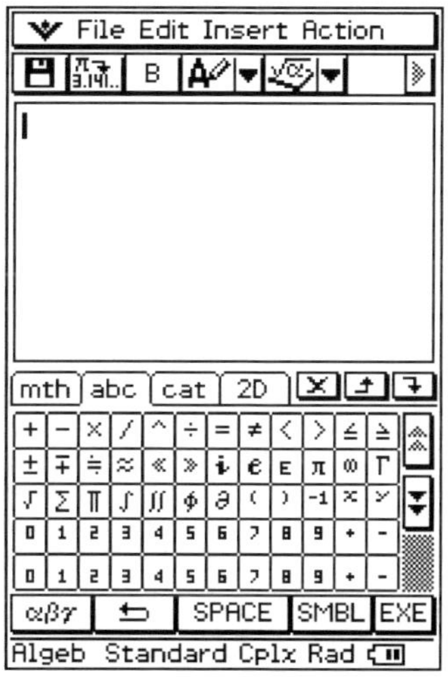

MATH layout

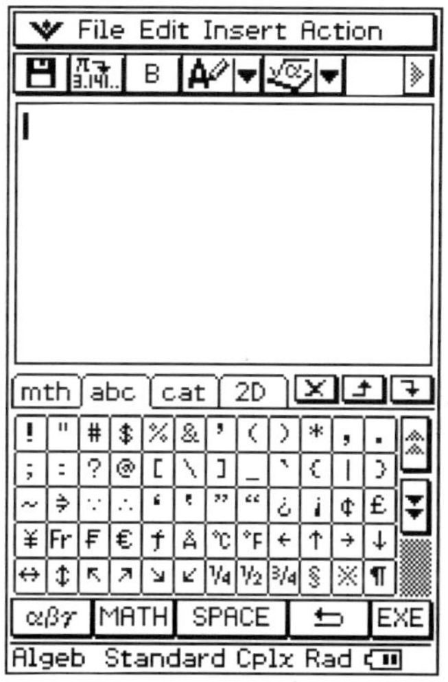

SMBL layout

Catalog (CAT)

The catalog (CAT) tab provides a one stop shop for most of the functions and operators found in other parts of the keyboard. Many exotic and powerful features exist nowhere else but in the vast catalog wilderness. This will be a quick tour – use the ClassPad manual as your field guide to identify any beasts that catch your eye.

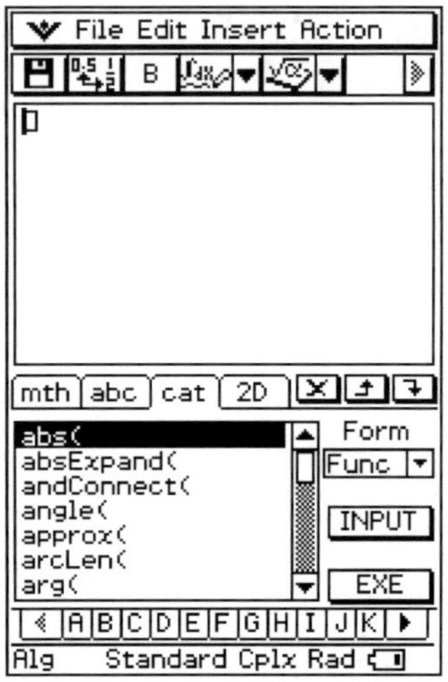

The catalog is just a big alphabetized list. We can only see a few things from where we stand, but the scroll bar just to the right of the list is eager to show us more. Pen taps on the scroll bar arrows are great for sneaking up to our prey when it's close, but they scream Carpal Tunnel before long. Tap on the shaded part of the scrollbar to extend your stride, or hold a scroll arrow down for a smooth glide. Dragging the scroll indicator sends us careening from spot to the next. Even more useful are the letters at the bottom of the catalog. Just press one of these letters to jump to the first thing that begins with that letter.

Once the function has been cornered, just selected it with a click, and tap the INPUT button. Or just tap on it twice. The trap is sprung, and our target will be deposited at the cursor, just as if we had typed it.

Whole new ecosystems lie beneath our feet, and we can dive into these using the Form drop down list. We were looking at the surface level – "Func" for functions. Press the arrow next to "Func" and we can see other layers: "Cmd" for program commands, "Sys" for system commands.

The elusive Factor is an exciting thing to find. We don't just want a picture of this animal – we want a real live one. In eActivity, switch the line mode from text edit to math evaluate.

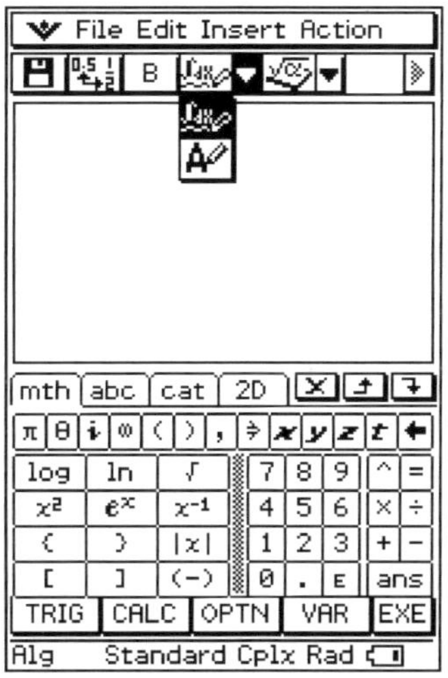

Make sure Form shows "Func", click on the 'F' letter – and it appears. Click INPUT to place it at the cursor, and try poking it with X^2-1. Finally, let go of the leash by pressing the "EXE" button.

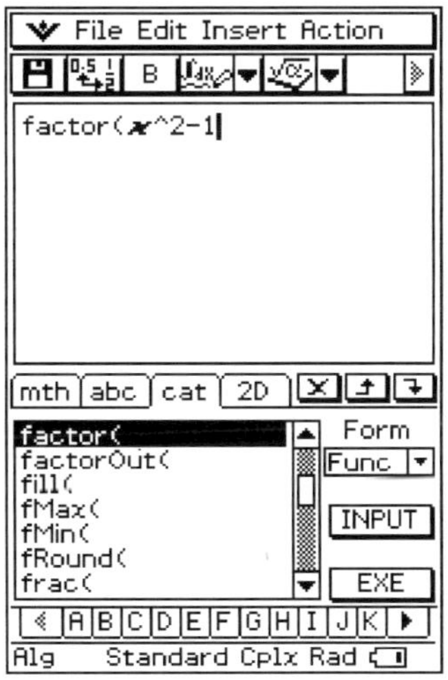

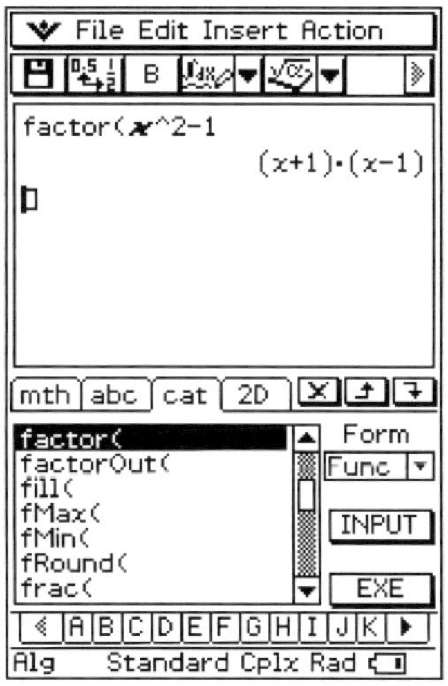

Natural Display Math (2D)

Modern calculators can be a powerful tool for using and manipulating expressions. But after we took the time to carefully enter an equation, often the results are disappointing. Instead of the simple equation we expected to see, we are faced with a barely recognizable sequence of characters.

Take, for example, the quadratic formula. If we typed it into a standard calculator, we'd see something like:

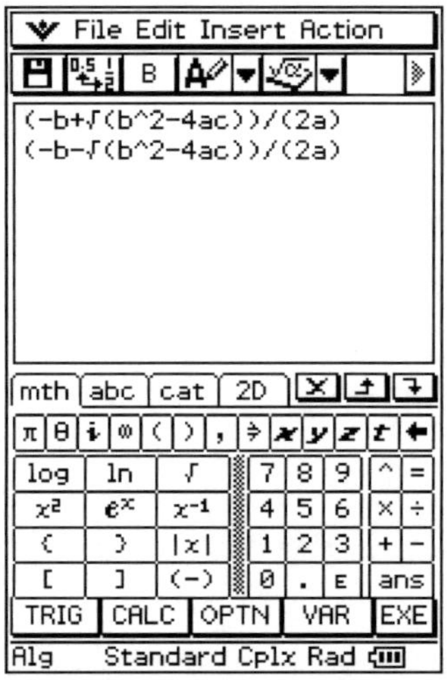

In contrast, the natural display features of the ClassPad allow you to see the quadratic formula as you would expect.

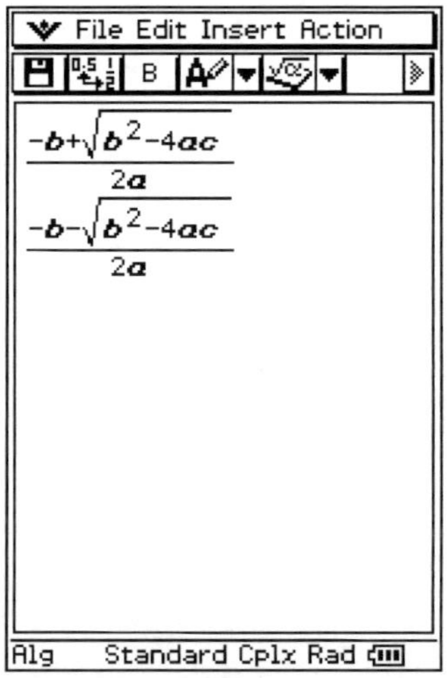

Instead of being crammed into one long sequence, the ClassPad allows the expression to extend up and down – giving it a natural '2D' look. This can make the difference between the whole expression fitting on the screen and part of it extending into the virtual realm beyond the screen's border.

All the tools for making natural-looking expressions are available on the 2D tab of the keyboard. For example, we could try the divide button.

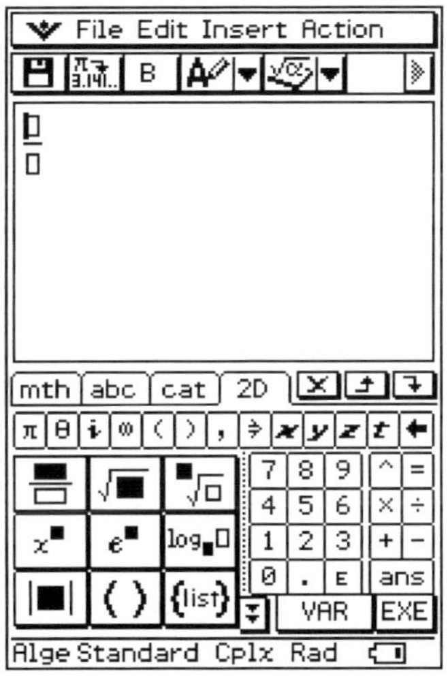

Notice there's empty boxes in the numerator and denominator to indicate that something belongs there but nothing has been put in yet. If we start typing away, the text flows into the numerator, where the blinking cursor resides.

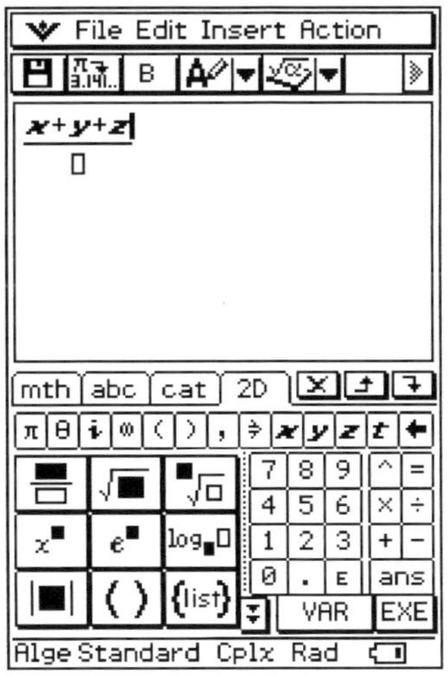

We can jump the gap between numerator and denominator by clicking with the pen on the denominator. However, the arrow pad is often the best choice for cursor maneuvers.

What if we really wanted that text in the denominator? That's not a problem with drag and drop. Just select the text by dragging your pen from one side of the numerator to the other – the numerator should be blackened.

Now press and hold on the numerator, and move the pen down to the empty box. The flashing cursor appears to show exactly where the text will end up. Once the pen is released, the text leaps to its new location.

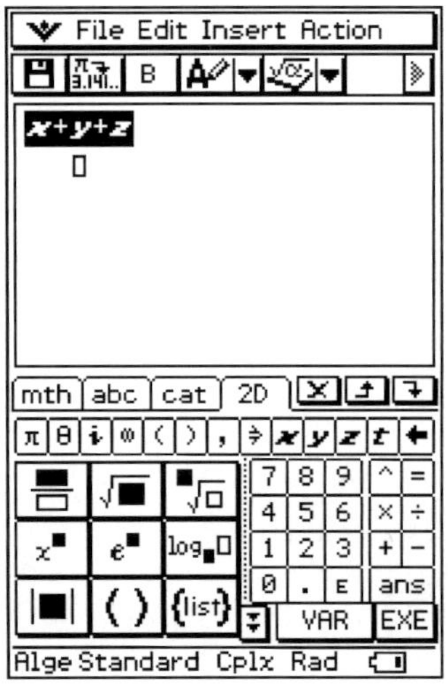

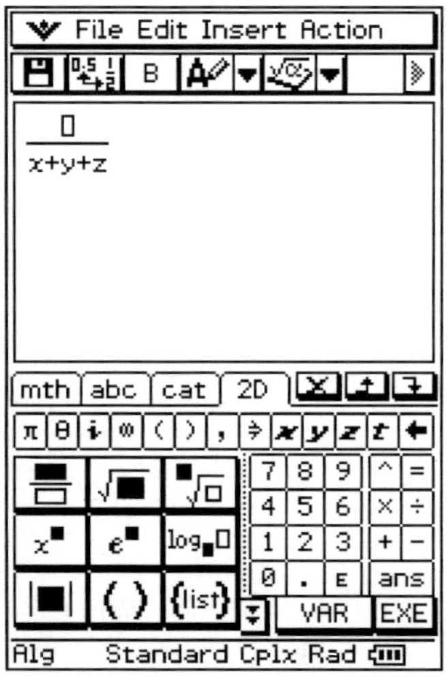

Another feature that reduces the tedium of building an expression is selection surrounding. Pressing a 2D key while something is selected causes the new math to appear around the selection. Let's say we've gone to the work of constructing the discriminant portion of the quadratic formula.

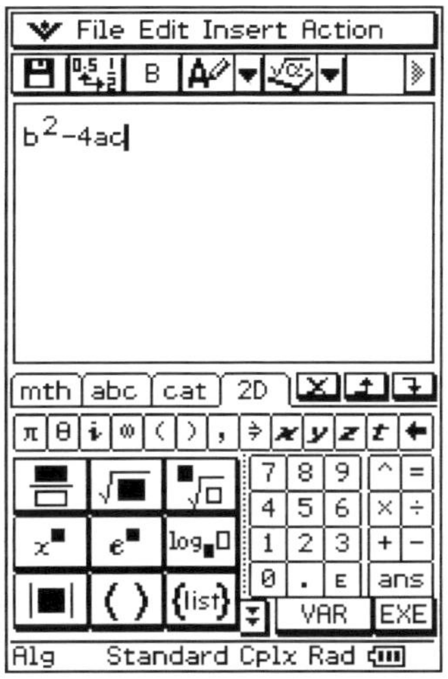

We can place the square root around the expression by simply dragging out pen over it to select it, then pressing the square root 2D button.

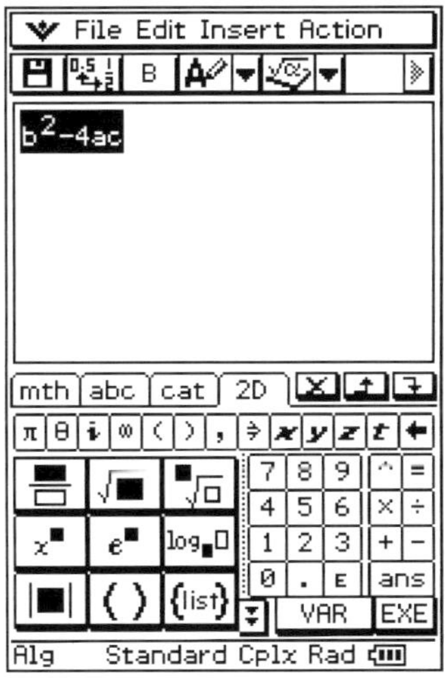

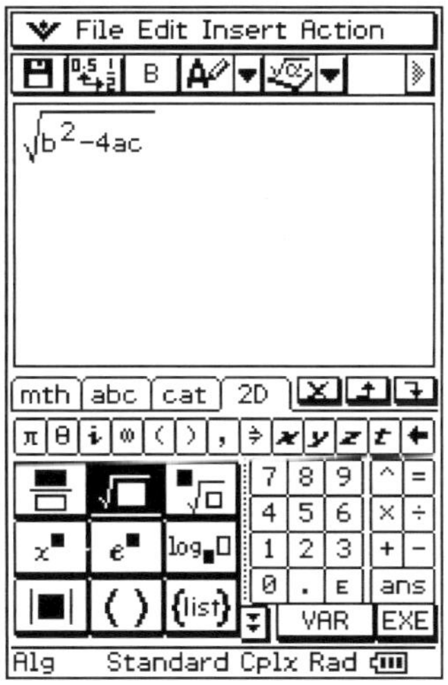

eActivity – Notebook

An eActivity is both a documentation tool, and a student notebook. Using it as a documentation tool you can create live examples with accompanying text, textbook expressions, graphs, and tables. The examples can be explored using the numerous modules on the ClassPad.

The eActivity below describes the law of motion equations in parametric form. A graph (parametric plot) is also attached. Clicking on the graph icon in the eActivity window brings up the graph in the lower window.

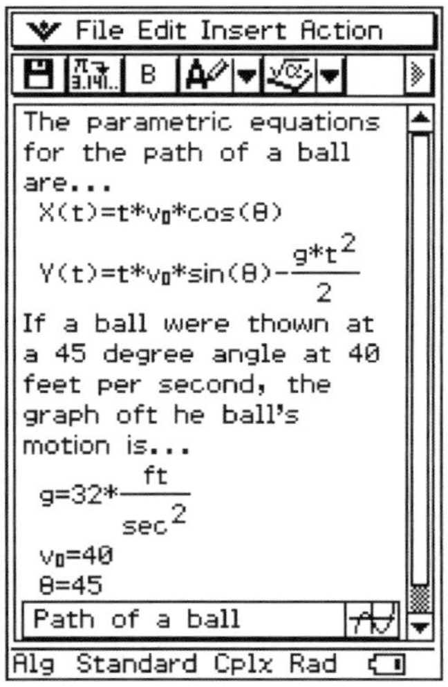

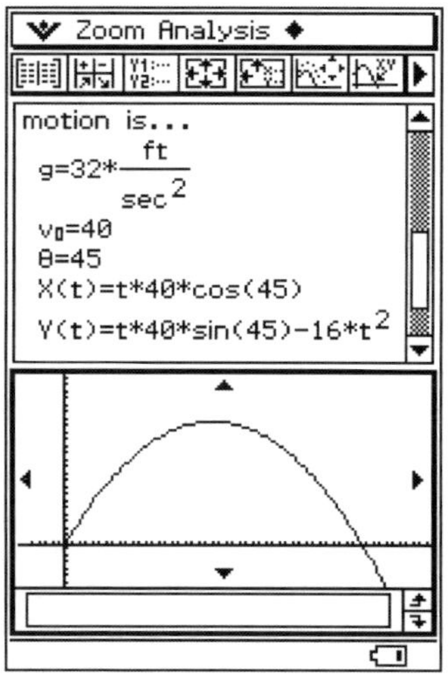

eActivity Toolbar

 OPEN/SAVE

Opens the storage manager, where you can open or save a file.

 FRACTION / DECIMAL

Switches the current strip from fraction display to decimal display.

 BOLD

Toggles the font in a text strip between bold and normal. Note that 2D math is always normal font. The bold font is useful for headings or titles.

 STRIP TYPE

This pull-down button contains the two icons representing the two different types of text, active math (calculator strip) or plain text.

 INSERT OBJECT

This pull-down button contains all the different types of objects that can be inserted into an eActivity.

eActivity Elements

The following is a list of the different items that can appear in the eActivity window.

TEXT STRIP

Text can be viewed and edited directly in the eActivity window. Text can contain multiple lines, as well as 2D math expressions. Individual text strips can scroll horizontally, but not vertically. Instead, the entire eActivity page scrolls vertically.

Even though text can contain 2D math expressions, it is not evaluated. Instead, the text is for information only.

You should notice that ***Text*** is the default strip type. Hence if you create some mathematical expression in the default strip and press enter, you will not have that mathematics evaluated, unless you change the strip type to ***Calculator***.

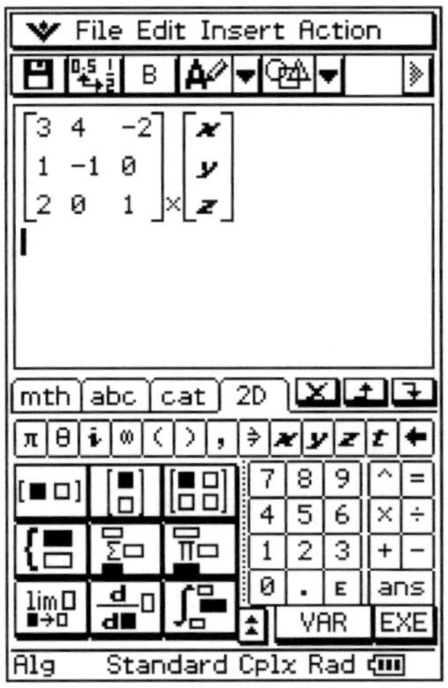

Text Strip – does not evaluate

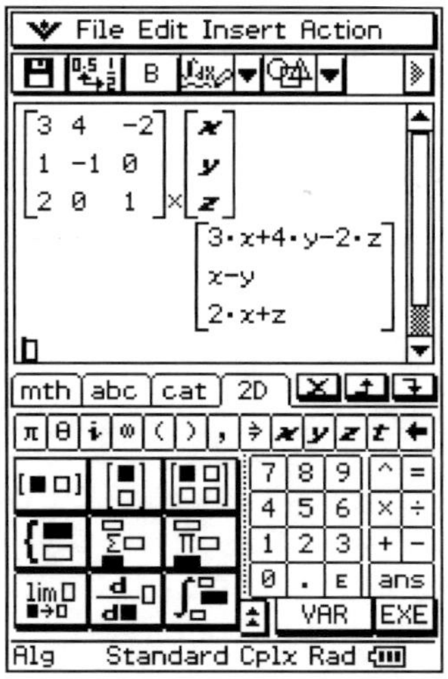

Calculator Strip – does evaluate

CALCULATOR STRIP

An eActivity can contain pairs of input/output expressions. Each input expression is evaluated and returns an output expression on the following row.

As long as you enter a valid expression into the Calculator Strip, then when you press the ENTER Key, ClassPad will create an output expression. You can edit the input expression, but not the output expression. However, you can select and copy and drag/drop from either expression. Both scroll independently in a horizontal direction, using the small arrow buttons which appear where needed.

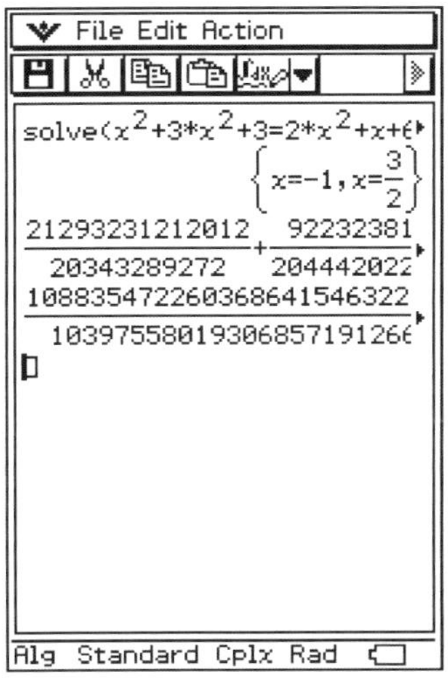

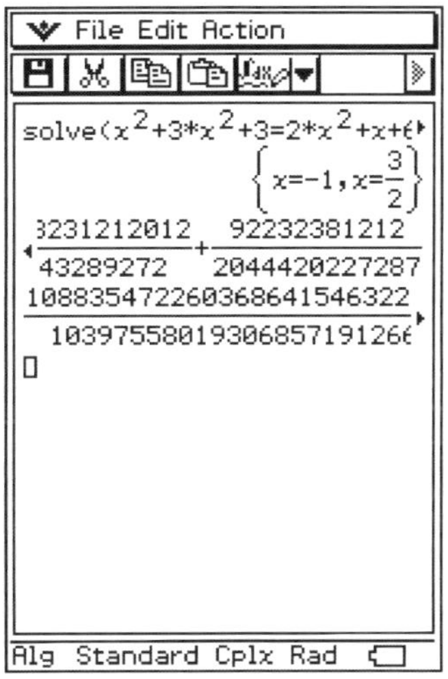

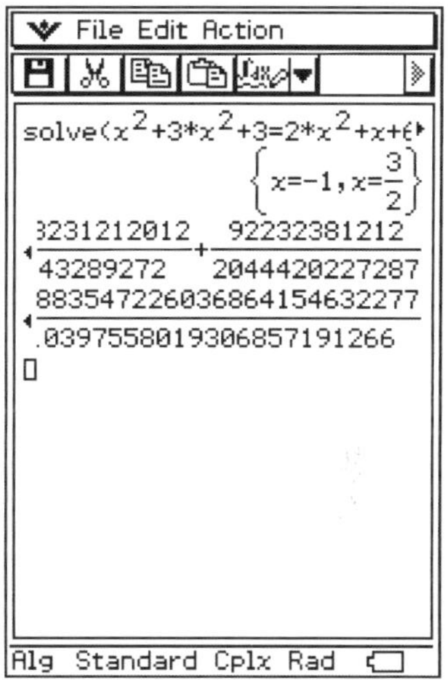

Calculator strips are evaluated from the top to the bottom. If you edit a calculator strip near the top, later strips that depend on it are recomputed.

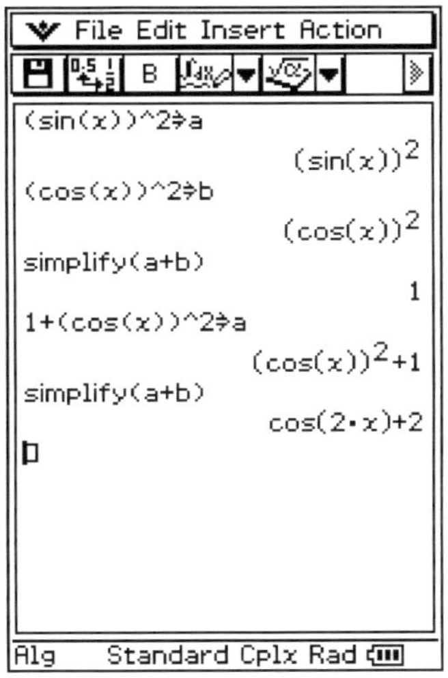

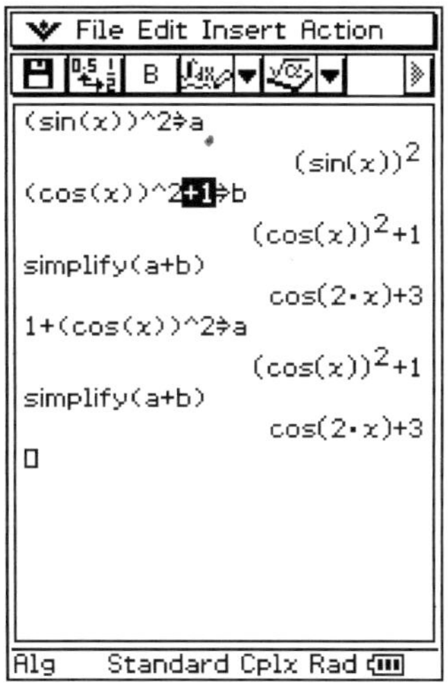

APPLICATION DATA STRIP

Data for other ClassPad applications can be embedded in an eActivity document.

An application strip displays the following elements in the eActivity.

Title – You can type in some text here

Expand Button – Click to expand application data into lower window

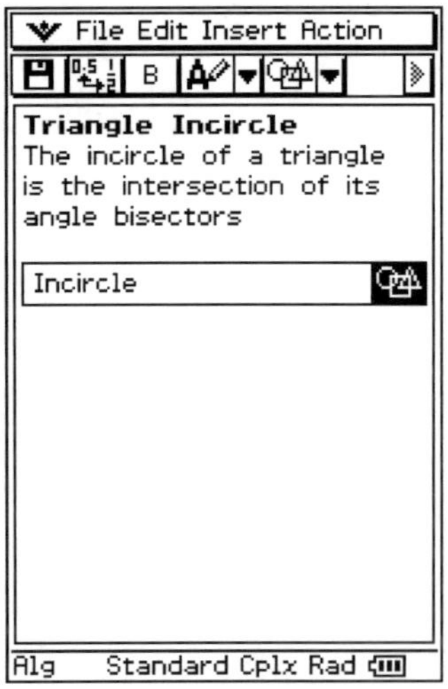

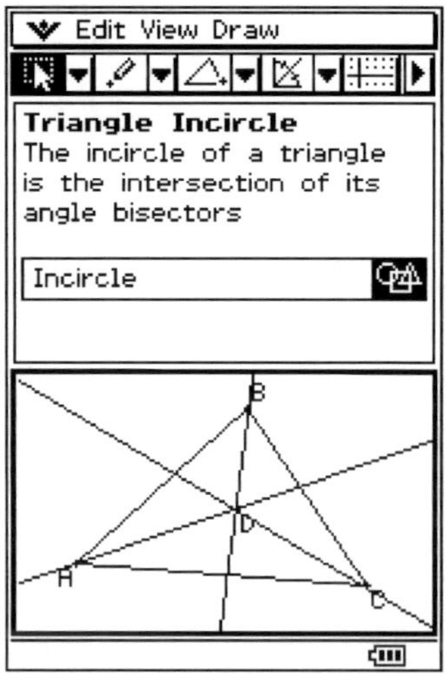

When an application data strip is expanded, it's expand button is displayed in inverse. This lets you know which data is currently expanded.

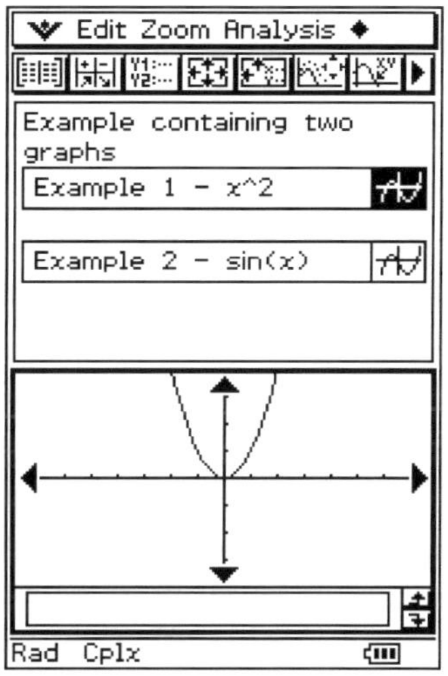

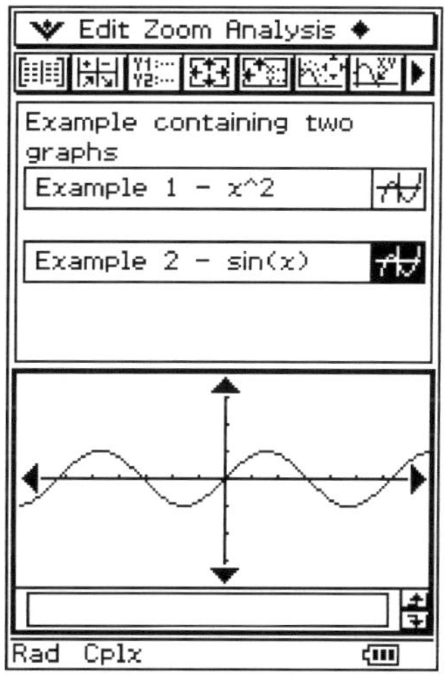

Moving Information Between Applications and Strips

An eActivity is like an active notebook or textbook. Active in the sense that the mathematics is live – change any input values and all the resulting computations change as a result. But also active in the that you can easily move information around within the notebook and between the applications embedded in the notebook.

We'll explore how to do that here:

Cut/Copy/Paste

You can cut, copy, or paste text or math expressions between blocks within the eActivity or between the eActivity and any of the other applications.

Depending on the application, you can cut, copy, and paste other types of data and paste them as mathematics inside the eActivity. For example, you can copy a line in the geometry application and paste it as an expression in eActivity text.

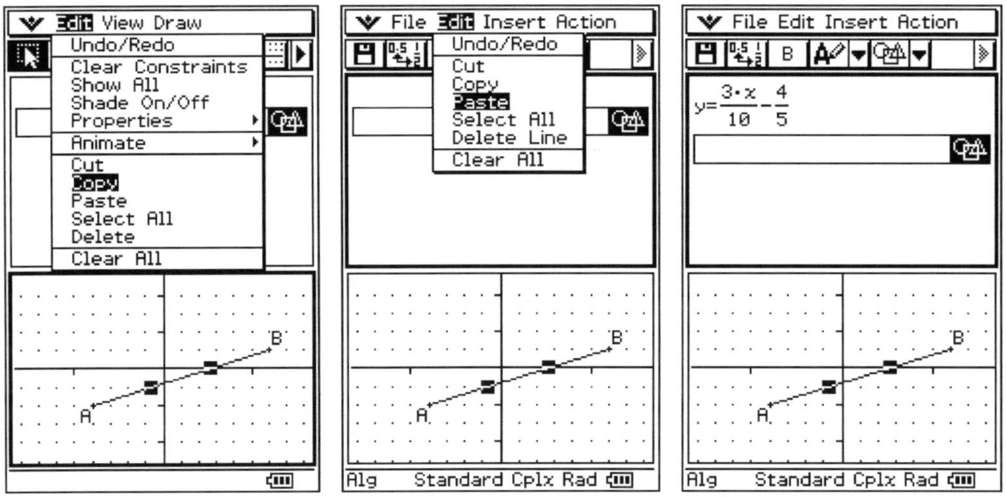

Drag and Drop

You can drag and drop text or math expression between the eActivity and any of the other applications. You can also drag and drop text and math expressions between blocks of text inside the eActivity.

Depending on the application, you can also drag and drop other types of data. In the example below, an expression is dragged onto the top of a graph window. A graph of that expression is created as a result of this action.

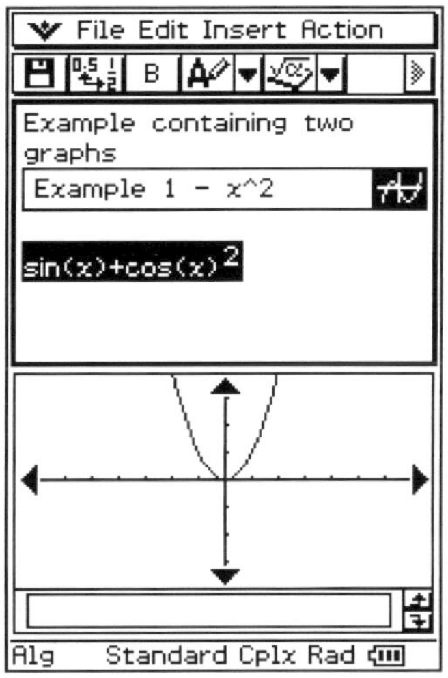

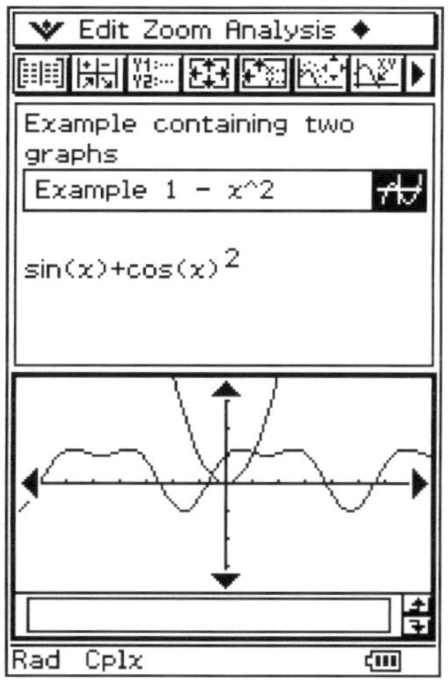

eActivity – Math System

As we have seen above, the eActivity is more than an individual application, it is a wrapper for all the other functionality which ClassPad offers. eActivity lets you pull together pieces of mathematics, geometry, graphs, tables and the like into a single document. It even lets you add a little text to explain what you are up to.

However, just viewed as a single application, eActivity is a powerful mathematics package in its own right. In this chapter we will explore these capabilities.

Much of the rest of the book will present examples where the mathematics of the eActivity application is combined with the other capabilities of the ClassPad, particularly with the interactive drawing and graphing of the geometry application.

Volumes & Surface Areas

Let's create a set of formulas for volume and surface area of cylinders, open or closed.

First, we need a title for our eActivity worksheet: let's call it Cylinders. Luckily, eActivity is by default in text mode, so it won't try to evaluate ***Cylinders*** with potentially confusing consequences. Let's be bold and make the title bold:

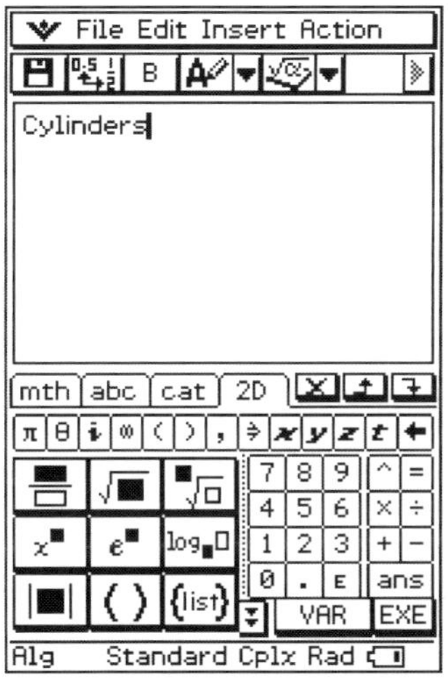

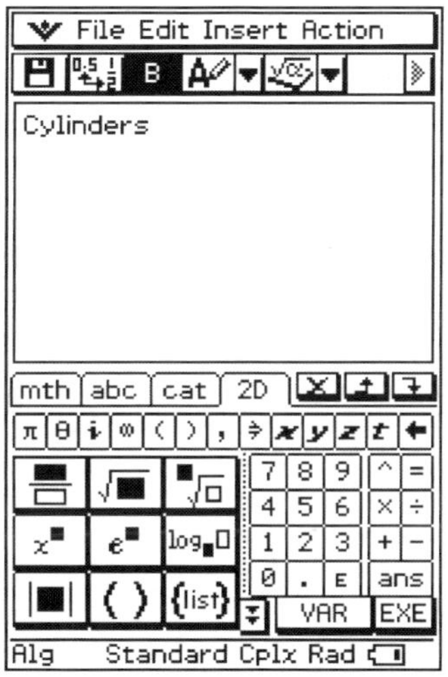

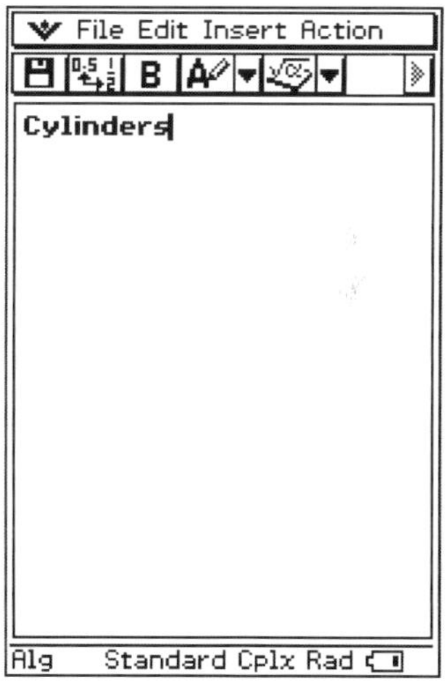

Now the inputs to a cylinder formula would be r and h, the radius and height of the cylinder.

We should use the single letter variables for these (found in the MATH / VAR tab of the keyboard) as this will make the formulas more succinct by allowing us to leave out multiplication symbols.

First, however, we need to make sure the next strip is a piece of mathematics, and not just a text field:

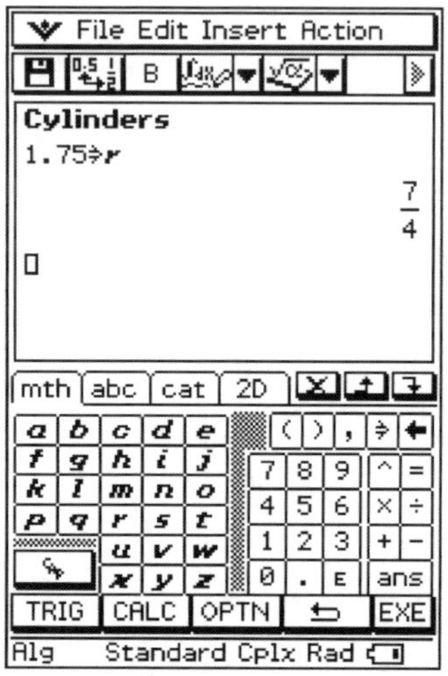

The double-stemmed arrow in the above screen shot is the assignment operator. It reads "store 1.75 in variable **r**".

We notice that ClassPad has converted this into a fraction. If this is not what we want, the fraction/decimal conversion button second from the left of the eActivity button bar lets us change this:

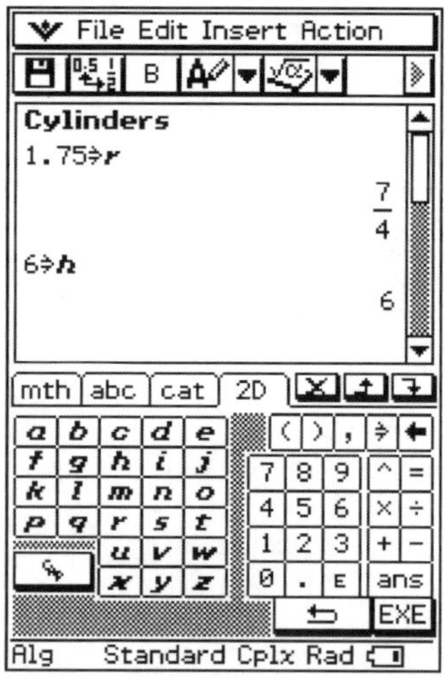

Back to text for a word of explanation, then our volume and surface area formulas:

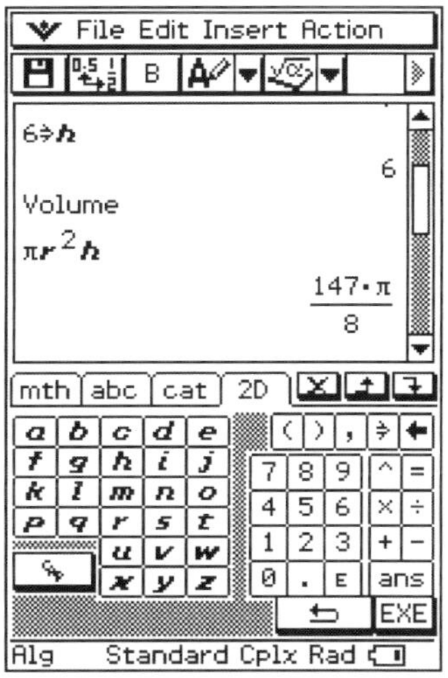

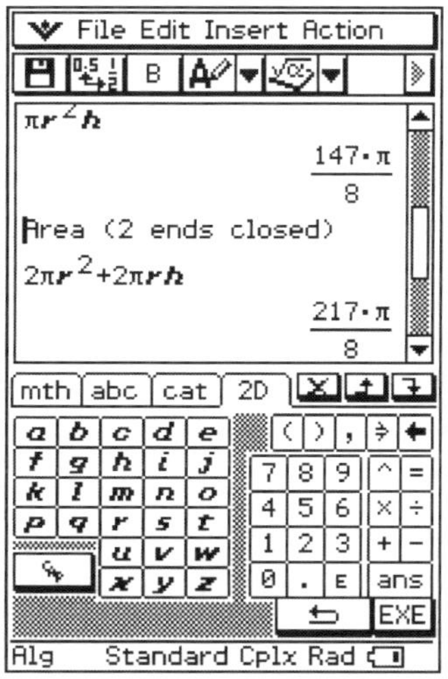

We can create a second area formula by dragging and dropping, then editing the text and equations we have dragged – first the text:

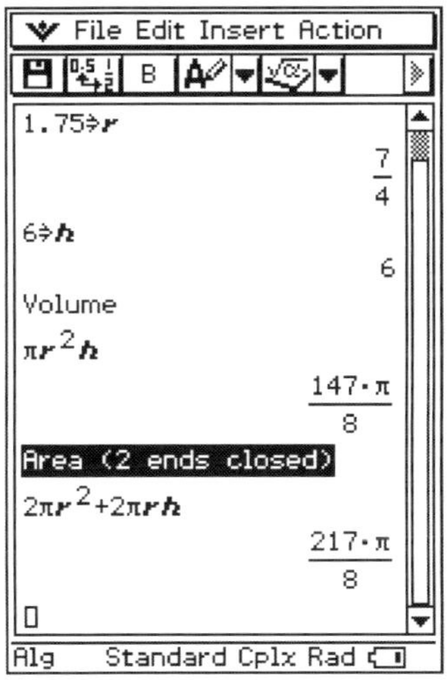

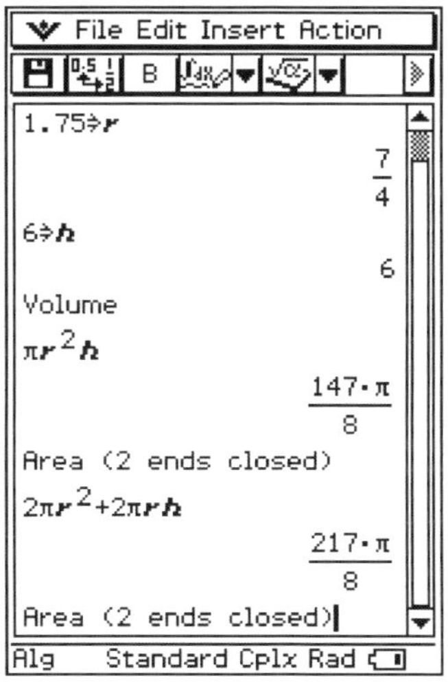

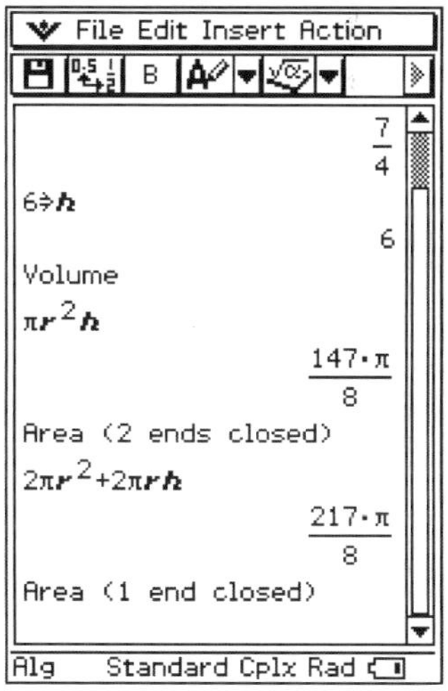

Then the equation – again, dragging and dropping allows you to grab the almost identical equation for the closed cylinder and modify it:

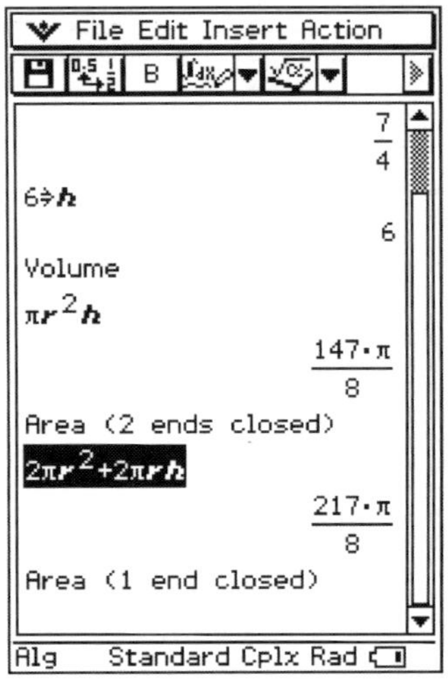

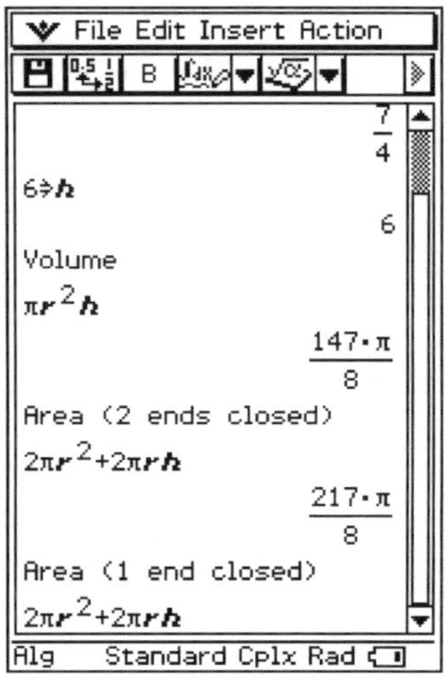

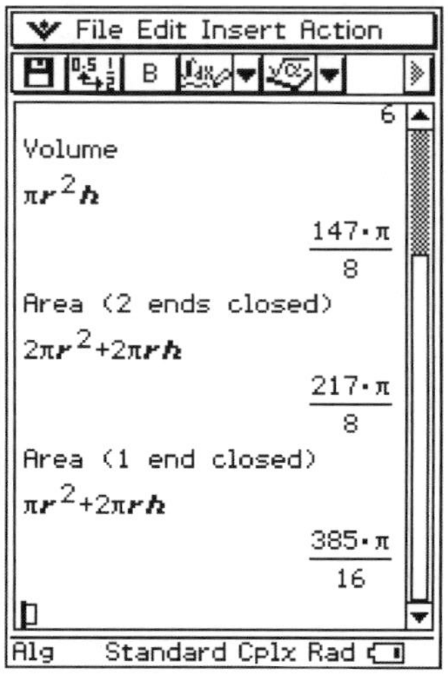

Now if you go back to the top of the page and change the value of **r**, and **h**, the corresponding values for the volume and surface areas will update automatically

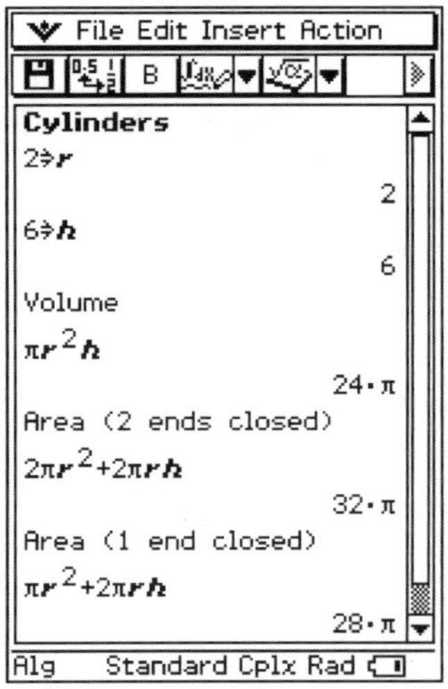

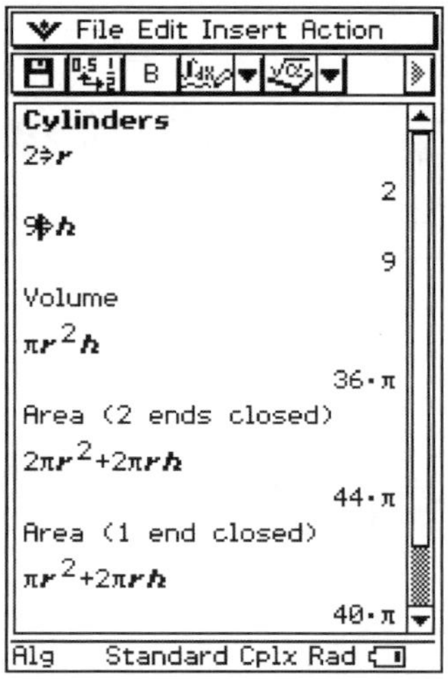

Minimum Surface Area Cylinder

Let's say we have a problem, where we are looking for the closed cylinder with minimal surface area which has a volume of 1:

We can look at the volume formula and see that it involves h and r^2. Hence if we make $h = \frac{1}{r^2}$ then the volume will be constant. But, the problem is that r is already set to a numeric value – let us write both in terms of a new unassigned variable x:

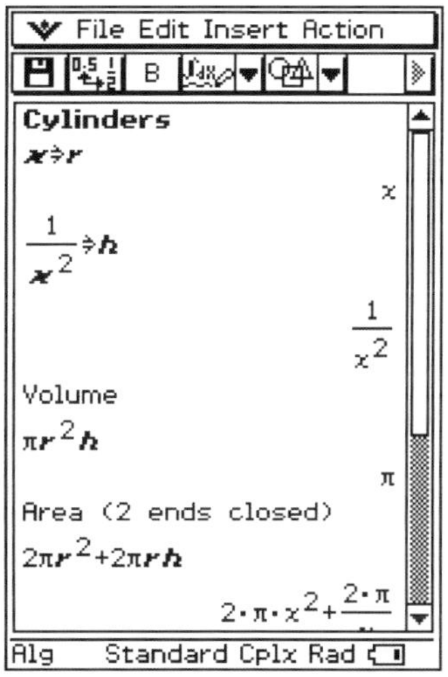

The volume, we see is constant, but we'd like it to be 1: so let's just divide h by π:

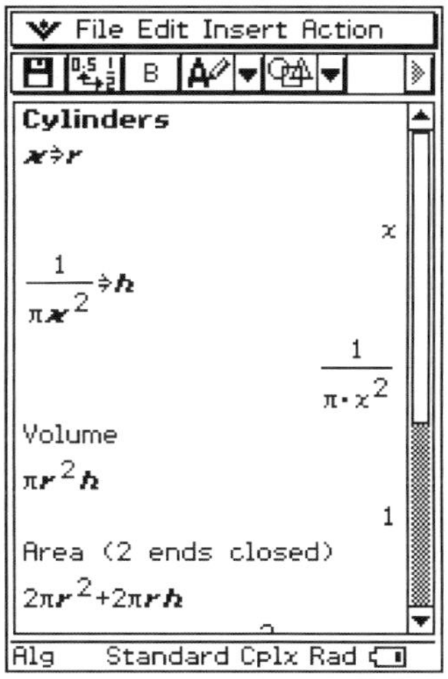

We can graph the area by creating a geometry strip and dragging and dropping the area formula into the strip. The graph does not look so good in half screen, but going to full screen gives a better picture. We can zoom in to locate the minimum:

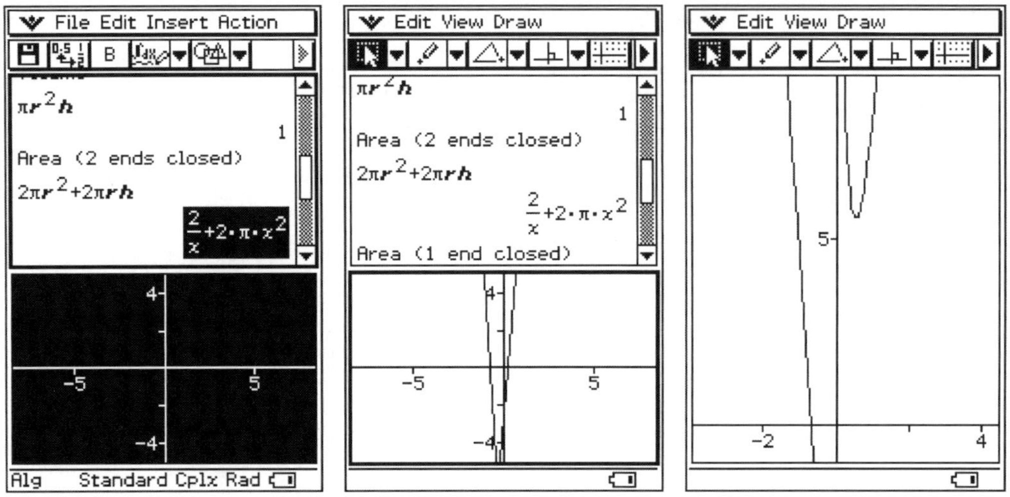

The reason the first graph did not initially pick out the minimum is because the asymptote finder in ClassPad's geometry has too course a granularity to distinguish the asymptote at this scale. Zooming in further on the minimum lets us pick out its approximate value. We can even examine a point on the curve to get the value more precisely:

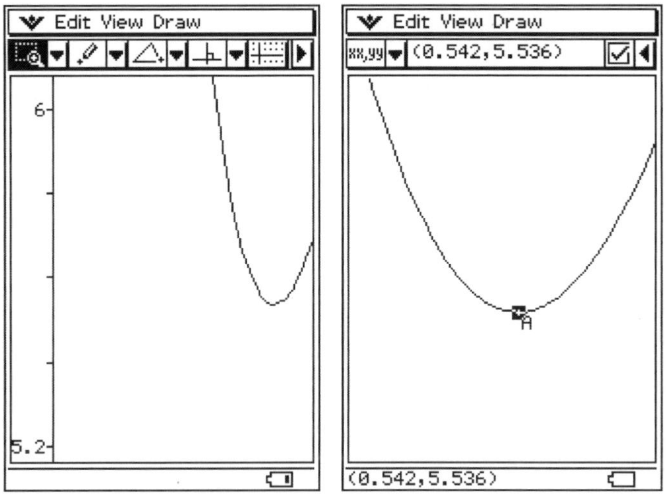

So we believe the minimum is approximately at x = 0.542. Let's go back into eActivity to check – we'll solve for the derivative of the equation equal to 0. We can simply drag and drop the area equation straight into the derivative:

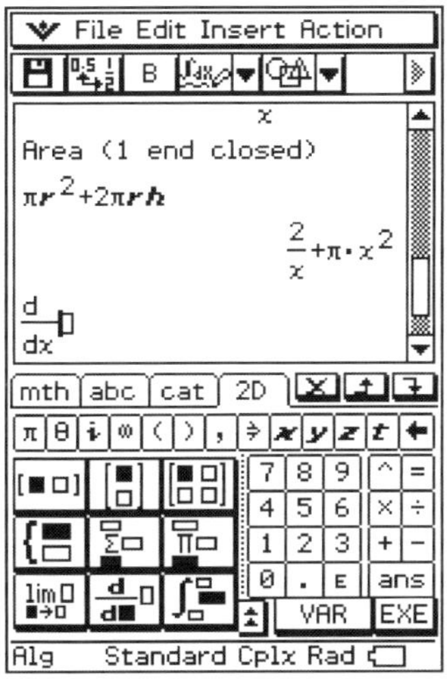

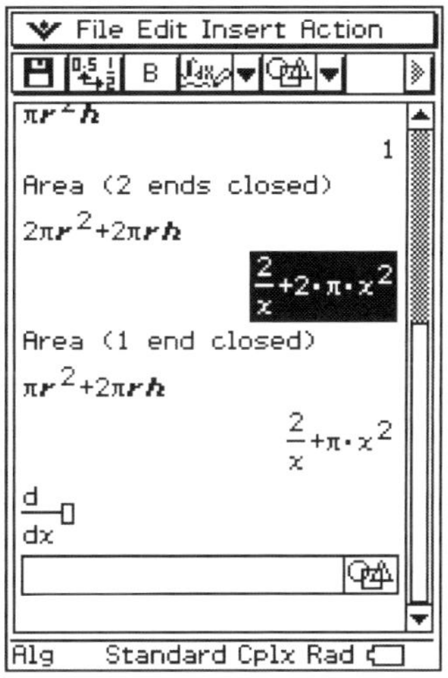

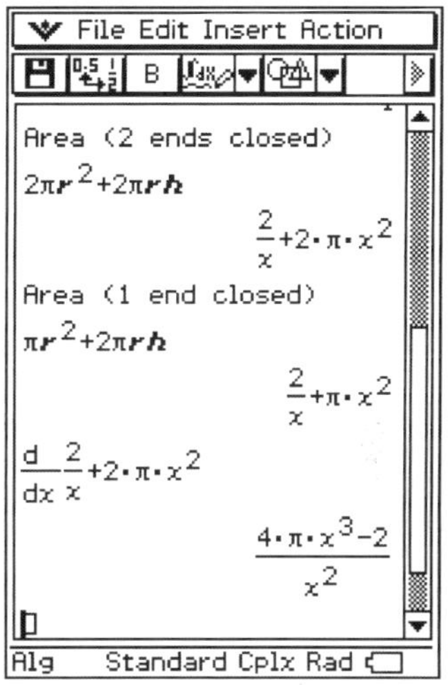

Then drag and drop the derivative into a solve statement (x is the default variable for solve and so does not need to be specified, also, by default if solve is invoked with an expression rather than an equation as its argument "expression=0" is assumed).

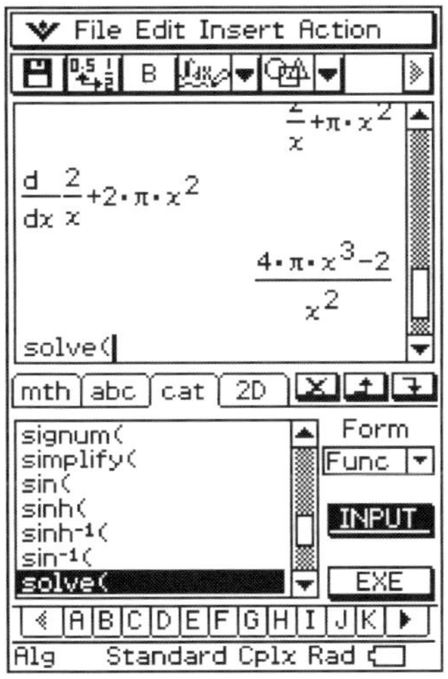

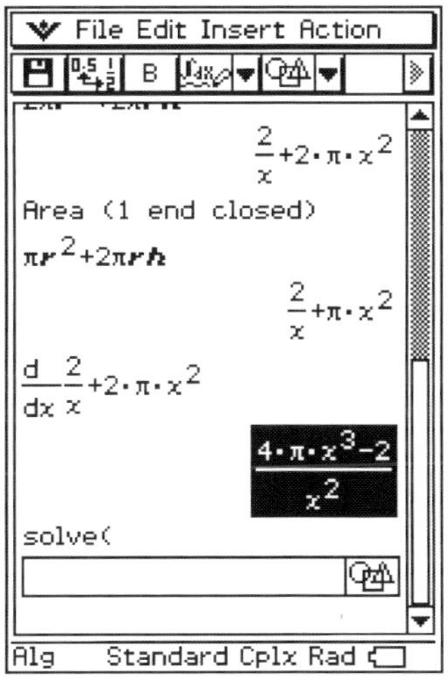

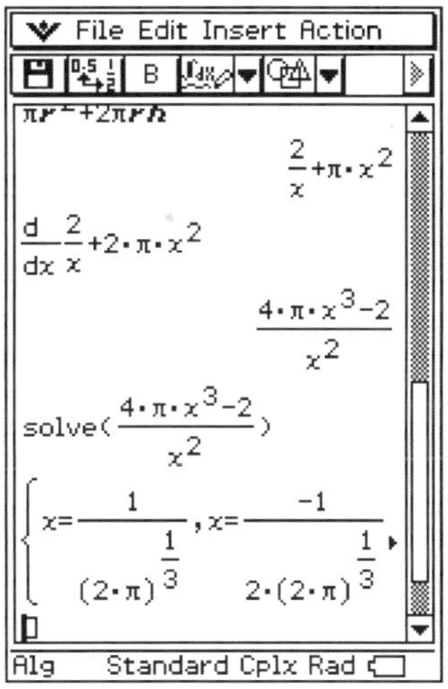

The negative value is not of much interest to us, but the positive one is. We can look at it numerically simply by pressing the numeric/exact button on the eActivity button bar:

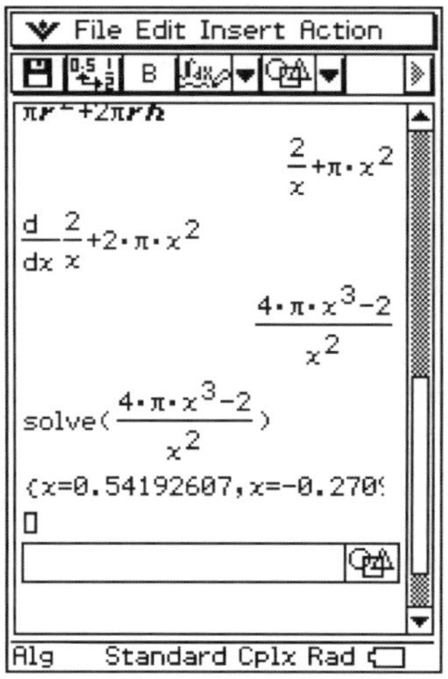

Which is pretty close to the approximation of 0.542, that we obtained graphically.

Clearing Variables

Before proceeding, a word of caution about variables – ***they persist.***

OK that was two words. But the point is, once you have defined a variable, it stays defined – and is accessible throughout the ClassPad. This is extremely useful if that is what you want, and extremely confusing if you have forgotten that you have defined x somewhere else and cannot understand why every time you type in x, the system responds with 2.

The **DelVar** command can be found on the Cmd sheet of the functions catalog on the soft keypad. It lets you remove the definitions one at a time. Alternatively you can invoke Clear_a_z to clear all lower case single letter variables at once.

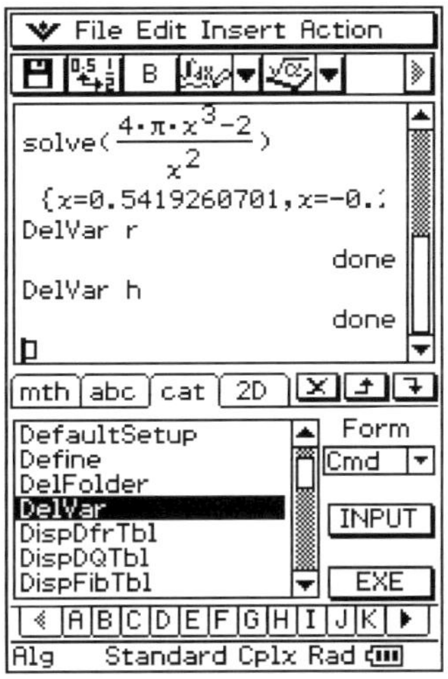

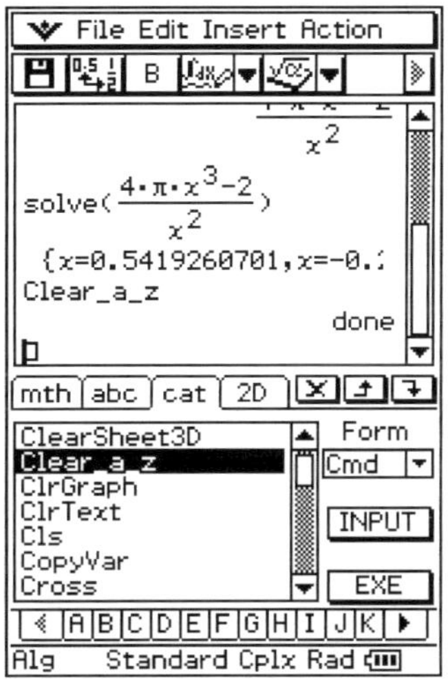

Simultaneous Equations

ClassPad lets you easily select pieces of expressions and drag and drop them to build other expressions. Let's explore these features in the context of solving a system of 3 linear equations by means of matrix inversion.

First, the 2D math input has a simultaneous equation template. By default it appears with space for 2 equations. We can lengthen it by clicking a second time on the template:

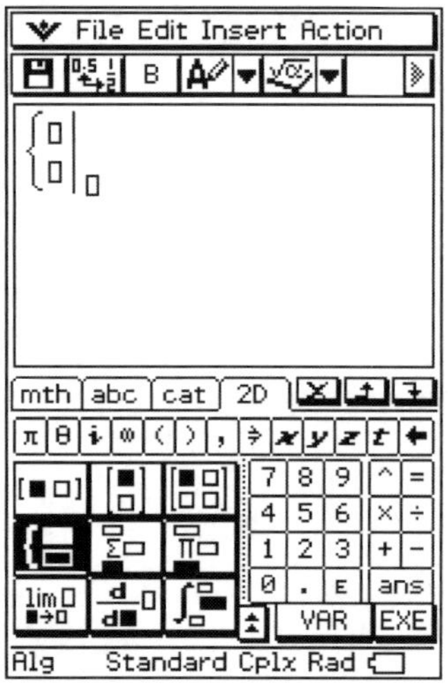

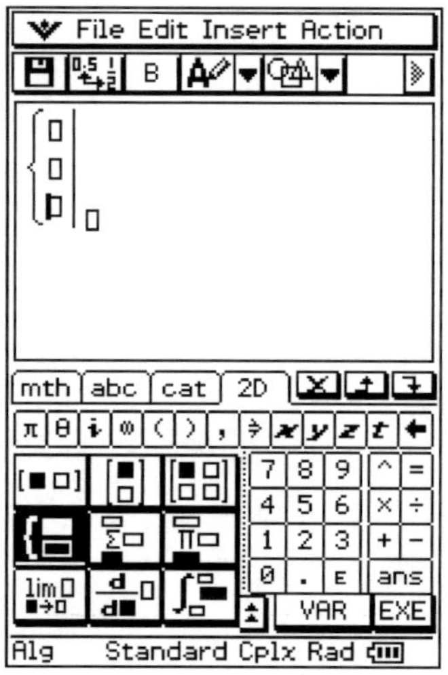

Now we can type in our equations and solve (remembering to make sure that we are in a Calculator Strip and not a Text Strip):

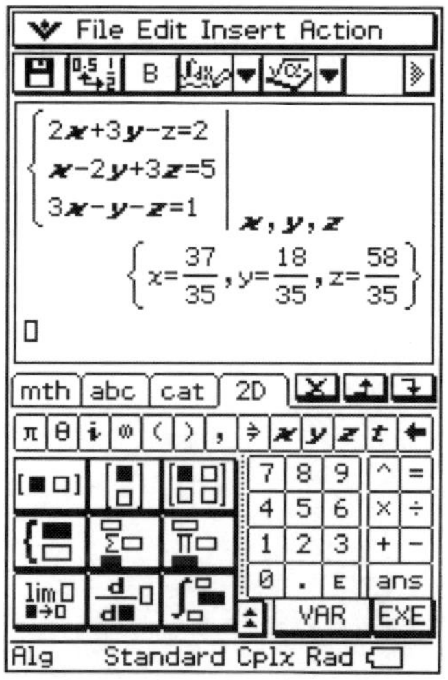

We have solved our equations, but not by matrix methods. To apply matrix methods, we need a matrix. Pressing the 2x2 matrix template gives us a 2x2 matrix. Pressing it again extends the matrix horizontally and vertically to give us a 3x3:

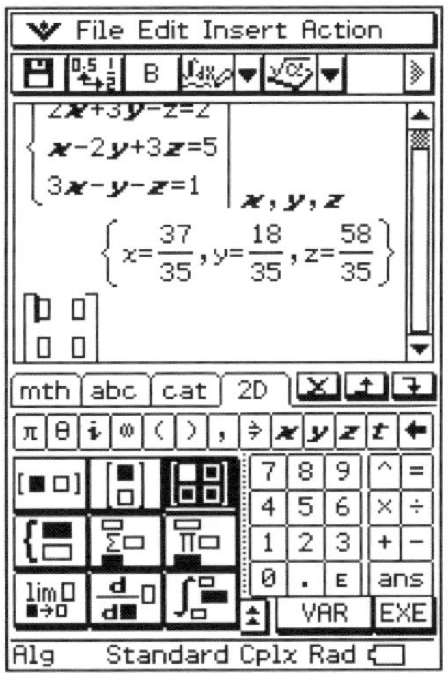

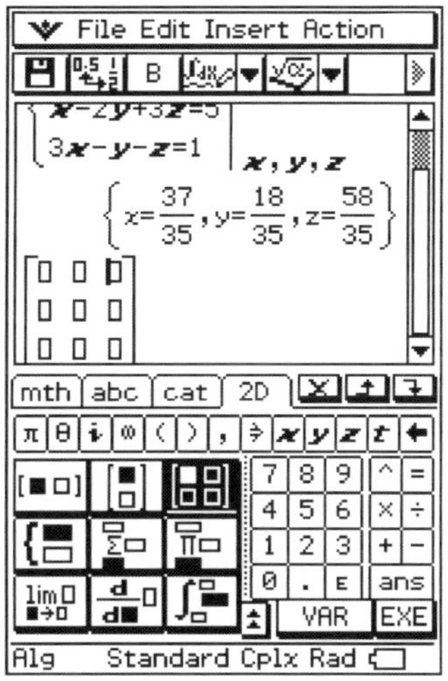

Now we can drag and drop coefficients from the equations into the matrix template:

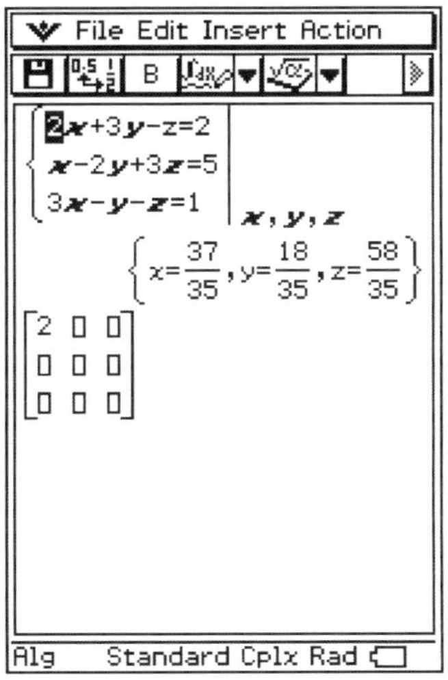

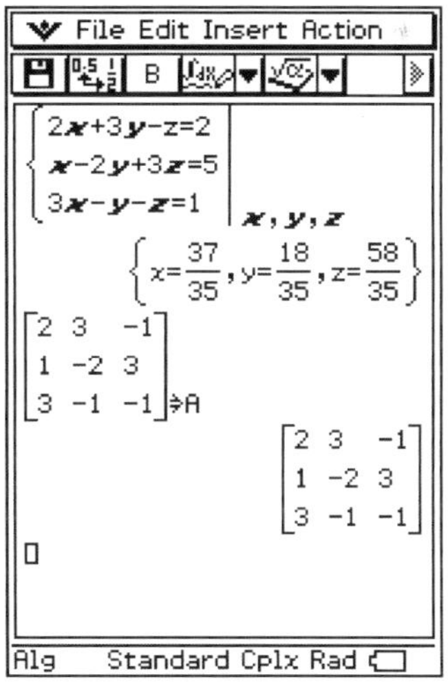

Note the ⇒ symbol (found to the left of the x,y,z,t on the soft keypad) is used to assign a value to a variable. In this case, we assign the matrix to variable A.

We will want to invert A then apply the inverse matrix to a vector consisting of the right hand sides of the equation. We create a column vector of length 2 with the Column Vector Key, then press it again to extend it to a vector of length 3.

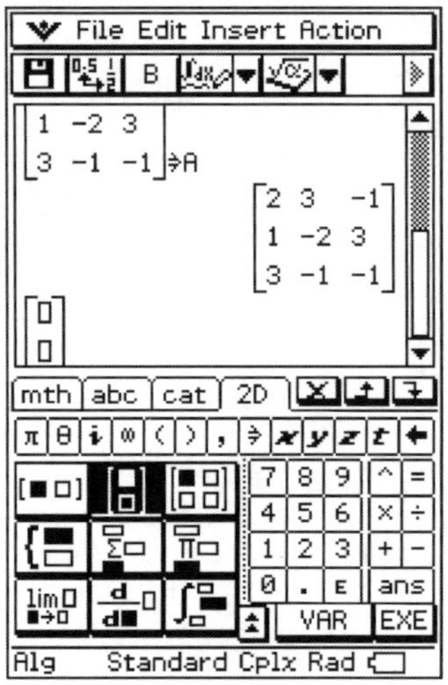

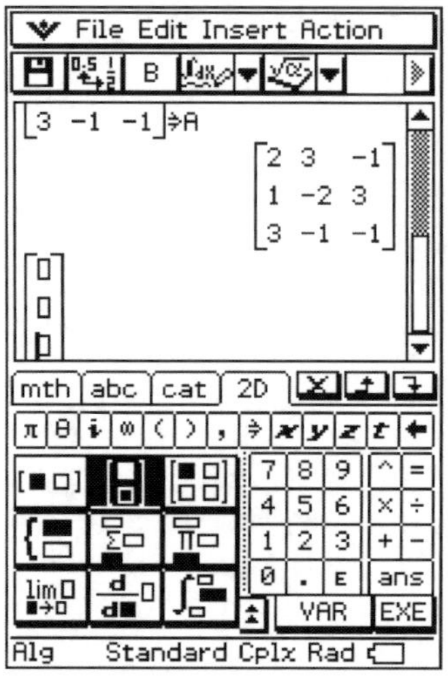

Then drag and drop in the coefficients:

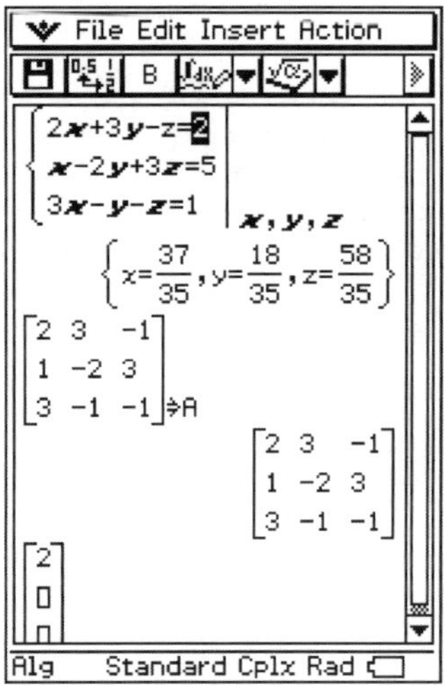

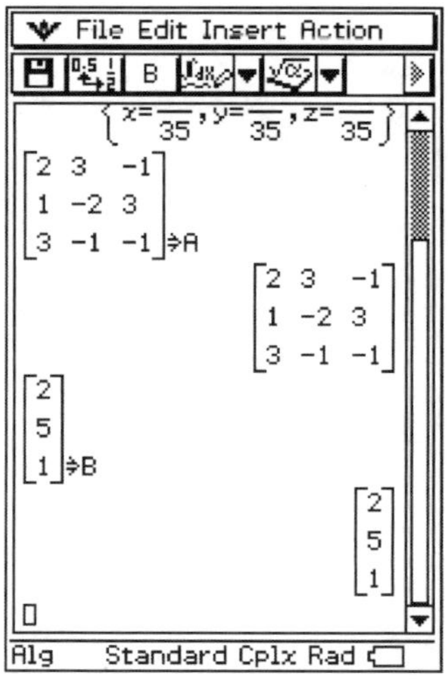

Now solving the equations is equivalent to inverting A and applying the inverted matrix to B:

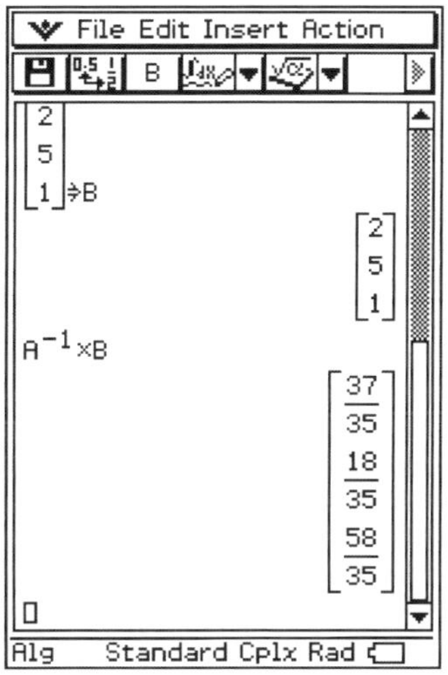

Which, as long as we dropped the correct coefficients into the correct matrix slots, should correspond to the solutions obtained automatically by ClassPad.

Variable Management

Before proceeding – remember to clear your variables. You can use the DelVar command as before. You cannot use Clear_a_z as this only clears the lower case variables and leaves the upper case ones alone.

If you have lost track of which variables are defined in your system, your only recourse is the **Variable Manager**. You invoke this from the System menu:

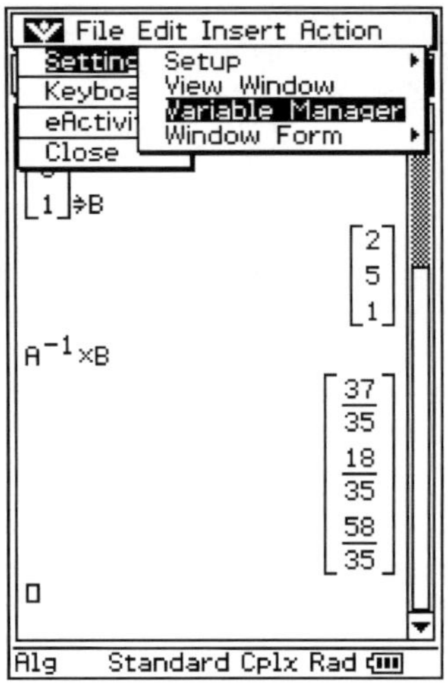

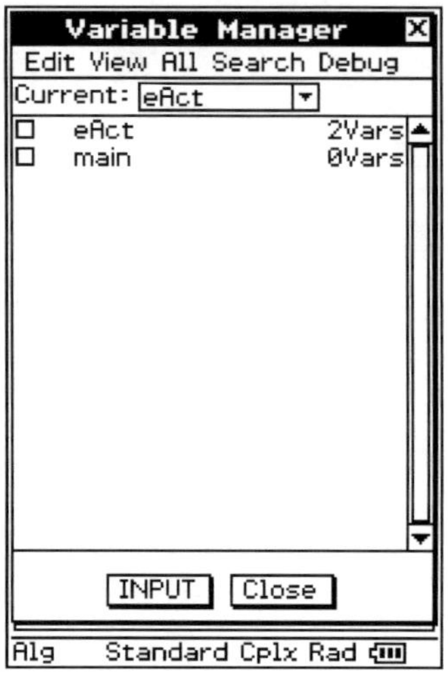

Double click on eAct to bring up a list of defined variables:

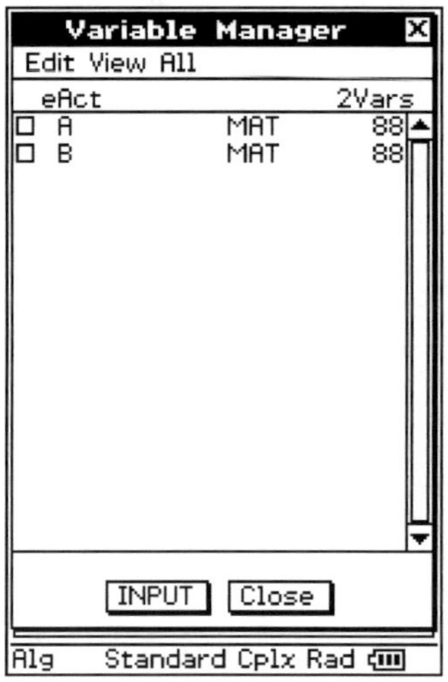

You can select the variables by clicking in the check box to their left, and delete them using **Edit/Delete**.

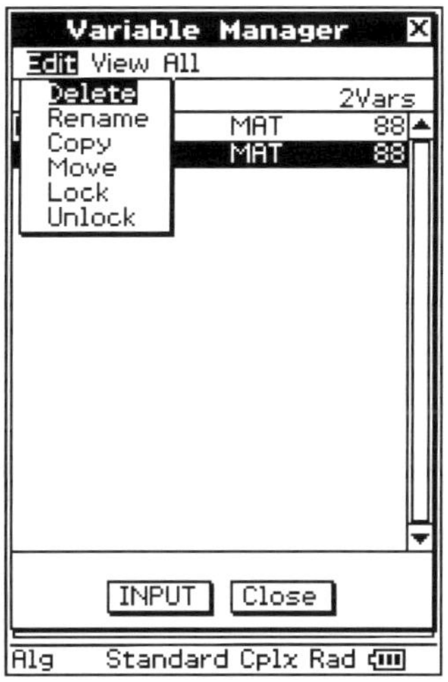

Characteristic Equations by Drag & Drop

Here comes more dragging and dropping, and we'll start with some name dropping:

> *eigenvalues, characteristic equations, identity matrix.*

Now we'll drag in a typically arbitrary mathematical assignment:

> Given an antisymmetric 2x2 matrix find the condition that the eigenvalues are identical.

We'd better start by defining our terms:

An antisymmetric 2x2 matrix looks like this:

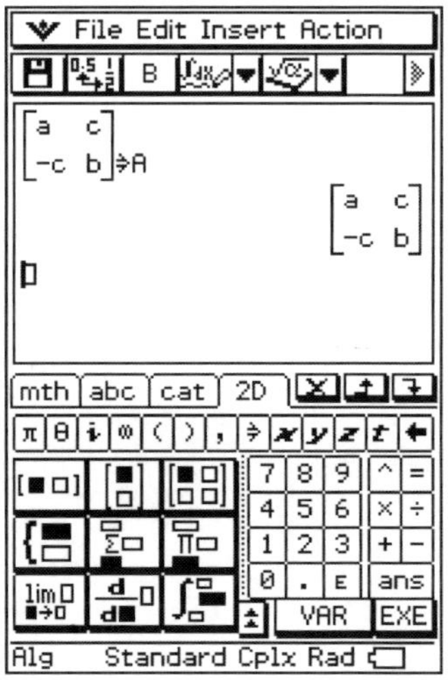

The identity matrix looks like this:

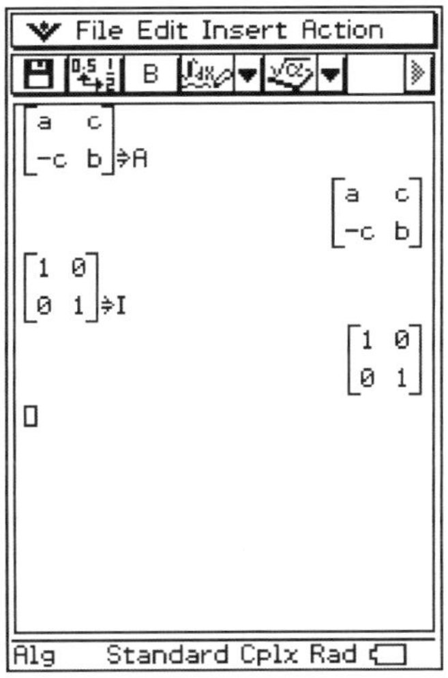

And an eigenvalue is a solution to the characteristic equation:

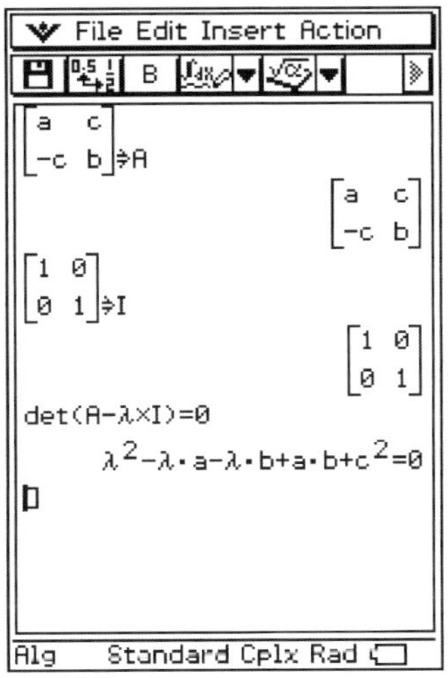

Eigenvalues are always denoted by the Greek letter symbol λ for some reason, so it's fortunate that ClassPad can accommodate Greek, otherwise our values might not be quite *eigen*.

Fortunate, also that it has a solve command:

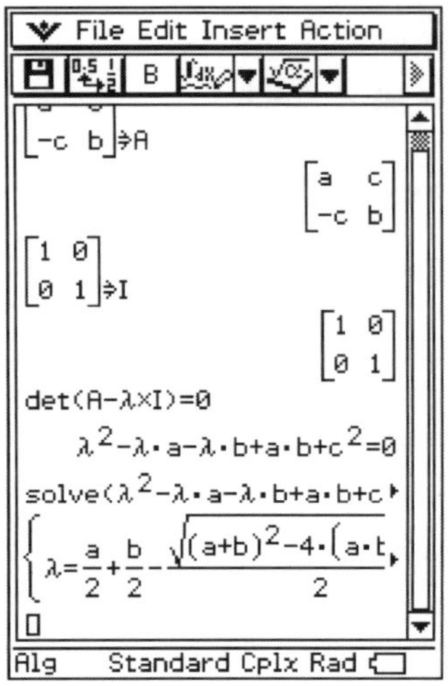

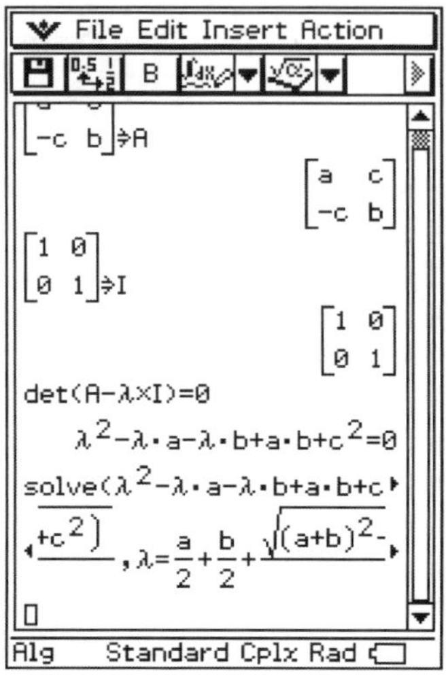

As one would expect, given that the original equation was quadratic, we have two solutions in the typical quadratic equation form, the solutions being the same when the expression under the square root is 0.

We can drag that expression out and solve for a, for example.

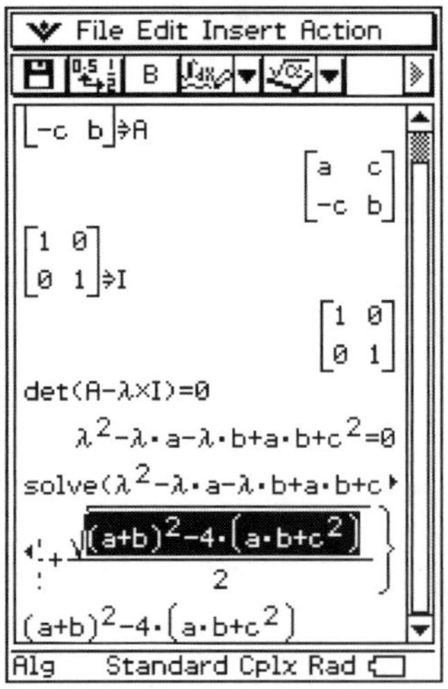

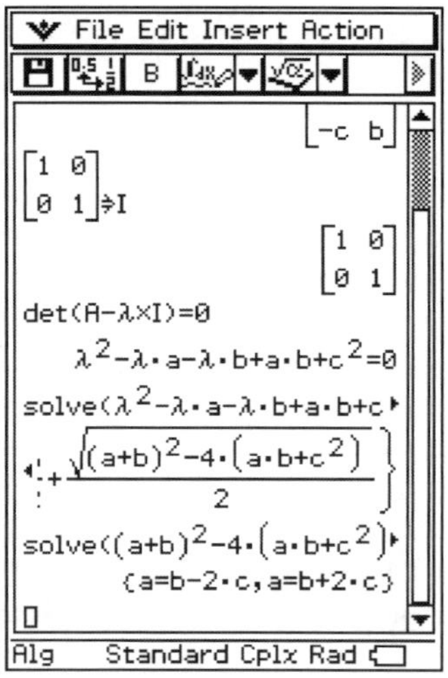

The ability to grab part of an expression and work with it is the key to getting the most out of your ClassPad. And we have the solutions to our problem: the condition for the eigenvectors to be identical is a=b-2c or a=b+2c.

Let's check using ClassPad's built in Eigenvalue function, we'll use b=3 and c=1. (The eigenvalue function is numeric only).

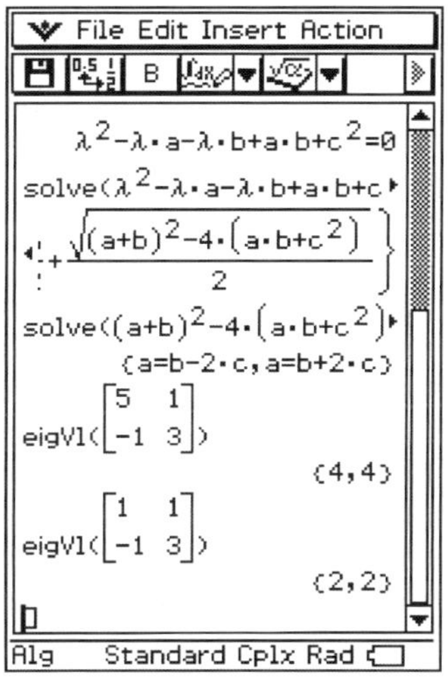

We see that the eigenvalues are indeed identical when a = 5 (b+2c) and when a = 1 (b-2c).

HotLinked Parabolic Mirrors

The third menu option on eActivity's **Insert** menu is **Geometry Link**. We'll use a parabolic mirror to focus in on this feature.

A geometry link is a direct two way link between an expression in eActivity and an object in geometry. When the geometric object moves, the expression changes. When the expression changes, the geometry object moves.

We'll build a model to allow us to study the reflection of a ray of light in a parabolic mirror. In order to do this we'll have to use geometry commands which we won't explain in detail until later on in the book. So if you are following along on your ClassPad, it might be better to stop and read just for this section. You can come back and make the model after you have gone through the geometry chapter (coming up next).

First, we insert a geometry link:

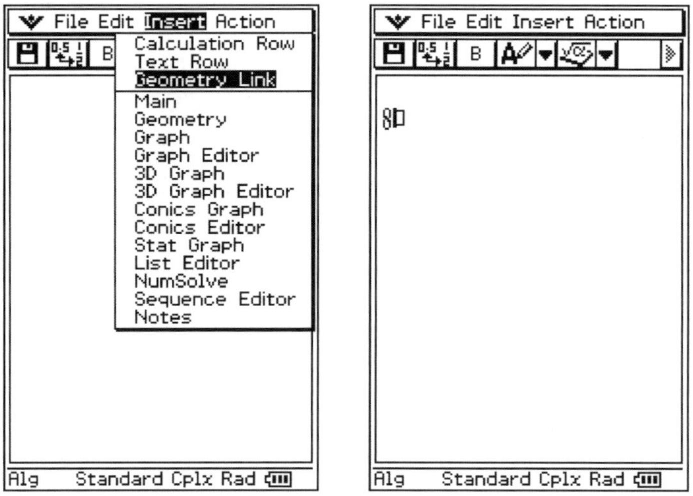

The Geometry Link looks like an ordinary Math Entry Field but has a small link symbol to its left.

We can enter our parabola's equation:

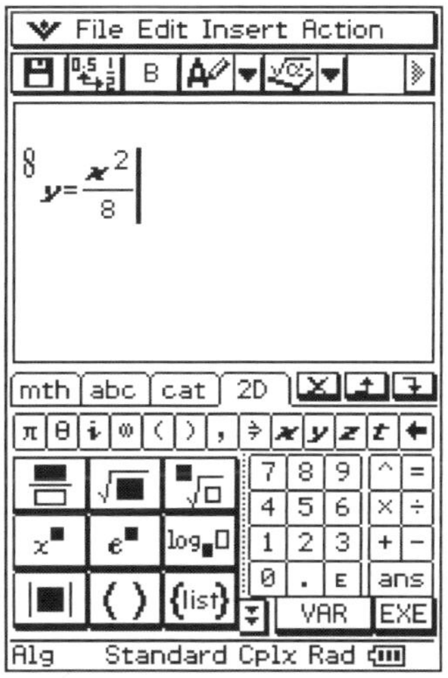

We now create a geometry strip and drag in the equation:

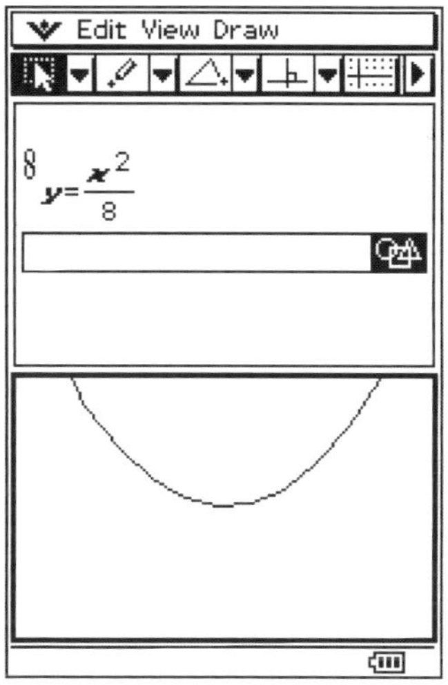

Now, if we change the equation in the Geometry Link, the geometry model is updated automatically:

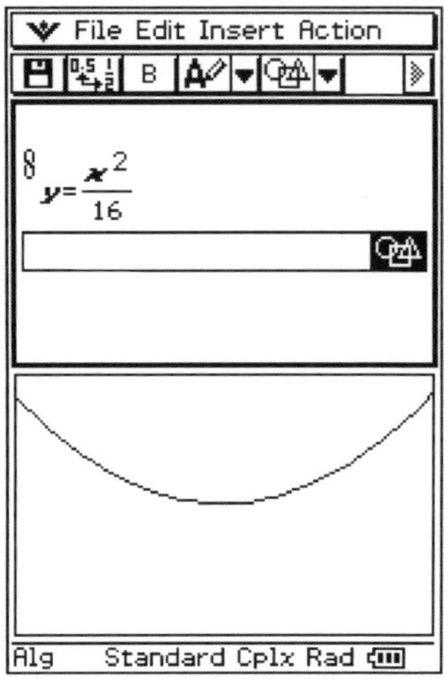

To study reflection in the parabola, we need to first construct its tangent. Then we create an infinite line through the tangent point, and set its direction so that it is vertical.

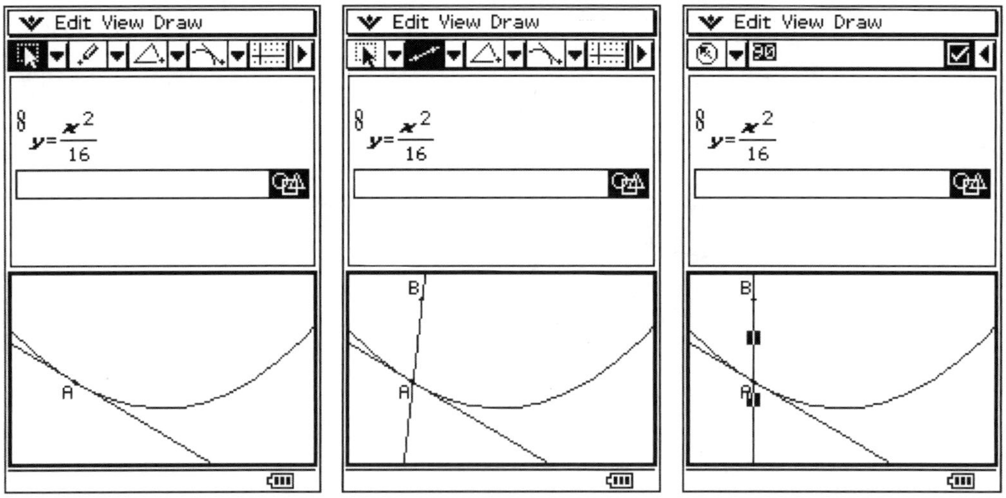

Now, we reflect the line AB in the tangent (switching on the axes gives a clearer view):

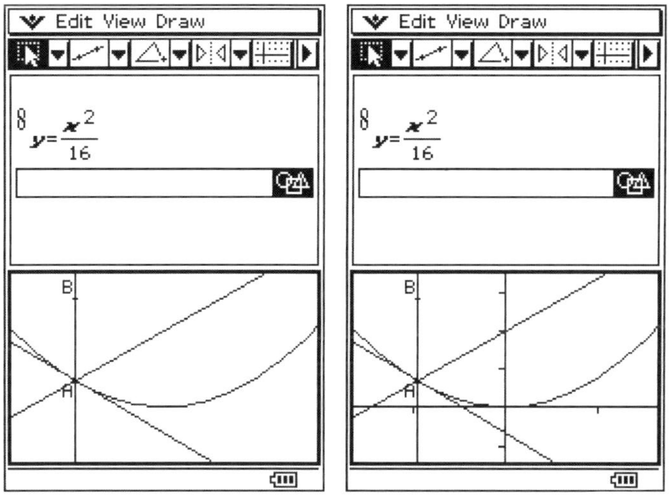

Now let's link the reflected line's equation back to eActivity:

First, we create another Geometry Link, then we drag the line into the link:

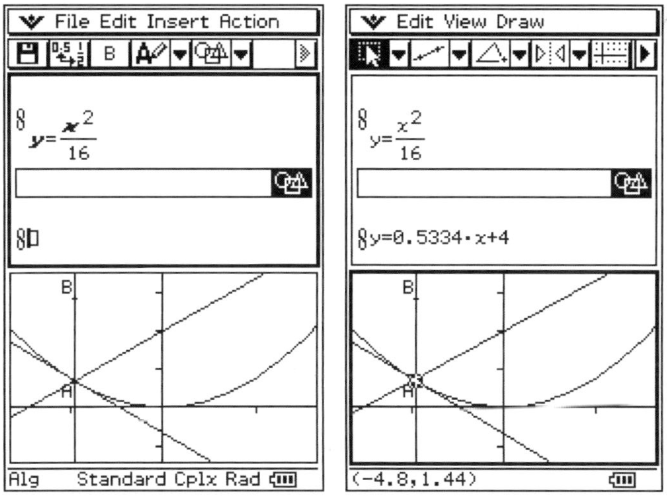

We can now drag point A round the parabola and watch its equation update in the eActivity window:

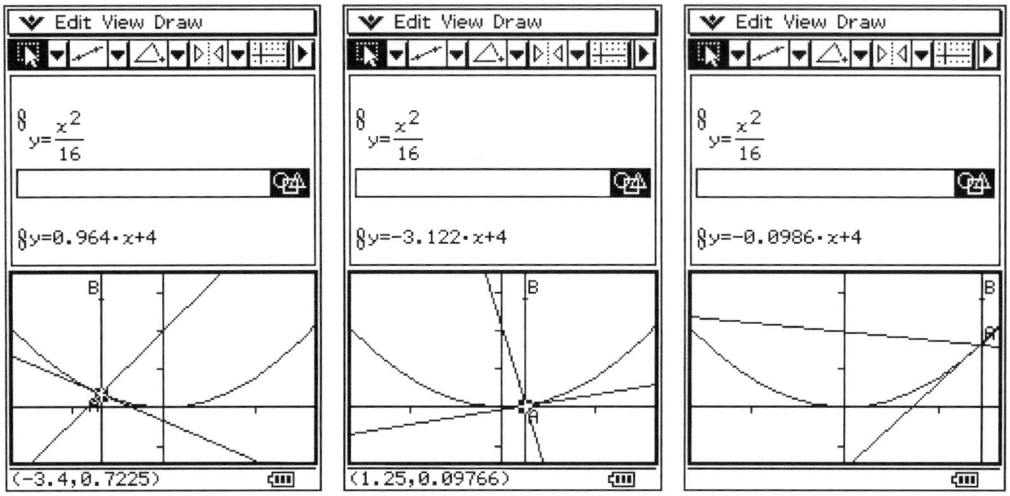

What do you notice about the equations. What property do the reflected lines all have?

We can experiment with changing the coefficient of the original parabola. The geometry and the line equation automatically update:

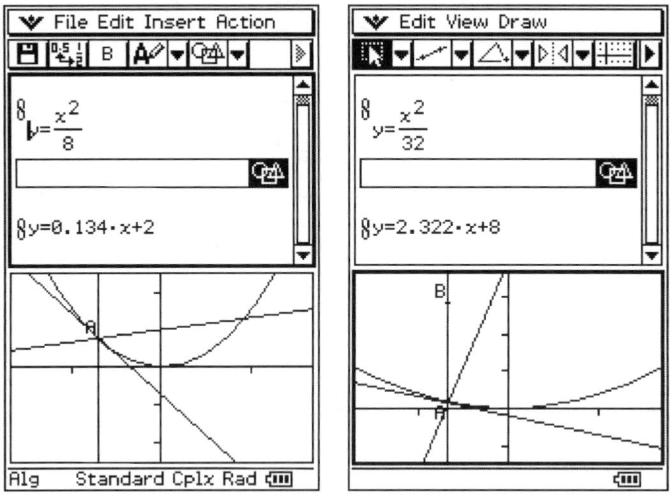

Can you identify a pattern in the line equations?

Geometry with ClassPad

One of the fascinating aspects of mathematics is that there are often several fundamentally different representations or ways of thinking about the same object. Nowhere is this more true than in geometry. A circle, for example, can be thought of as the locus of all points distance 3 from a center point. Alternatively it may be thought of as the solution set of an equation like this:

$$x^2 + y^2 = 9$$

or again as the multivalued function:

$$y = \pm\sqrt{9 - x^2}$$

or as the image of the unit circle centered at the origin under a dilation of scale 3, or as the limiting case of regular polygons with increasing numbers of sides.

The beauty of the ClassPad is that it is designed to let you think about geometry in many different ways at the same time. It does not force you to change your problem to fit its mental model - it is prepared to change its mental model to fit your problem. The different representations (or mental models) available in ClassPad are:

- Coordinate - geometrical objects are described in terms of Cartesian Coordinates.
- Algebraic - geometry is described in terms of algebraic equations.
- Constructive - we study figures which can be constructed by combining various ruler and compass style operations.
- Transformational - we look at the image of geometrical figures under the influence of transformations such as translation, rotation, reflection, dilation.
- Constraint based - we define geometry in terms of relationships - such as specified lengths and angles.

In the following sections, we will explore these different representations in the ClassPad, and we will particularly look at the interplay between different representations.

The basic techniques for operating the geometry system are shared between these different approaches, and we will discuss these here:

Selection

A pen is just like a mouse, right? In fact it's more direct, because you are manipulating the pen on the same surface that the drawing lives on, whereas the mouse sits on a table a couple of feet and 90 degrees away from the screen?

Well, yes, but a mouse has a critical advantage over a pen, which is that you can see a cursor indicating where the mouse is before you click the button. With a pen, you don't see any effect on the screen until you have pressed down, and then it might be too late!

For this reason, the ClassPad geometry tends to wait till you lift the pen up before accepting a selection. However, in order to give you feedback on what you are about to select, it will be highlighted. This gives you a chance to move the pen off one object and onto another (or off all the objects) before the selection is accepted.

On a computer, the typical selection model goes like this:

	No Shift	**Shift**
Mouse click on object	select object	select object
Mouse click on another object	replace selection with new object	add new object to selection
Mouse down on already selected object	prepares entire selection to do some action – if followed by a drag, drags entire object	removes object from selection

The ClassPad does not have a Shift key! And even if it did, would you really want to be manipulating the shift key with one thumb while trying to do fine selection on the screen with a pen in the other?

Instead, ClassPad has 2 different shift modes, on the top two buttons of the **Select / View** dropdown.

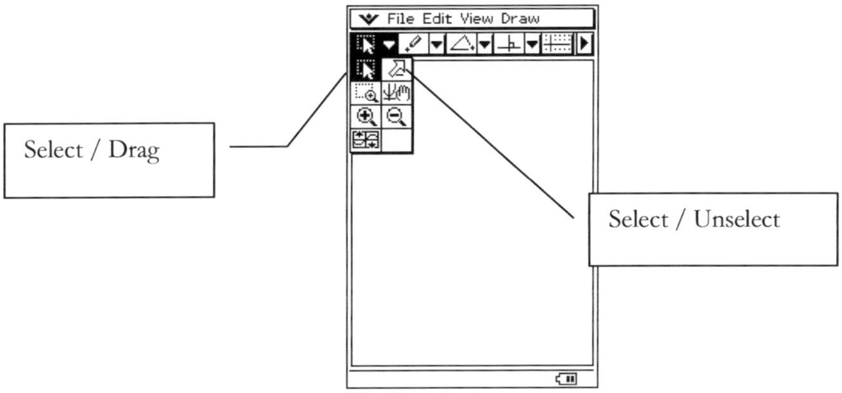

	Select / Drag Mode	**Select / Unselect Mode**
Press on object and lift	Selects the object	Selects the object
Press & lift on an unselected object	Adds the object to the selection	Adds the object to the selection
Press on a selected object & drag	Draws a selection box around the current selection and drags the whole selection	Nothing
Press on an existing object & lift	Nothing	Unselects the object
Press on empty space	Unselects everything	Nothing

Zooming / Panning

The Geometry Module has the usual sort of zooming options arrayed out on the **Select / View** drop down:

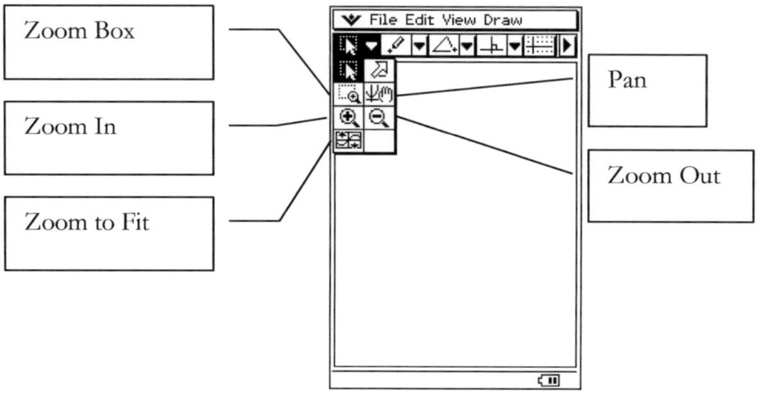

It also has some zooming options concealed on the hard keys:

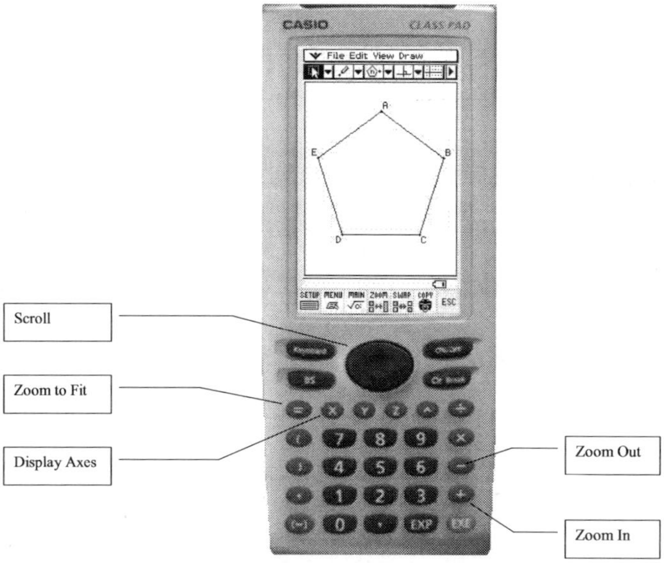

Although you should be aware that these only take effect if the Geometry Window is currently the input focus. (That is, if it was the last Window you clicked in.)

Measuring

A geometrical figure is of little use if you can't measure anything on it. The critical part of the Geometry User Interface for measuring things is effectively hidden behind the Button Bar. To reveal it you need to click on the Tab Arrow at the extreme right of the button bar:

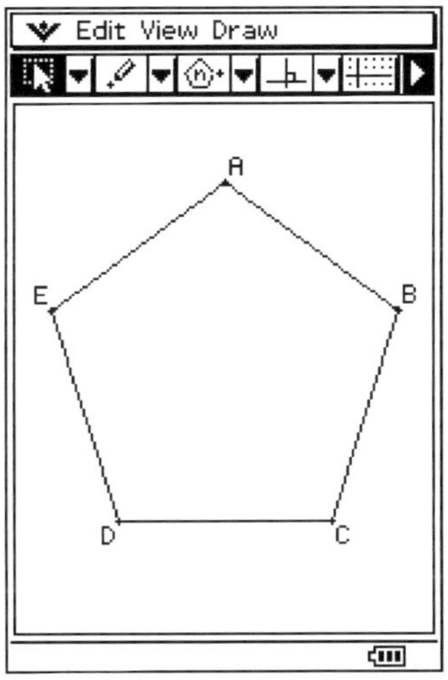

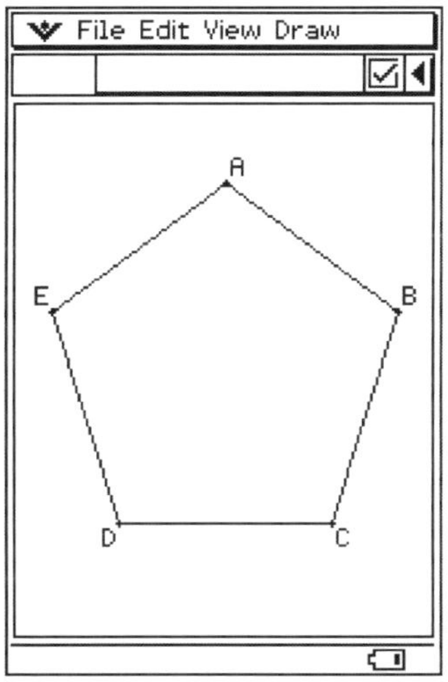

If an object is selected, then the measure box shows some measured attribute of that object, length in the example below:

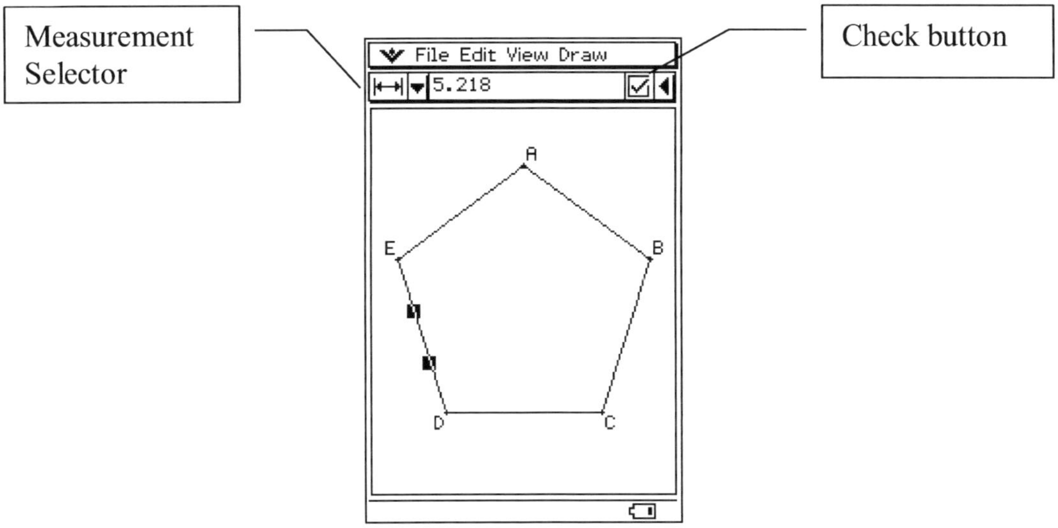

The drop down button to the left of the measure box is the measurement selector button. This lets you choose which attribute of the object to display (we'll change to a display of slope):

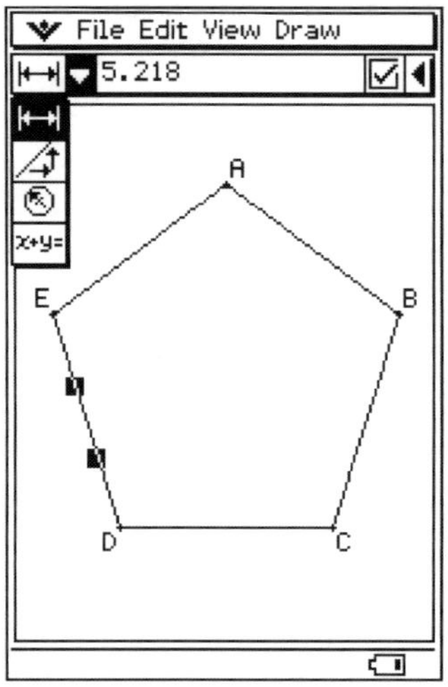

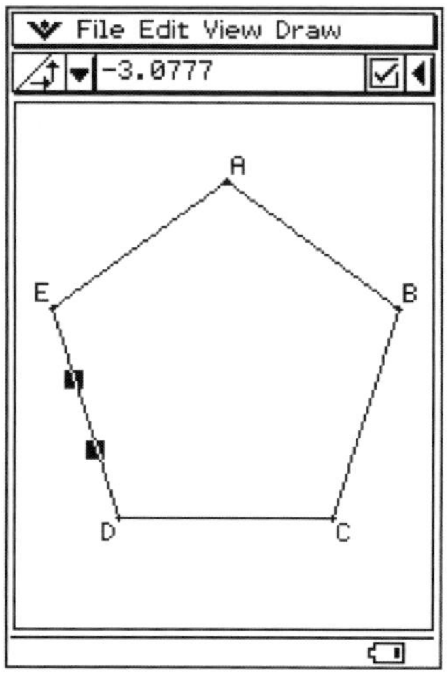

If two objects are selected, then the measurement box displays an attribute of the relationship between the objects (angle in the example below).

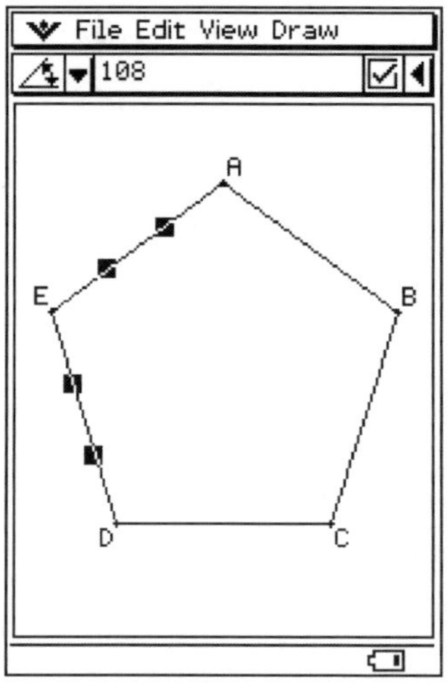

The check box is used to constrain the model so that it maintains the value in the measurement box. Its use is the topic covered in the section ***Constraint Geometry***.

Constructive Geometry

To do constructive geometry on the ClassPad we use the geometry creation tools, which are available from the geometry drop down button, or from the Draw menu.

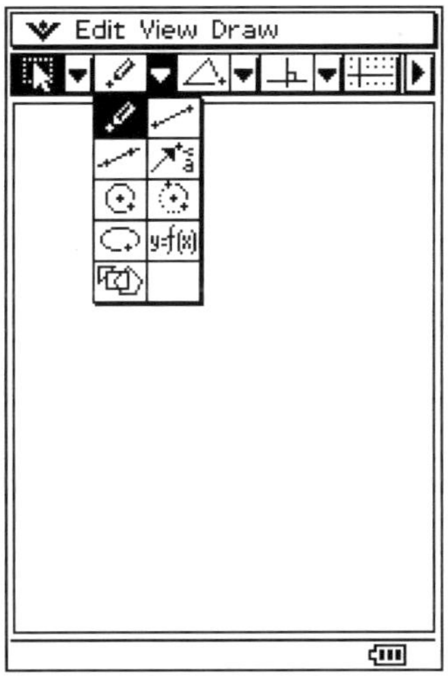

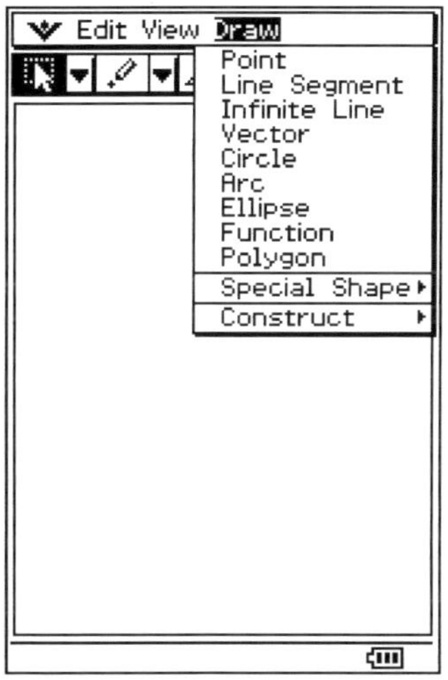

The drawing commands are these:

 Point

 Line segment

 Infinite line

 Vector

 Circle

 Arc

 Ellipse

Function

 Polygon

We can also use the secondary constructions. These are found in the top half of the Construction / Transformation drop down button or the Construct menu.

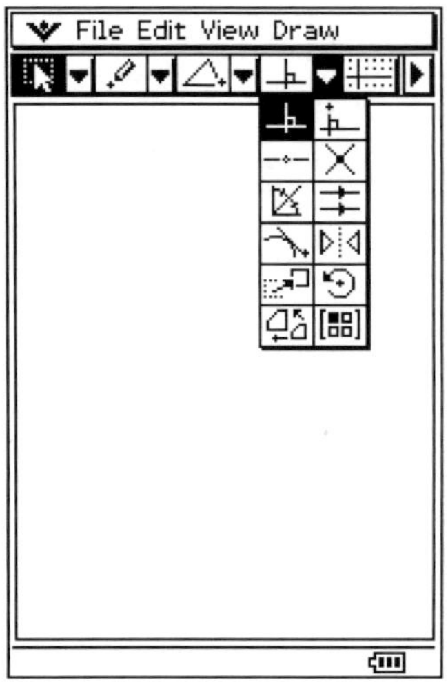

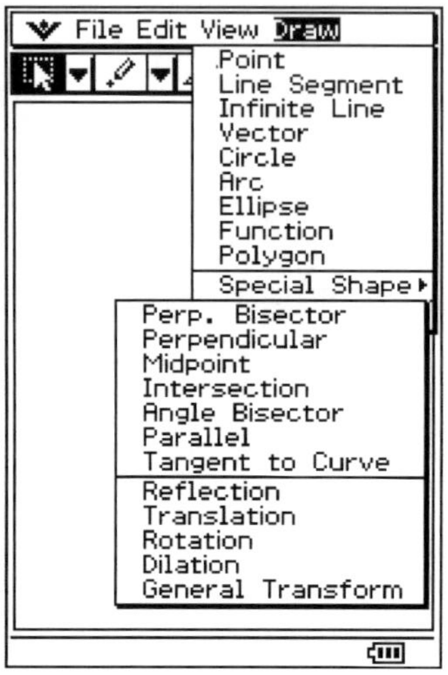

Available constructions are:

 Perpendicular bisector (of a segment)

 Perpendicular line (creates a line perpendicular to a given line through a given point)

 Midpoint (of a segment or between 2 points)

 Intersection (of a pair of lines, or circles or between a line and circle)

 Angle Bisector (of a pair of segments)

 Parallel line (creates a line parallel to a given line through a given point).

 Tangent (of a curve or function)

Let's look at some geometric examples

Centroid

We are going to look at the centroid of the triangle whose vertices are (0,0), (1,6), and (5,3). The centroid is the point of intersection of the triangle's medians. (A median is the line joining a vertex of the triangle with the midpoint of the opposite side.)

In most of our geometry examples, we will want to do some mathematics. It is always best, therefore, to start geometry up as a module within the eActivity, rather than starting it up on its own from the Menu.

First we need to create our triangle - do this with the Integer grid turned on so that it is easy to snap the vertices to the desired coordinates:

You should also turn on the axes, so you can see where the origin is.

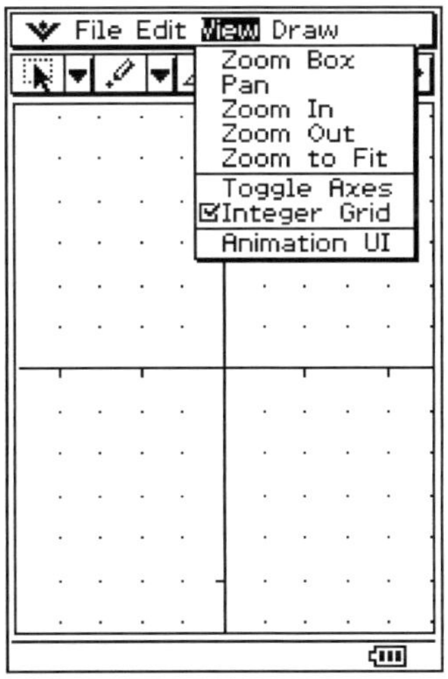

Although the triangle tool on the special shapes drop down button (second from the right) allows you to create a triangle in a single click, or a click and drag – it does not let you position its vertices accurately. The best tool to use for that is the line segment tool.

When using the line segment tool, you press down the pen and drag it to the location where you wish to place the first end point, then lift up. Then press down again to place the second point. The reason for this somewhat cumbersome design is to give you feedback on where you have positioned the line before it is too late.

Make the first vertex lie on the origin. Put vertex B at the point (1,6) and vertex C at the point (5,3).

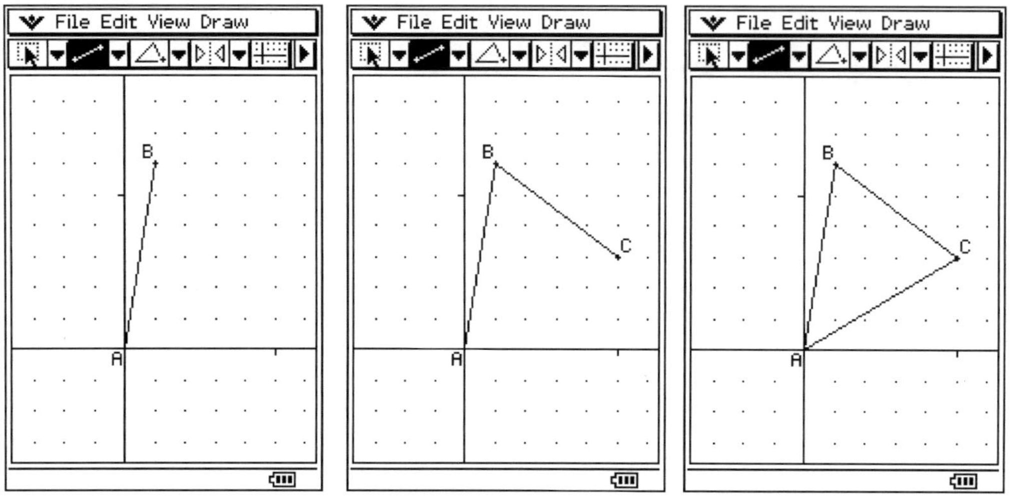

We'll start by constructing the median between A and BC. To do this, we first need to construct the midpoint of the segment BC. We do this by selecting the segment, then clicking the midpoint button.

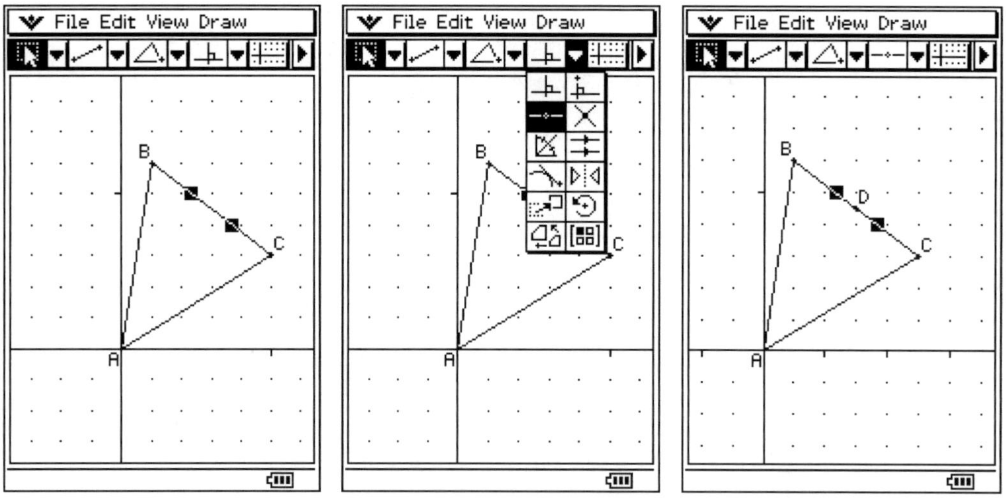

We then draw a segment from A to the new point D. This is the median. We can repeat the same process to construct the median between B and the line AC.

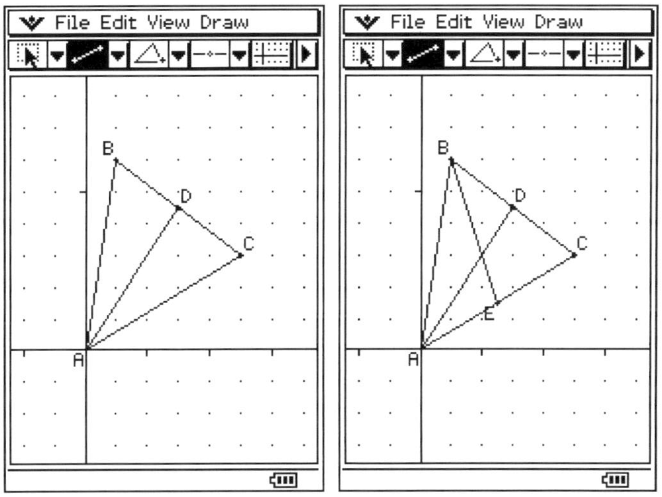

We construct the intersection point of these two medians by selecting the segments, then clicking on the intersection button:

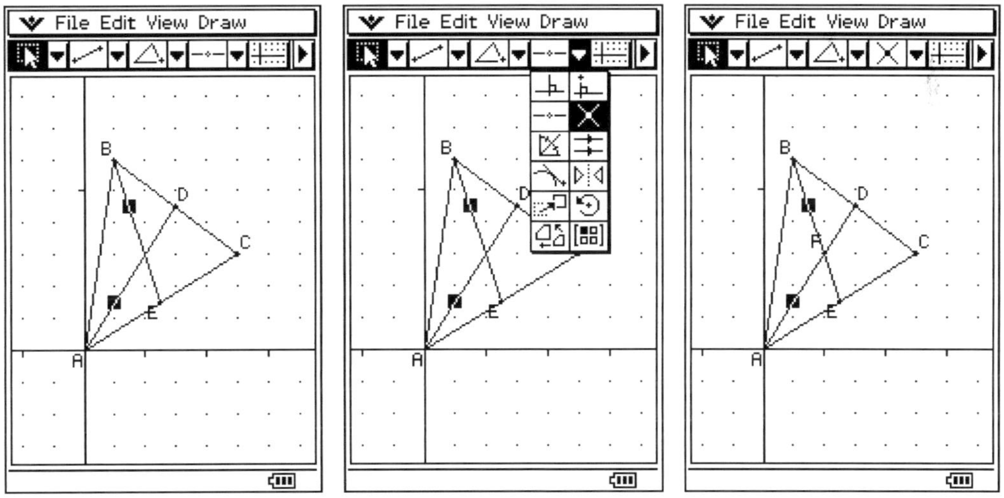

The point F at the intersection of two of the medians is called the Centroid of the triangle. One interesting property to examine is how far along one of the medians this point is.

Let's examine the coordinates of point F: we can do this by selecting the point and bringing up the measurement box (by clicking the tab arrow at the extreme right of the button bar).

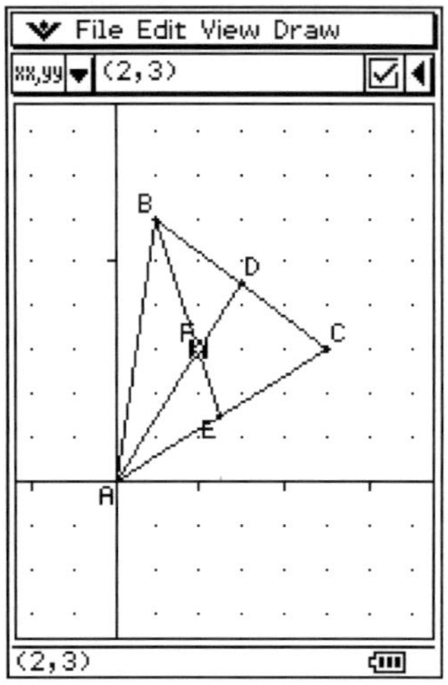

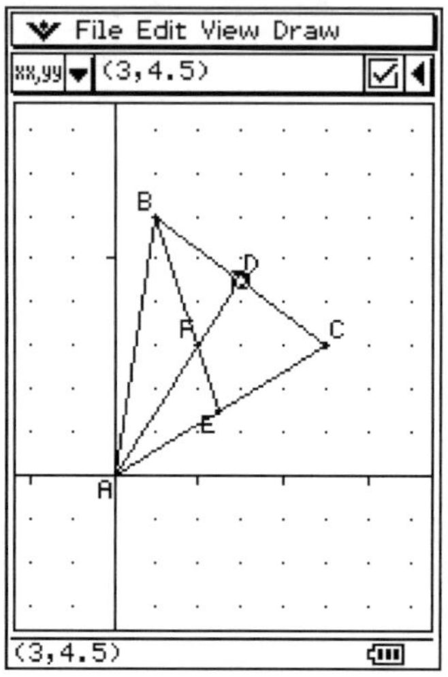

F has coordinates (2,3). We can select point D to find its coordinates: (3,4.5). How far along the line AD is the point F?

Now, how far along the line BE is F? We can find this by measuring the distance from B to F and comparing it with the distance from B to E: selecting two points puts the distance between them in the measure box.

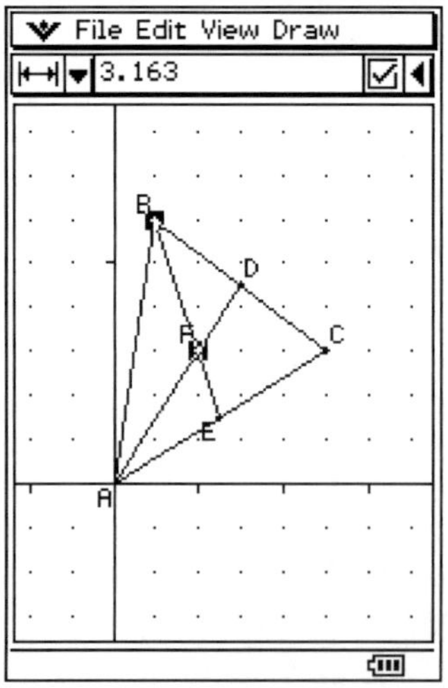

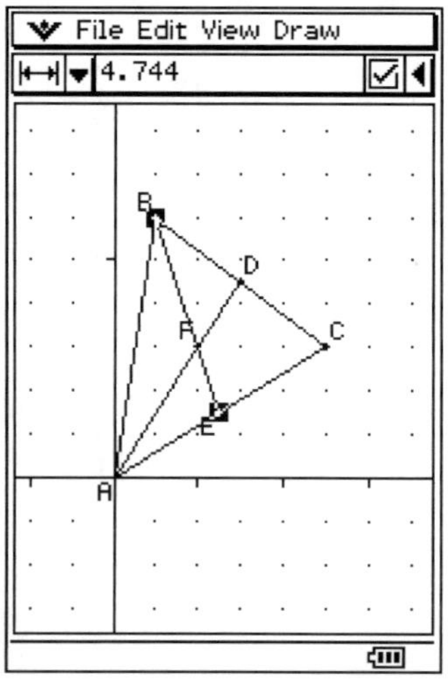

We can calculate this ratio, and display it as a decimal by pressing the fraction / decimal button:

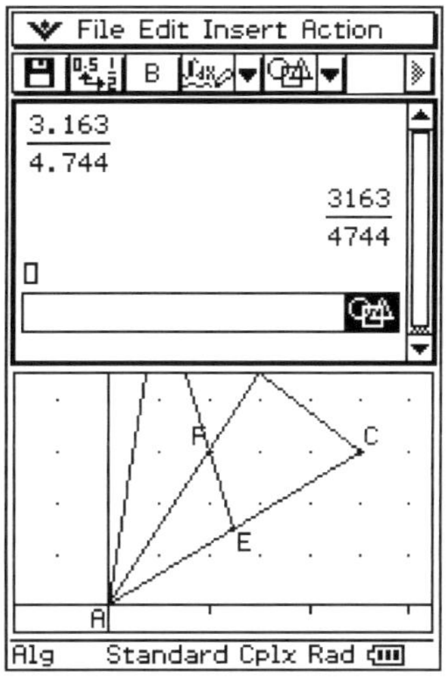

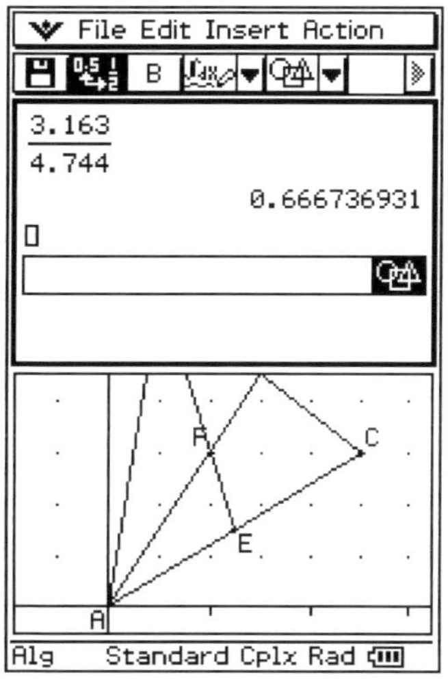

Another interesting way of looking at the centroid is to relate its coordinates to those of the vertices of the triangle.

If we drag a point and drop it into a math edit control (for example a math line in eActivity), the coordinates of the point are dropped in column vector form. For example dragging A into the eActivity window gives the vector $\begin{bmatrix} 0 \\ 0 \end{bmatrix}$.

Drag in A, and B and C and add them:

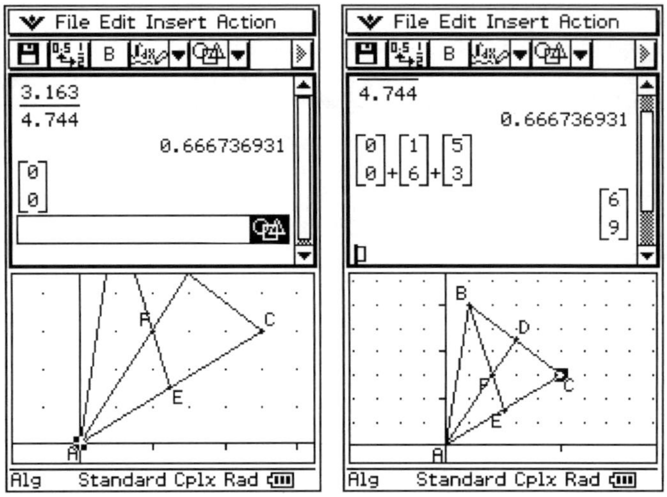

Can you see a relationship between F and the sum of the vectors A, B, C?

Try dragging A to the point (-3,-3). Where does F now lie? Does the relationship above still hold?

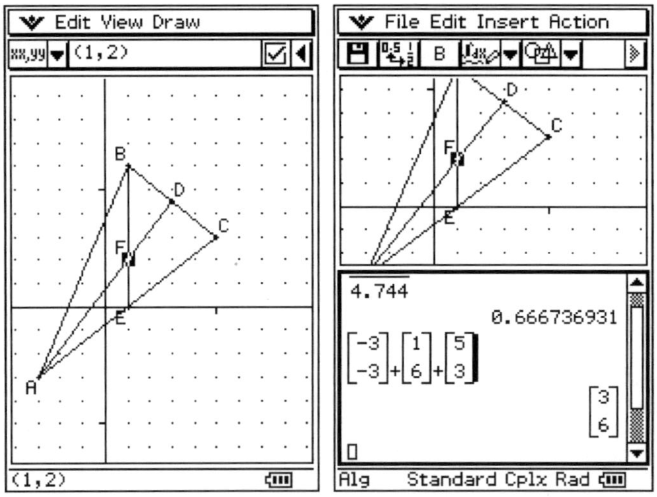

We have drawn 2 medians. However there is a third median, from C to the midpoint of AB. Let's construct it now - the same way as before: first we select AB and construct its midpoint then we draw a line segment from the midpoint to C.

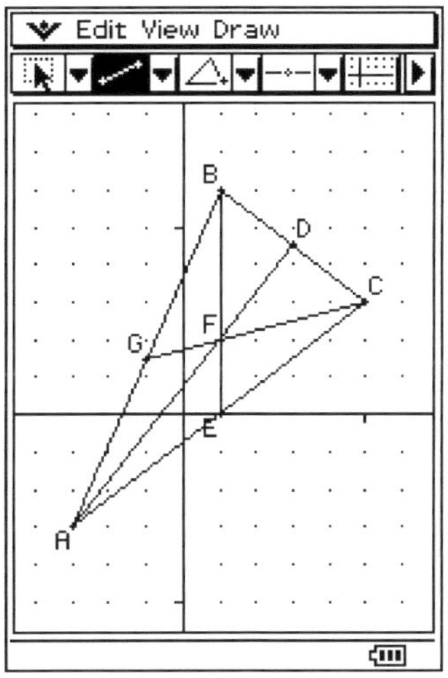

We notice that the new median appears to go right through point F, the intersection of the previous two medians. Is this just an optical illusion or does it actually run through the point? There are a couple of ways to answer this.

One would be to zoom way in and see if the three lines keep looking like they intersect:

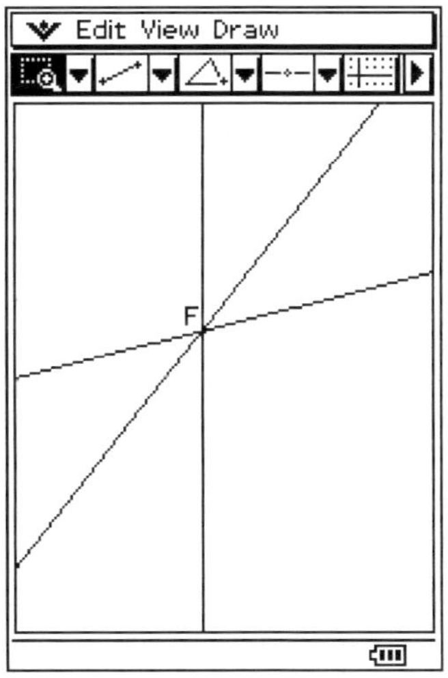

Another is to select the point F and the new median at the same time, and look at the measure box. We see the distance between the point and the line is displayed as zero. We can change the display to display incidence. "Yes" in the measure box says the point really does lie on the line.

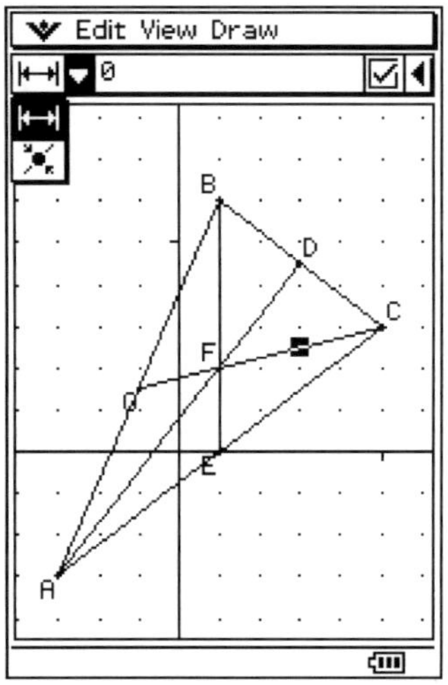

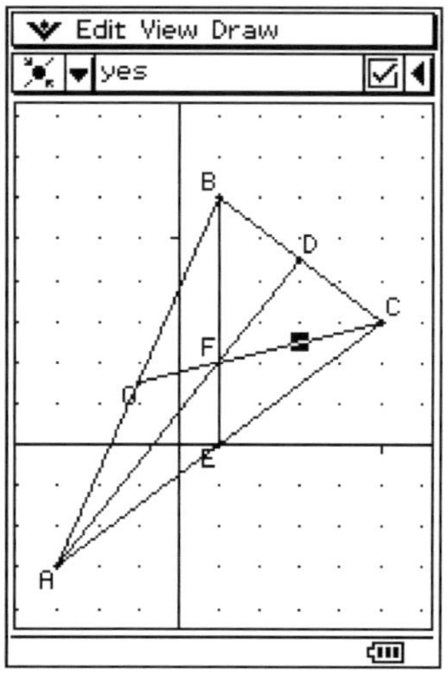

Now try dragging the vertices of the triangle to other locations and observe that the medians keep in touch:

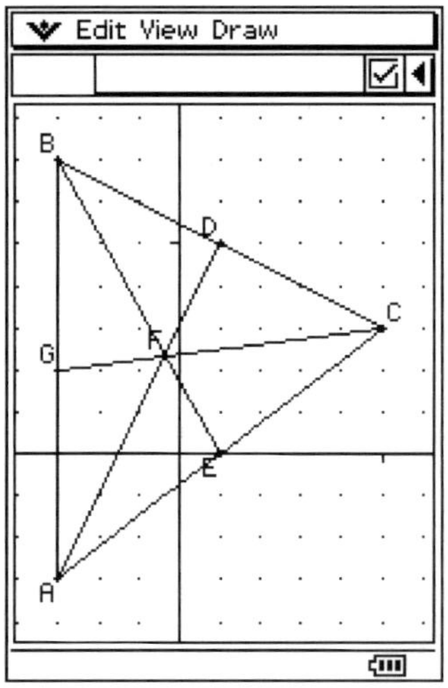

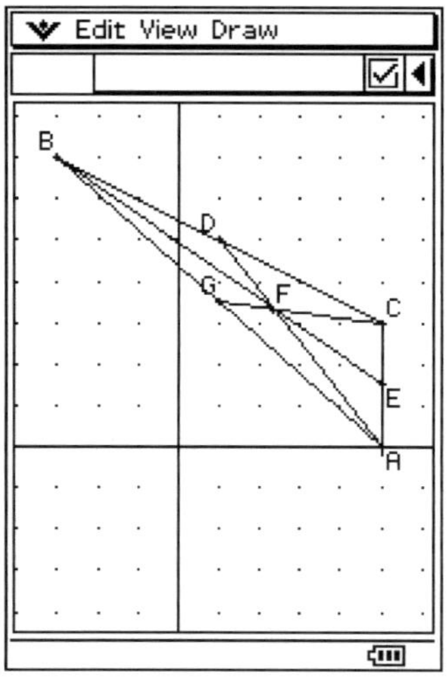

Further Exploration

The centroid happens to be the center of gravity of a triangle (assuming it is made of a uniform substance). Try cutting a triangle out of cardboard and constructing the centroid. Now try balancing the triangle on the point of a pencil. The balance point should lie at the centroid.

Another interesting investigation using the centroid figure is to compare the areas of the 6 small triangles which the original triangle is broken up into.

To display the area of a triangle, we select the three end points and go to the measure box. The measure box has the options of perimeter and area.. (By the way, you can shade a triangle by selecting its end points and clicking **Shade On/Off** in the **Edit** menu.

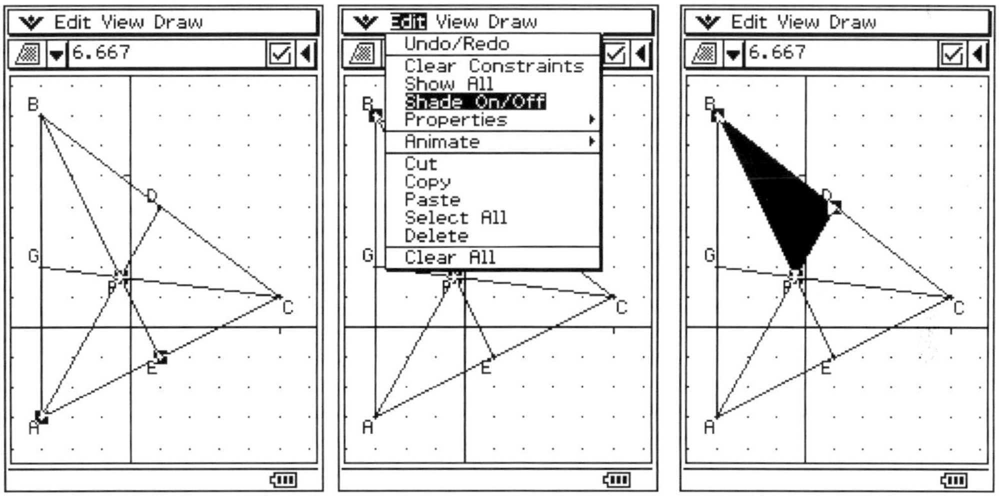

What do you notice about the areas of the small triangles? Compare them with the area of the big triangle.

Other Triangle Centers

The centroid is not the only "center" of a triangle. There are other interesting centers of a triangle. These include:

- The circumcenter (point of intersection of the perpendicular bisectors of the sides).
- The incenter (point of intersection of the angle bisectors)
- The ortho-center (point of intersection of the altitudes).

Circumcenter

To construct a perpendicular bisector, you first construct the midpoint of the side as above. Then select the midpoint along with the side and use the perpendicular tool:

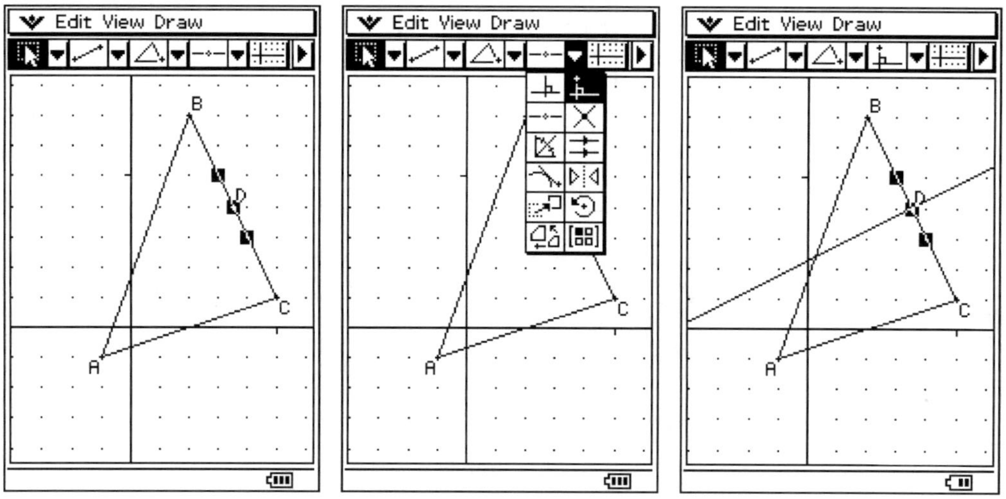

When you have created the intersection of the perpendicular bisectors, draw a circle centered at this through one of the vertices of the triangle.

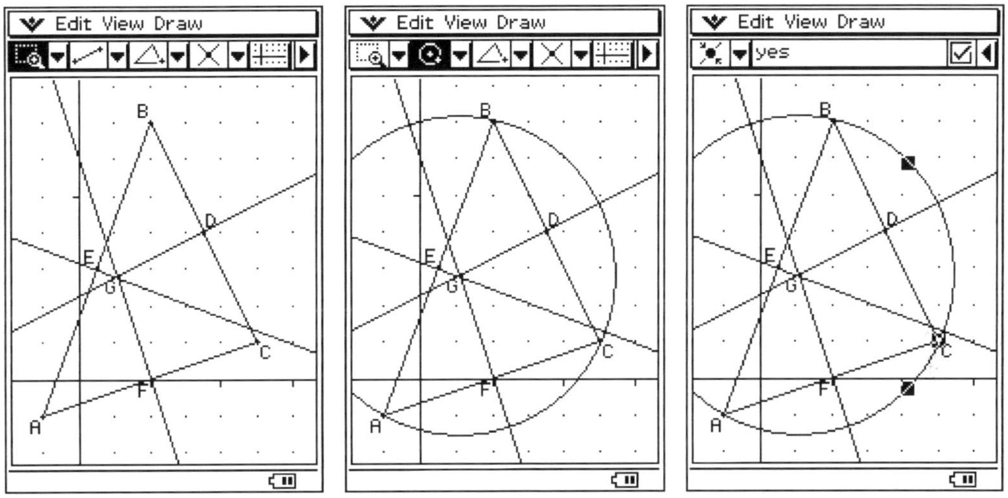

You should see that this circle goes through the other vertices of the triangle. Again you can check it with the incidence measure.

Incenter

To construct an angle bisector, you just select the appropriate pair of edges and click the angle bisector button from the Construction drop down.

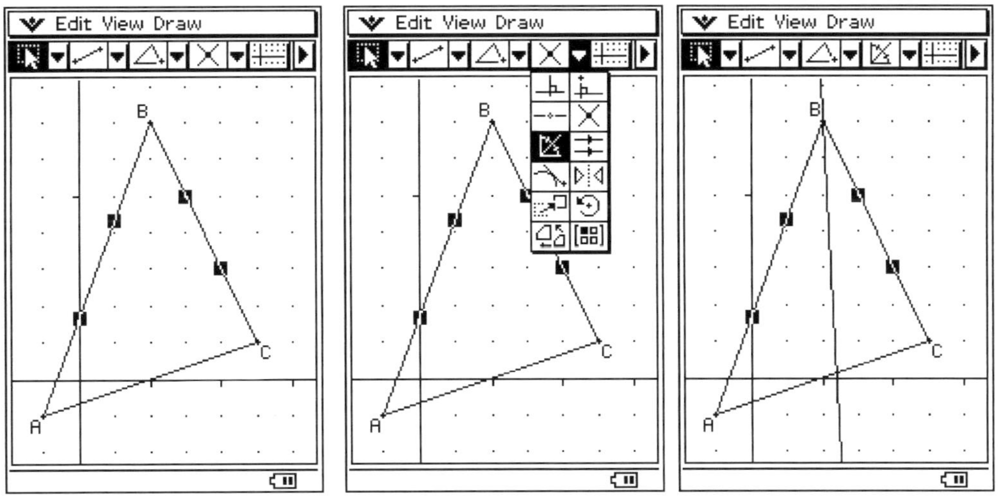

When you have the intersection of the angle bisectors, create a perpendicular to one of the sides through this center.

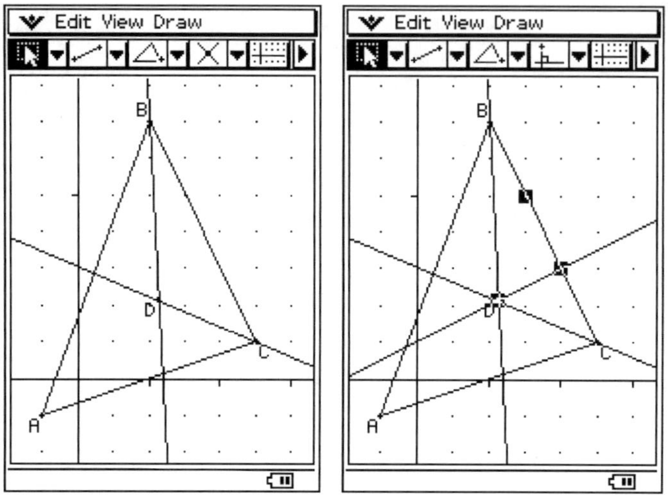

Now construct the intersection point between this perpendicular and the side, and draw the circle centered at D through this point.

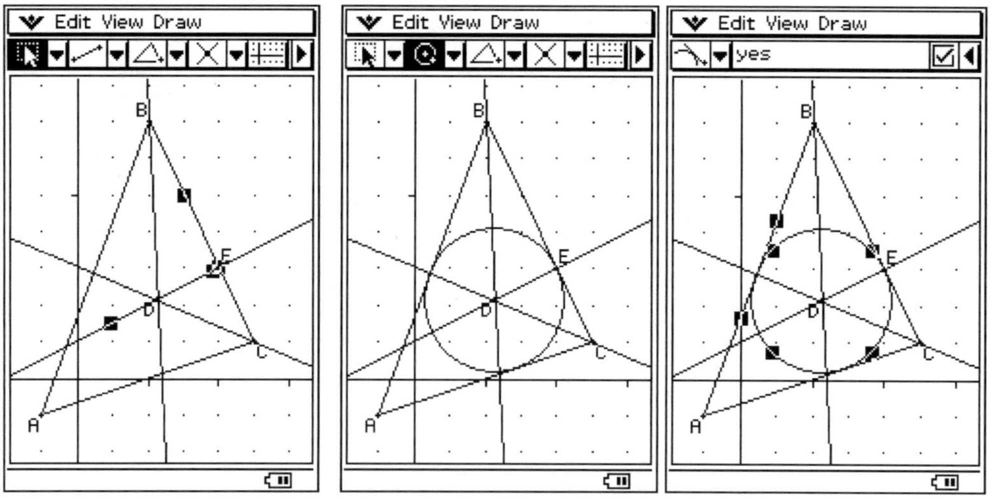

Observe that this circle is tangential to the three sides. You can check this by selecting a side and the circle and looking at the tangent option in the measure box.

OrthoCenter

The altitude of a triangle is the line through a vertex perpendicular to the opposite side. Constructing altitudes is easy: select the vertex and the opposite side and use the perpendicular tool:

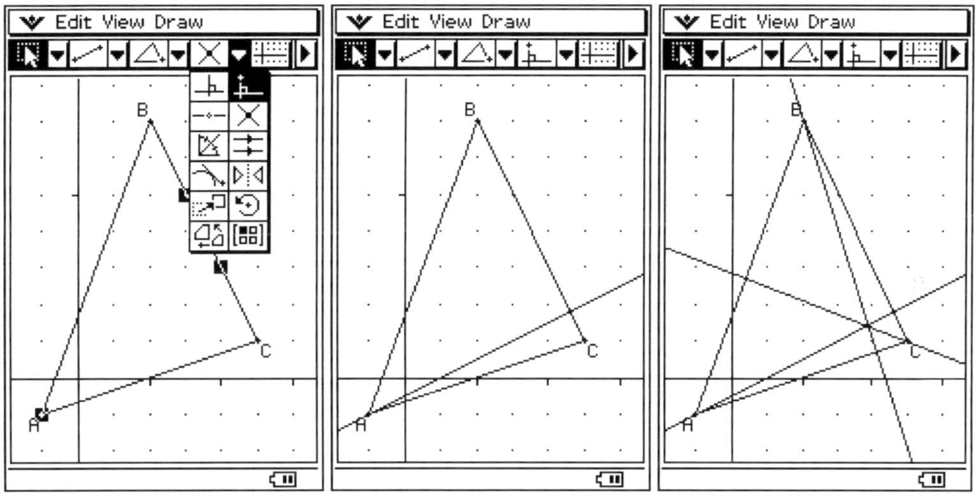

Constructing a Regular Hexagon

One of the easiest compass / straight edge constructions is the one for a regular hexagon. We will see how to do this on ClassPad.

First we create a circle AB. Then we create a second circle, center B passing through A. To center the circle at B you must hold down the pen and drag it over B until B is highlighted. To make it pass through A you must make sure that A is highlighted when you are locating the radius point.

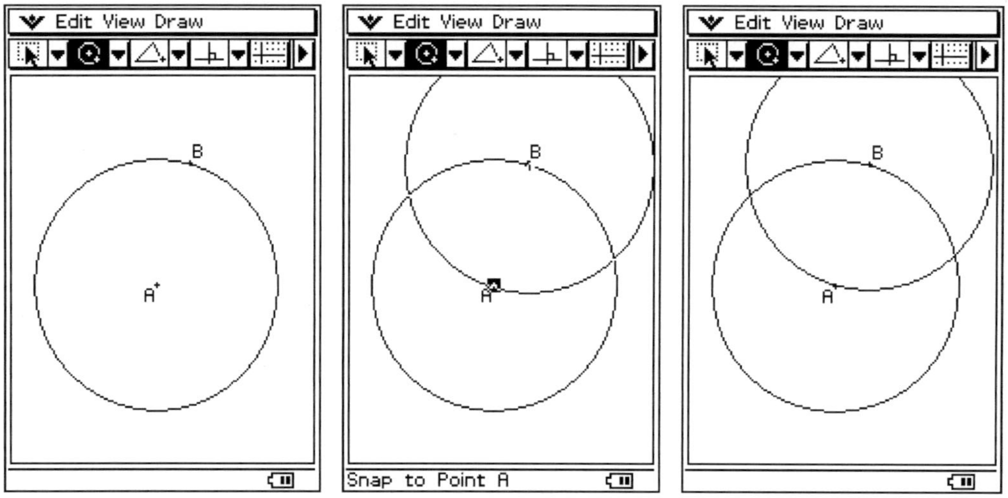

Next we want to construct a circle whose center lies at the intersection of the two circles we have already created. We could create the intersection points by selecting both circles and using the intersection tool. However, a shortcut is simply to draw the circle, but making sure that when you are positioning the center that the two circles are highlighted. Note that the Status line at the bottom of the screen reads **Snap to Intersection** when you have the point located correctly.

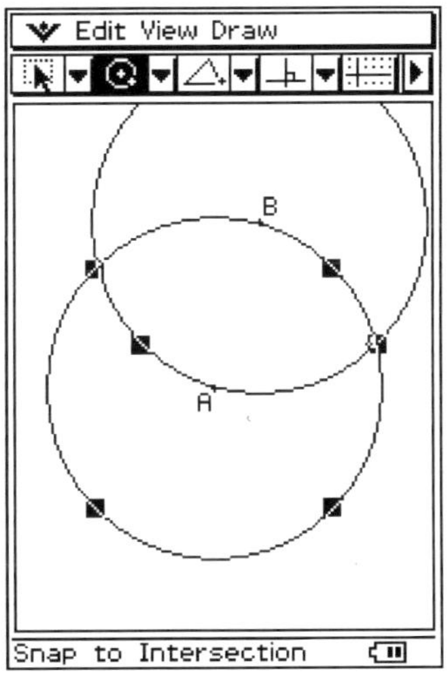

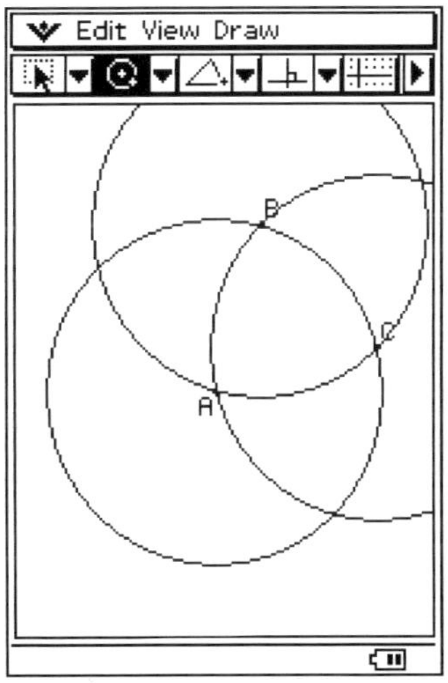

We now keep repeating this procedure creating a circle centered at the intersection between the original circle and the latest circle until we find ourselves back at the start:

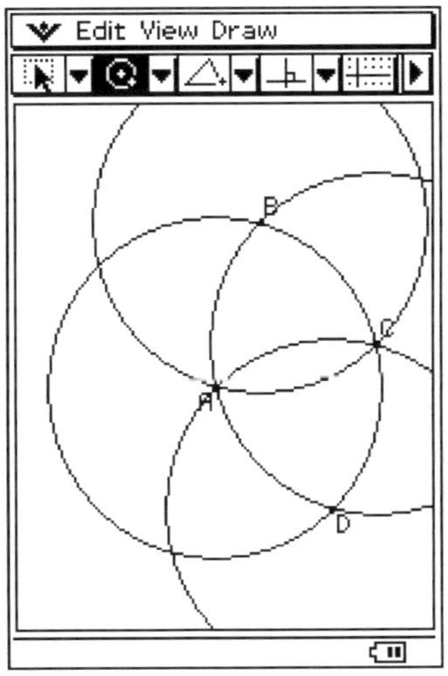

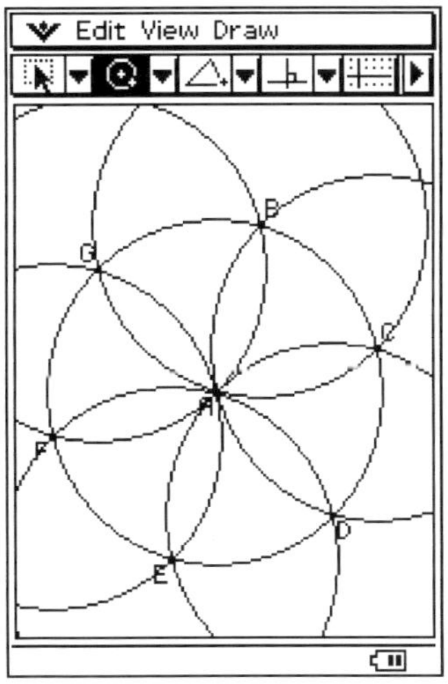

We can now construct the hexagon by joining up the circumference points we have created.

The drawing is a little cluttered. Perhaps we should make the hexagon stand out a little. We can do this by selecting it and choosing **Thicker** from the **Edit/Properties** menu:

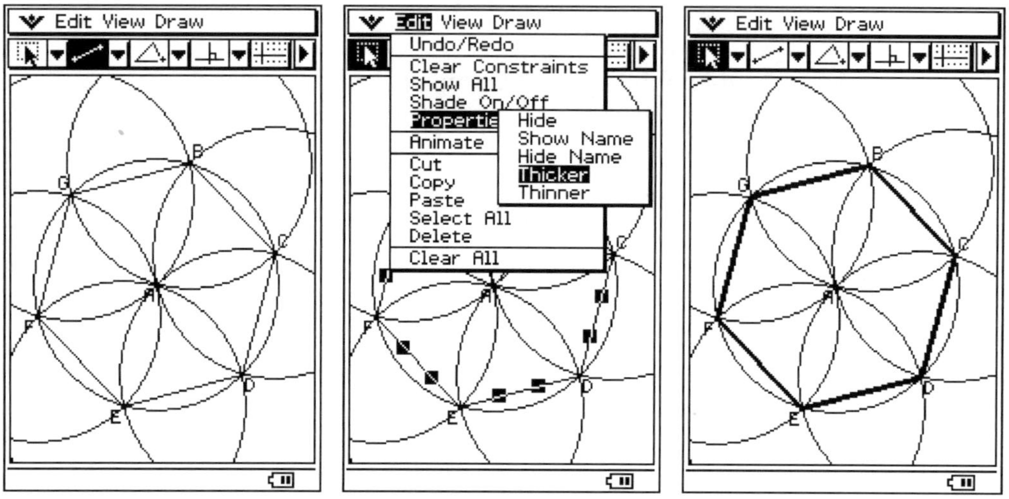

Alternatively, we could shade the hexagon, which would make it stand out more, but obliterate the pretty petal pattern underneath.

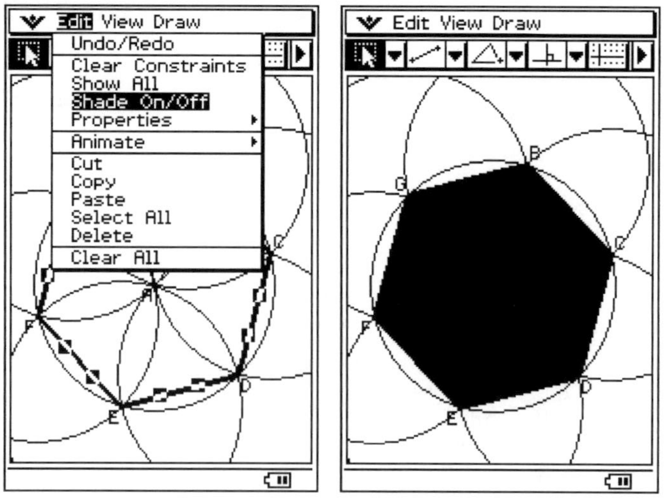

Further Exploration

There are ruler and compass constructions for many figures. Below, we show the construction for an equilateral triangle:

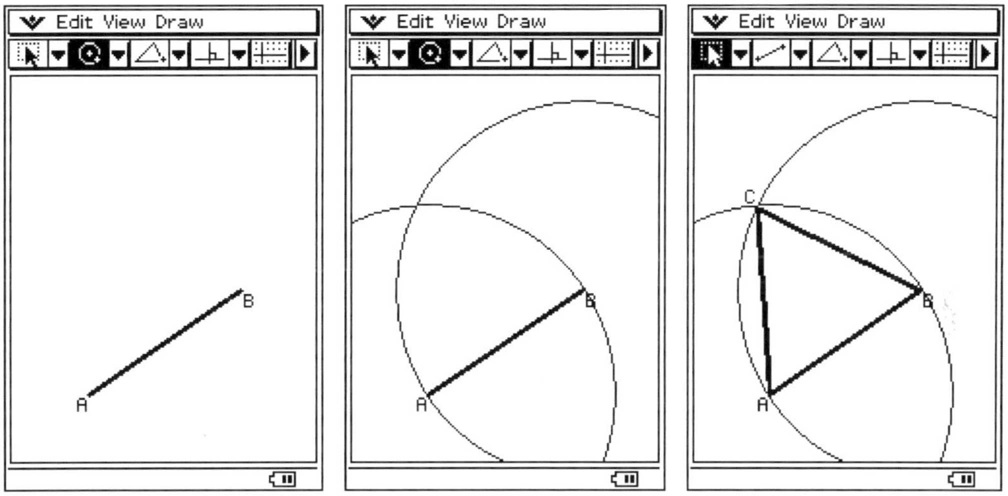

What other figures can you construct with ruler and compass? Look up some old fashioned geometry books to give you a start.. This is how the Greeks did geometry 2500 years ago, and it still hasn't quite gone out of fashion in some circles.

Another Angle on Circles

In this section we investigate angles subtended by chords on a circle. Grandiose word that - "subtend" - but the picture is simple: a chord is simply a line segment both of whose end points lie on the circumference of a circle, an angle subtended by the chord is just the angle of a triangle whose vertices are the two ends of the chord and a third point on the circumference.

Which angle, you ask - a triangle has three. Of course it's the one opposite the chord. Here is the picture: the angle BDC is an angle subtended by the chord BC. To measure the angle, we just need to select the two line segments BD and DC and bring up the measurement box (by clicking on the tab arrow at the extreme right of the toolbar).

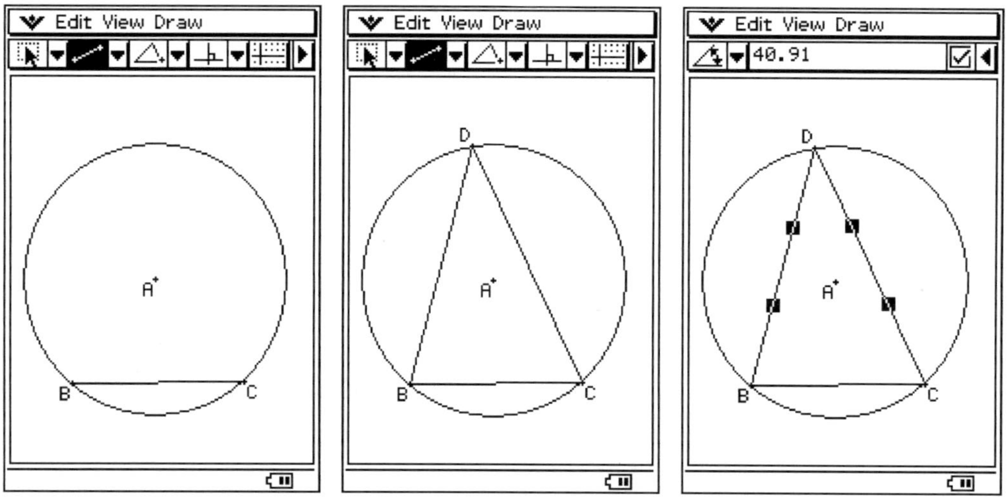

Try drawing another angle on the same chord, and look at its value:

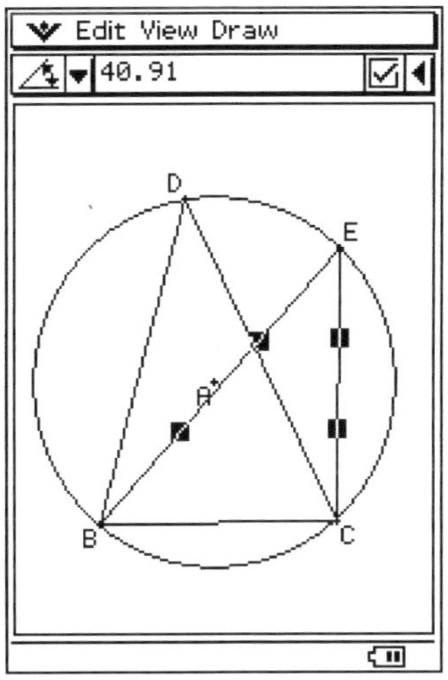

You'll see that it is identical. Now, we would like to drag the point D around the circle and check that the angle stays the same all the way round. However, if we try this, we find that the circle changes radius to conform to the location we have dragged D:

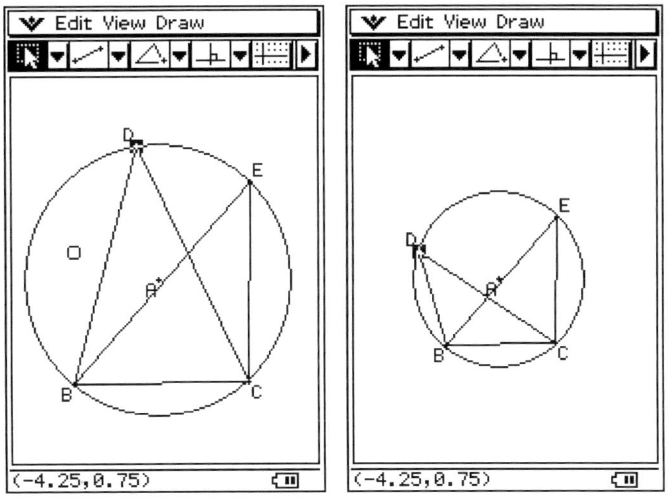

Not only that, the chord BC has changed!

This is a characteristic of the ClassPad geometry: if you drag a point, the ClassPad will do its best to make the geometry conform to the location of the point. We need to tell

ClassPad that the circle is to stay fixed. First, we should Undo the drag we have just done to get back where we started:

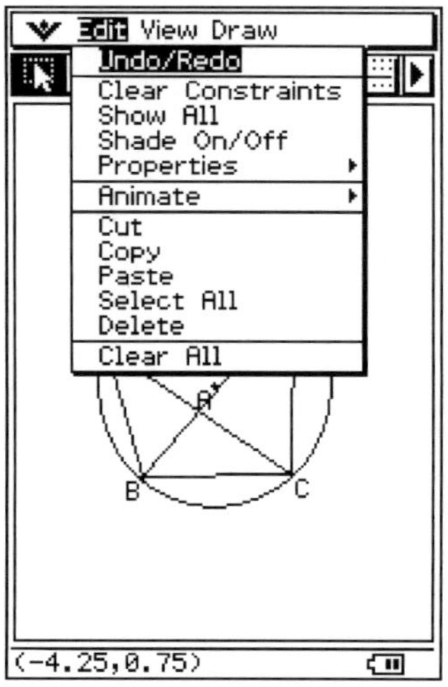

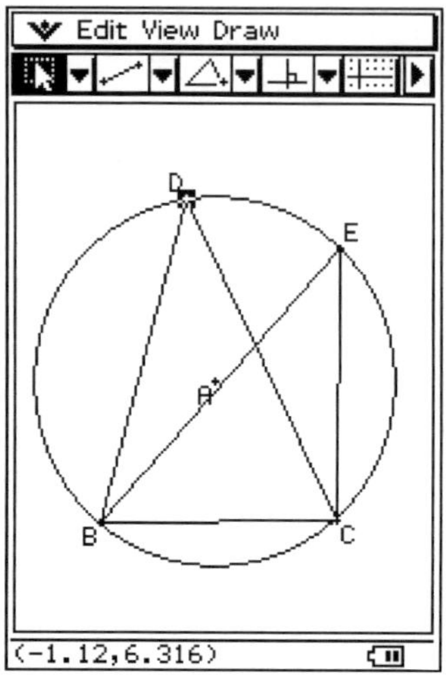

Now we should fix the circle by first selecting its center A and bringing up the measure box. You should see a display of the coordinates of the point. You can fix the location by clicking the check box to the right of the measure box.

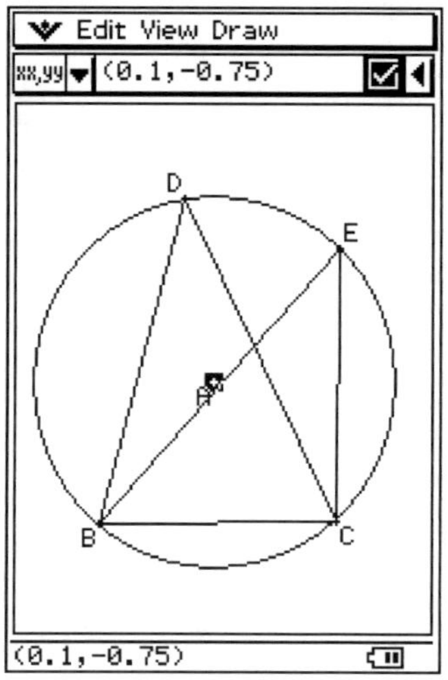

After fixing the location of the center, you should fix the radius of the circle. You can do this by selecting the circle (make sure you first unselect the center) and clicking the check box with the radius displayed.

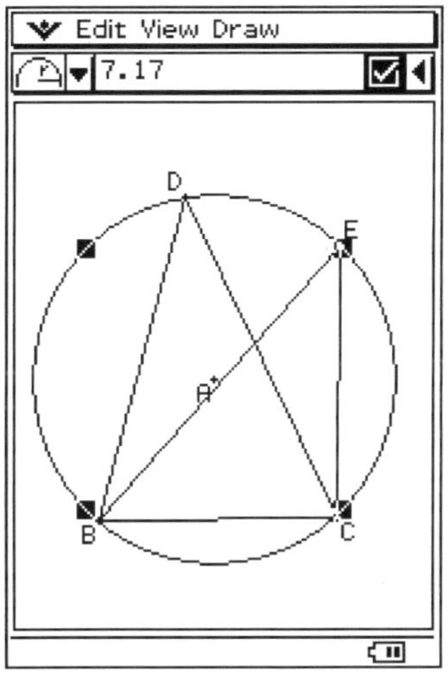

This time, we see that when we drag it point D snaps back to lie on the circumference of the circle.

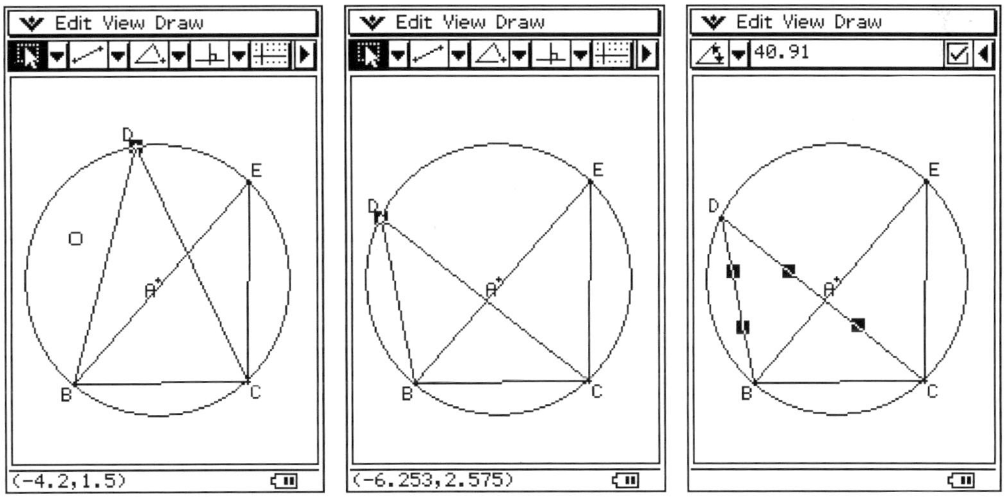

Angles in the middle

So we have seen that the angle subtended at the circumference by a chord is the same no matter where on the circumference the point is placed. The question is, why?

And the answer involves looking at another angle: the angle subtended by the chord at the center:

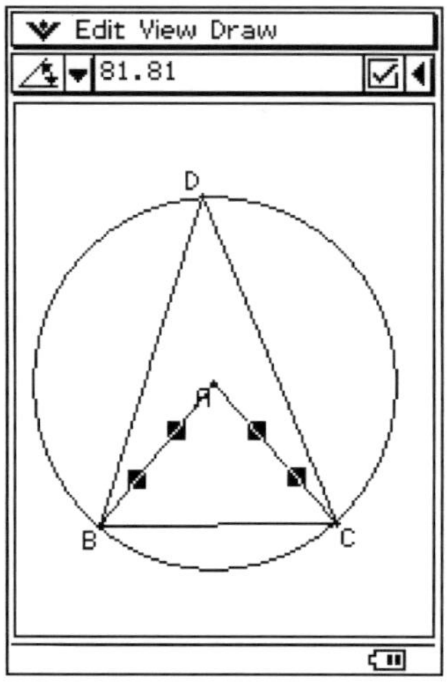

What is the relationship between this angle and the angle subtended at the circumference?

Try dragging the location of the chord BC to verify that this relationship persists:

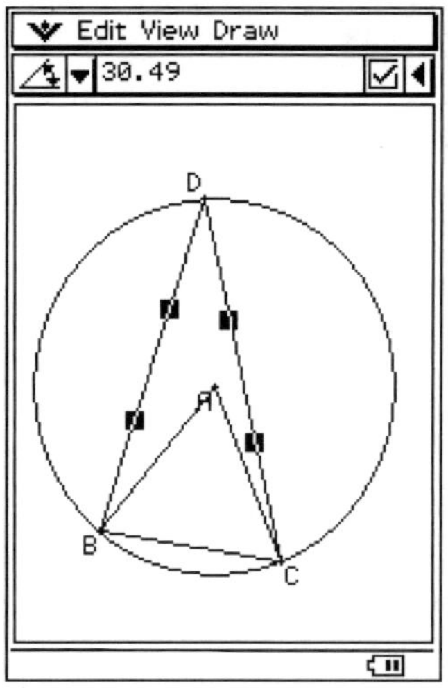

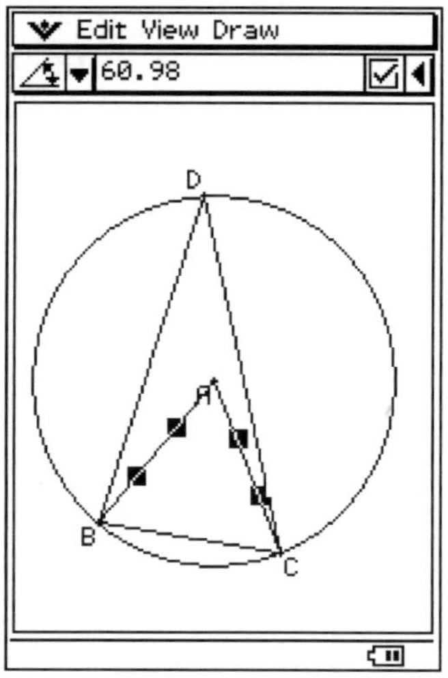

If we can prove to ourselves that the angle at the circumference is always half the angle at the center, then we will have proved that every angle subtended by this chord at the circumference is the same.

The proof is easy using ClassPad to handle some of the algebra. First let's compare the angles ABC and ACB on our drawing:

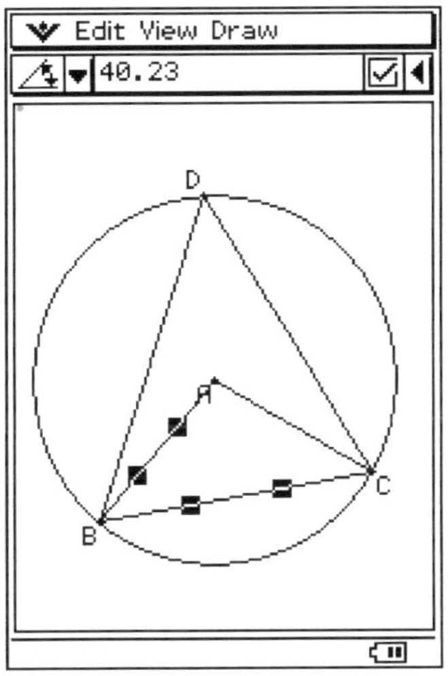

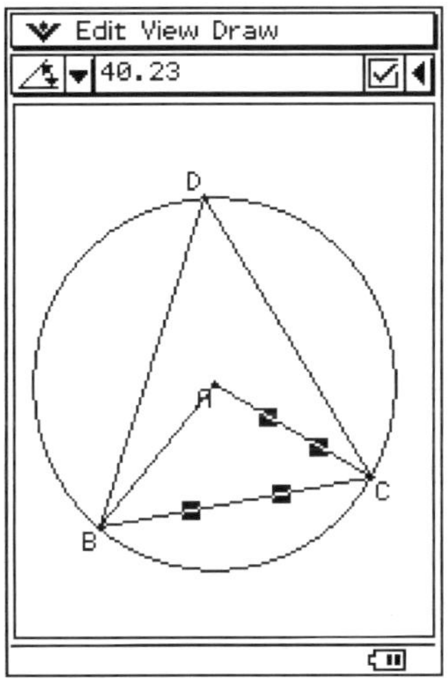

We see that they are the same. Why? (Think about the lengths of AC and AB).

Now let's add one more line

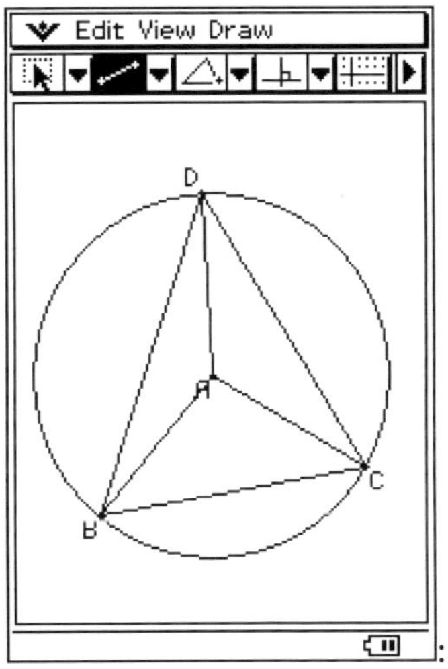

Can you find two more pairs of identical angles in this drawing?

Lets give names to these angles: let's call ABC (and ACB) **r**, let's call DBA (and BDA) **s**. and let's call ADC (and ACD) **t**.

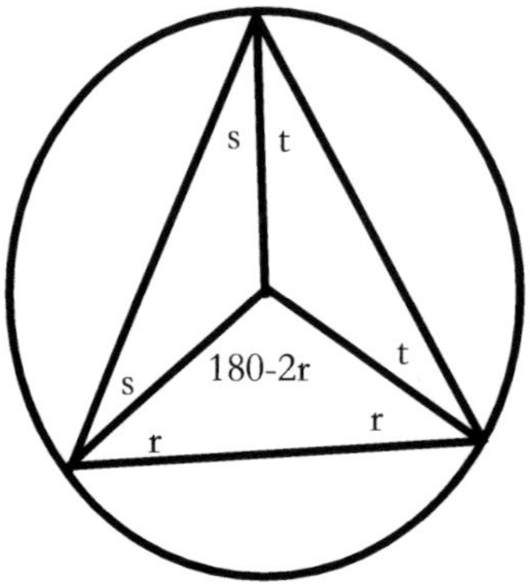

Let's call the angle at the center **u**

We know that the angle at the center is 180-2r (the angles of a triangle add up to 180). Typing this into a math entry box in the eActivity gives a correct, but unhelpful result:

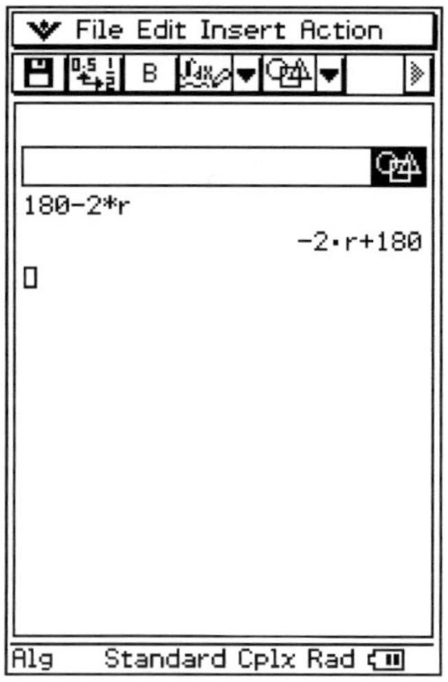

Now we also know that the angles of the large triangle add up to 180. We can get ClassPad to solve for r:

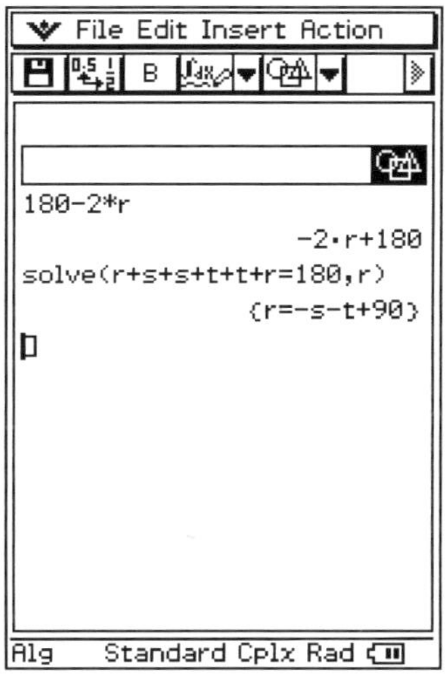

We can now put the value for r into the above equation. This can be done by selecting the right side of the solution and copying then pasting into the first equation:

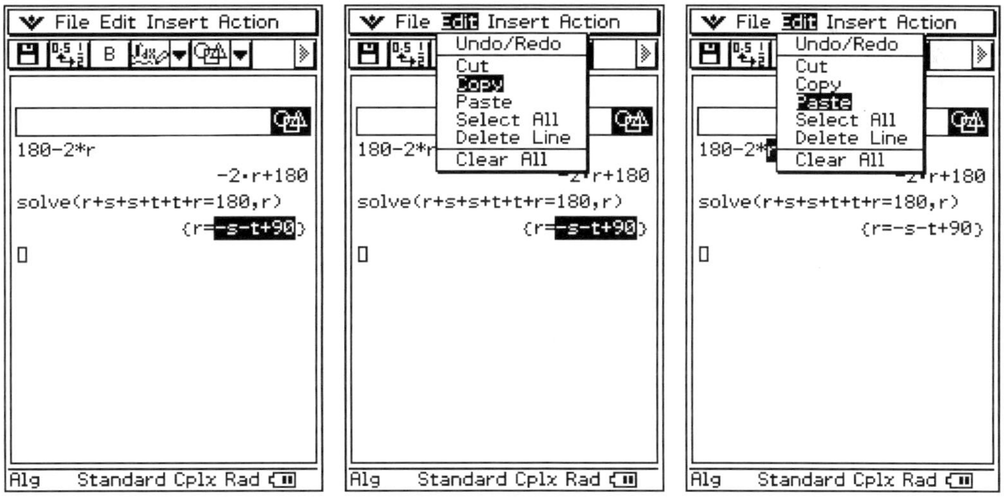

(remember to put in parentheses so that the expression will read correctly)

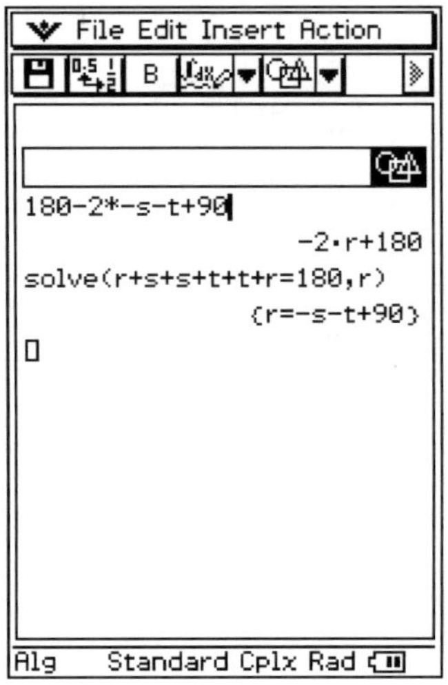

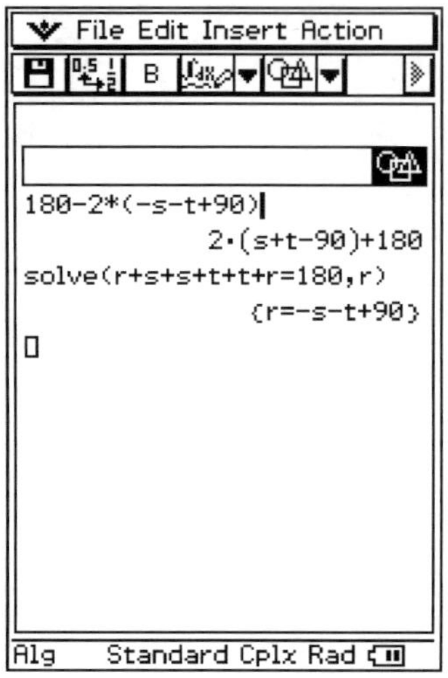

Now this expression is still not all that helpful. But let's simplify it: put the cursor at the beginning of the first input line and then select Simplify from the **Action/Transformations** menu.

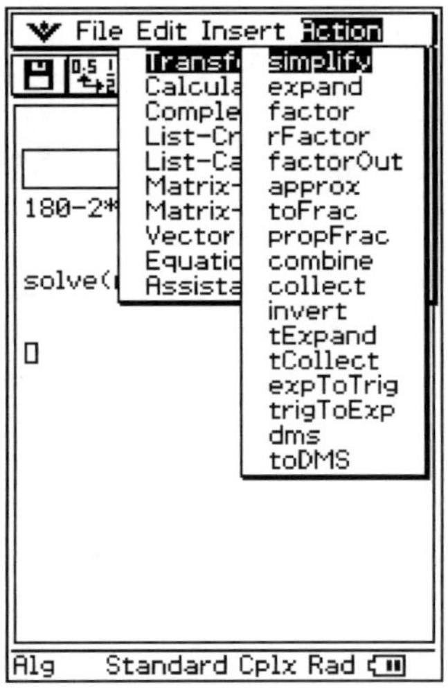

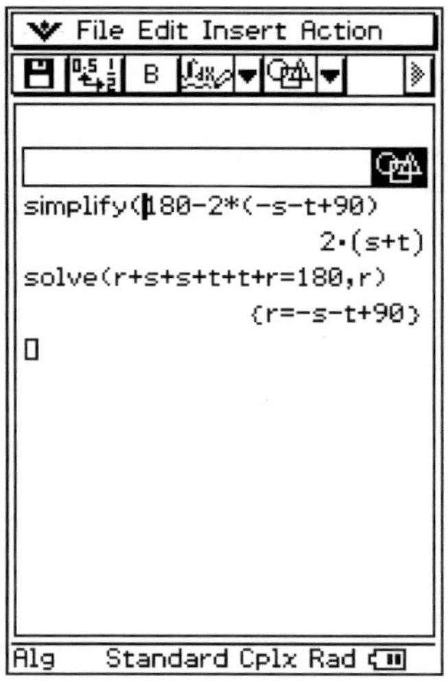

Press Enter and we see that the angle at the center is 2(s+t). But s+t is the angle at the circumference. So the angle at the center is shown to be twice the angle at the circumference. Not just once or twice, but once and for all!

Further Exploration

ANGLES IN A SEMI-CIRCLE

What if AB is not just a chord, but in fact a diameter?

First, how do we draw a diameter of a circle?

One way is to start by drawing a chord, then specifying that the center of the circle lies on the chord. To do this, select the chord and the center and bring up the incidence measurement.. This will likely say **no.**

To specify incidence, click on the check box. If ClassPad is able to make the circle center lie on the chord, **no** will change to **yes** and the drawing will update accordingly.

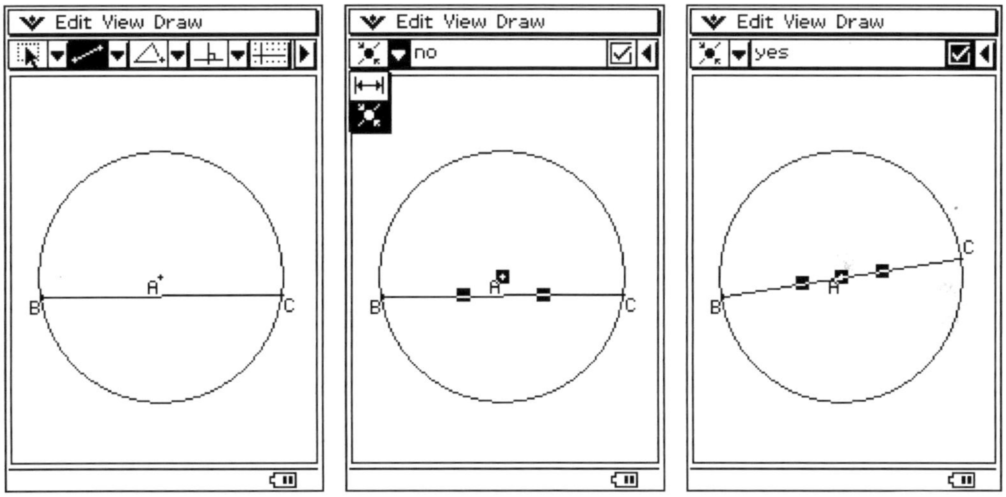

If the chord goes through the center of the circle, what angle does it subtend at the center? 180 degrees, right?

So the angle subtended at the circumference should be …

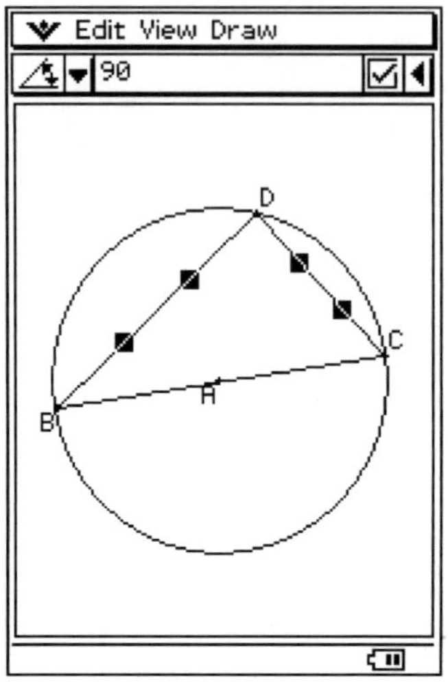

Intersection of two circles

Finding the intersection of two circles is easy enough in ClassPad's geometry module: simply select the circles and click on the intersection tool. However, we are going to see how we can approach the problem algebraically using eActivity's math capabilities and (hopefully) gain some insight into how the algebraic representation relates to the purely geometric:

Let's be specific, and give ourselves the following problem:

> Find the points of intersection between the circle center (4,-3) radius 5 and the circle center (-6,2) radius 10

Turning on the integer grid will help you position the circles exactly. Otherwise, you can sketch them and use the measurement box to locate them.

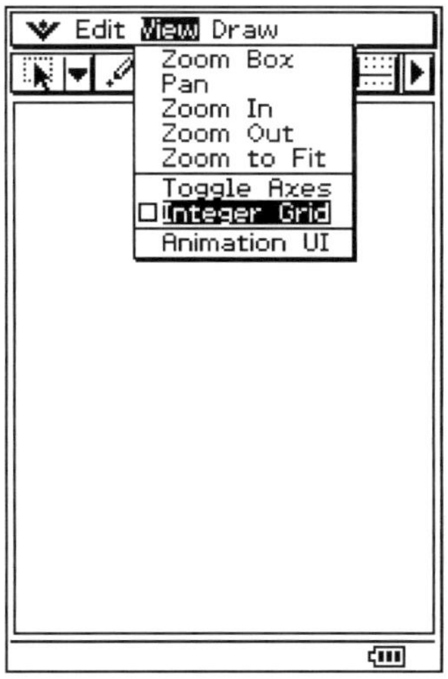

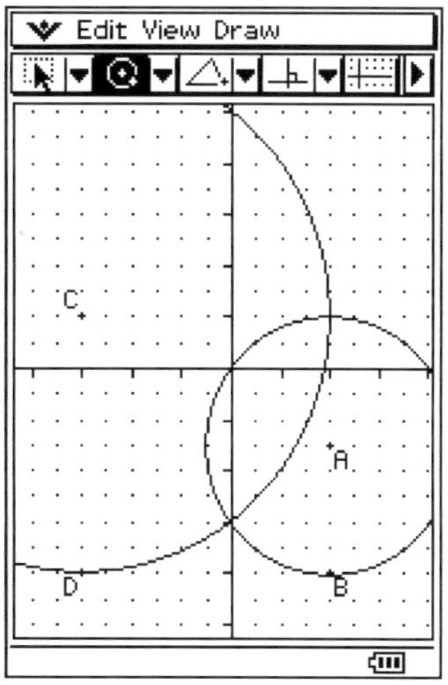

To solve the problem algebraically, we need the equations of the circles. We can get them by dragging into empty math entry regions in the eActivity application:

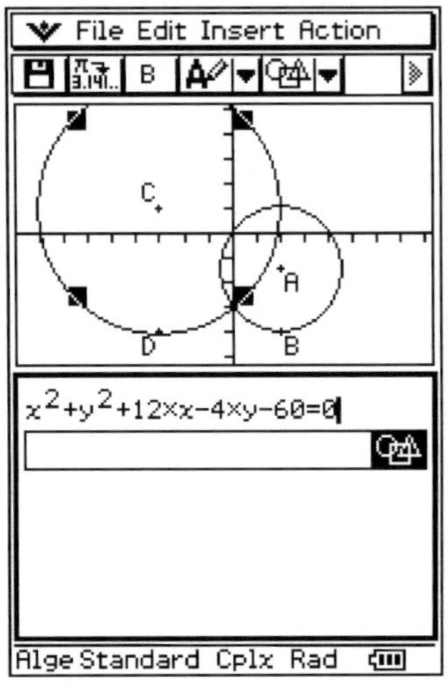

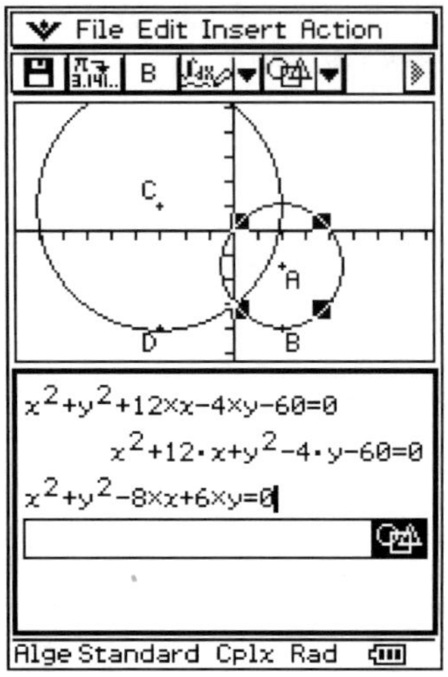

Now we need to solve both of these quadratic equations simultaneously. However, we notice if we subtract the two equations, the squared term in both x and y should disappear. Let's do this: we can drag and drop or copy and paste the first equation into a new math entry box. We then subtract the left hand side of the second expression (being careful to put it in parentheses).

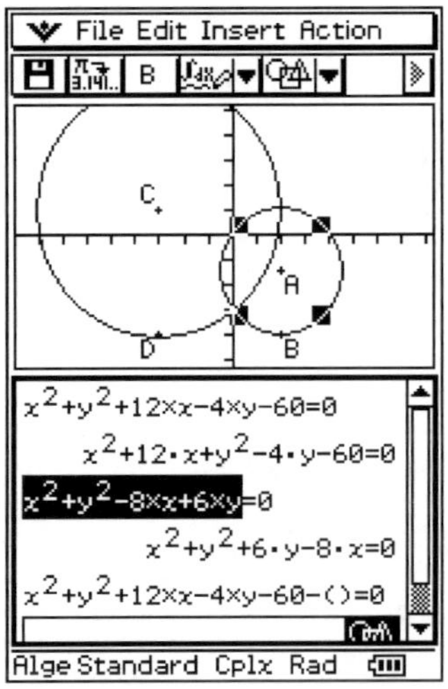

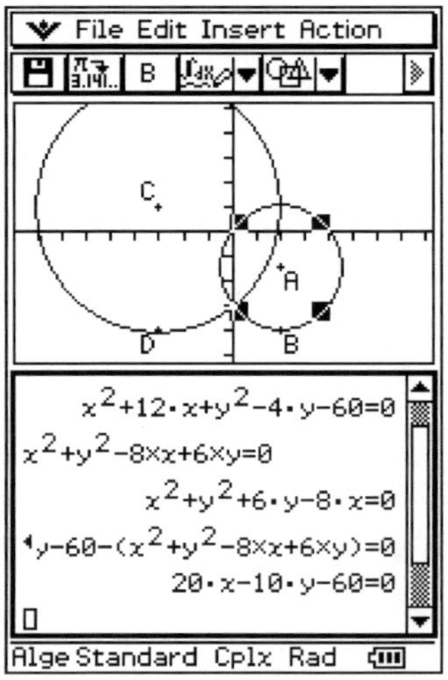

The solution is a linear expression. Solving for y puts it in the form y=mx+c.

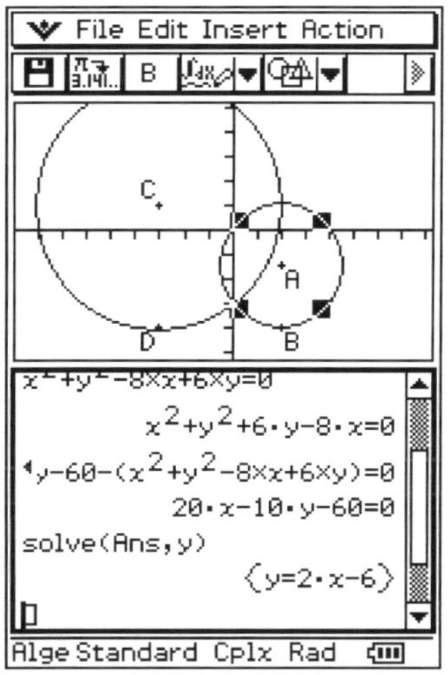

This is the equation of a line, so why not look at it in the geometry window. We can do this simply by dragging and dropping the linear equation (not the curly brackets, though).

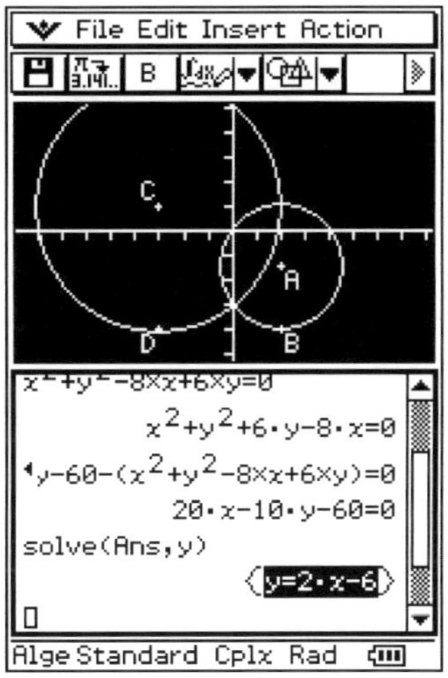

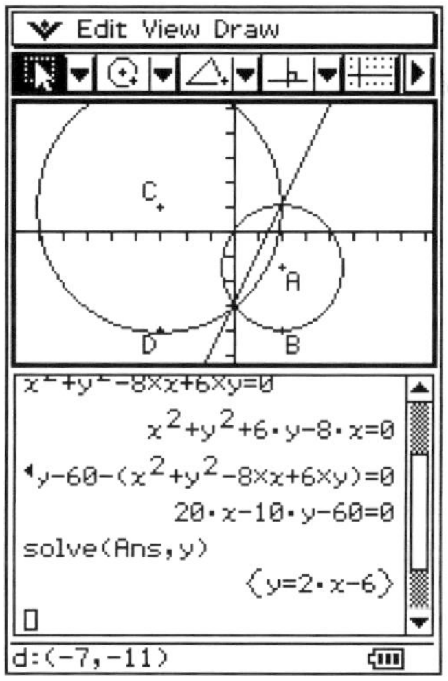

We see that this is the line through the two intersection points. So to find the intersection, we need simply to simultaneously solve one of the original circle equations and the equation of the line. This can be done by substituting the value for y into the equation for the circle.

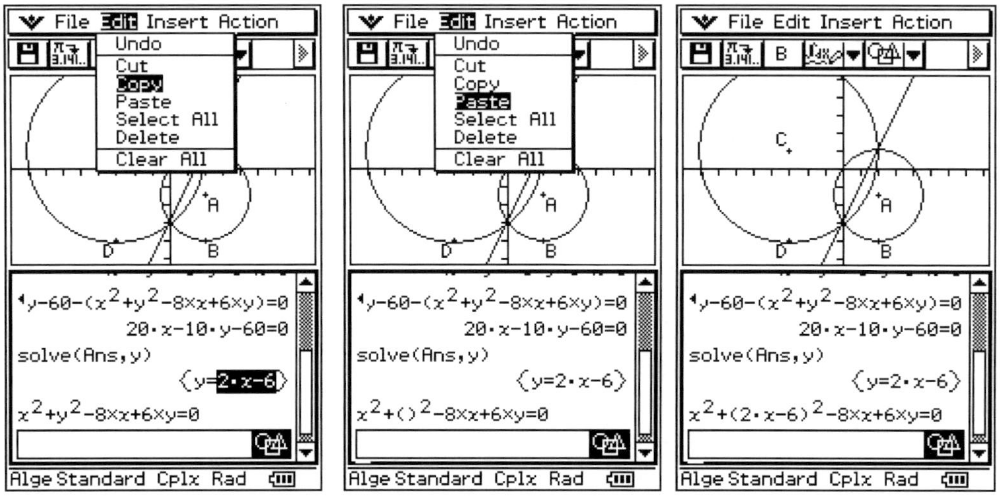

You need to paste twice: once into the y squared term and once into the y term. Remember your brackets each time! Now the expression is too long to fit nicely on the page. But it can be simplified by expanding out the brackets:

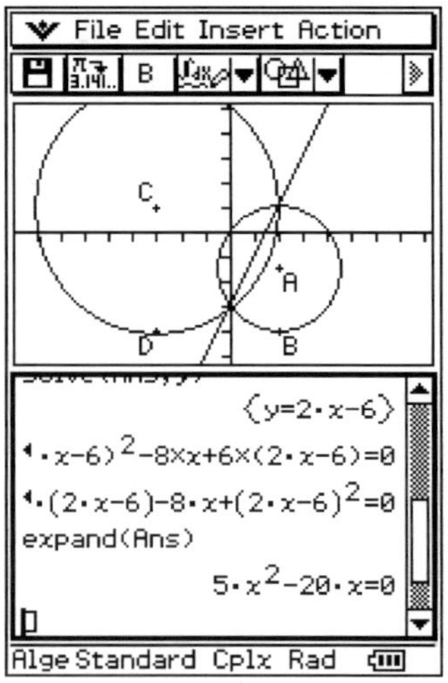

Solving for x will give us the x coordinates of our two intersection points. These can be substituted into the line equation to give the corresponding y coordinates

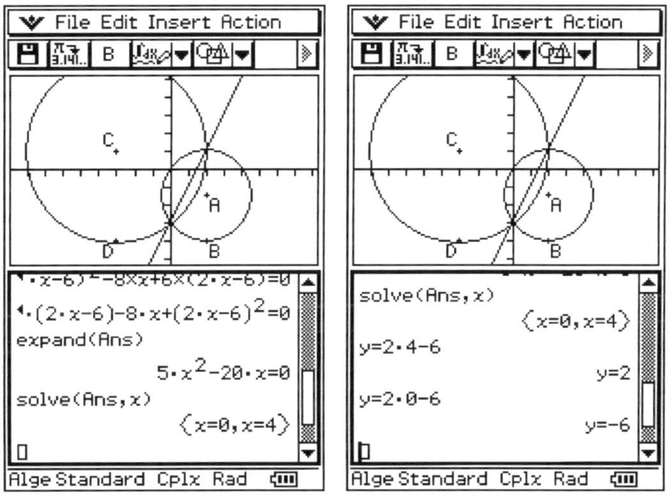

We can check our work geometrically: simply select the two circles and invoke the intersection construction. Intersection points E and F are created. We can drag these into the eActivity and verify that the values are the same as those found algebraically.

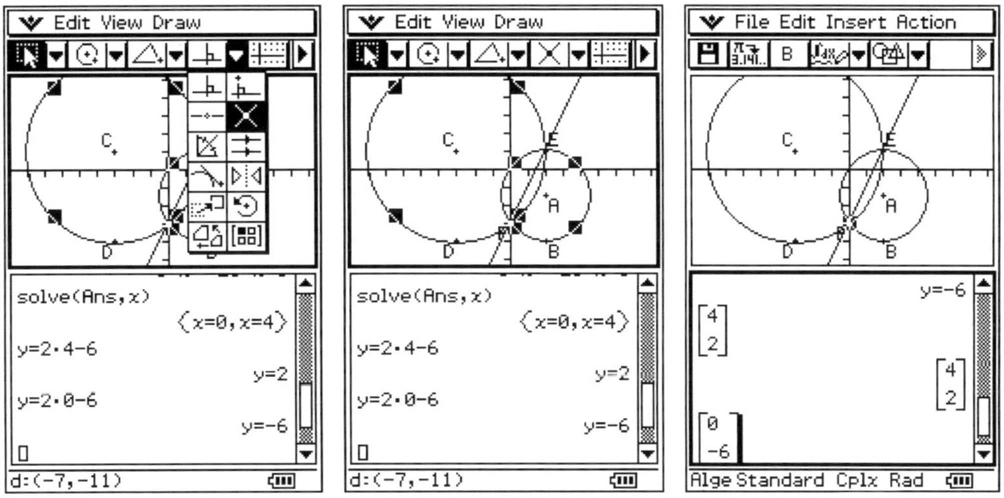

Archimedes Frog

Archimedes thought about a frog which kept jumping half the distance between its position and the edge of a pond. He found it paradoxical that the frog could keep jumping forever, but never make it out of the pond. Mathematically, he was considering an infinite series, whose sum is finite. Let's look first at the geometry..

We'll make our frog start at point A and make point B be the edge of the pond. We'll make the distance between A and B 1. Now we can use the Midpoint construction to create the successively smaller jumps: first the midpoint of A and B gives C, then the midpoint of C and B gives D and so on..

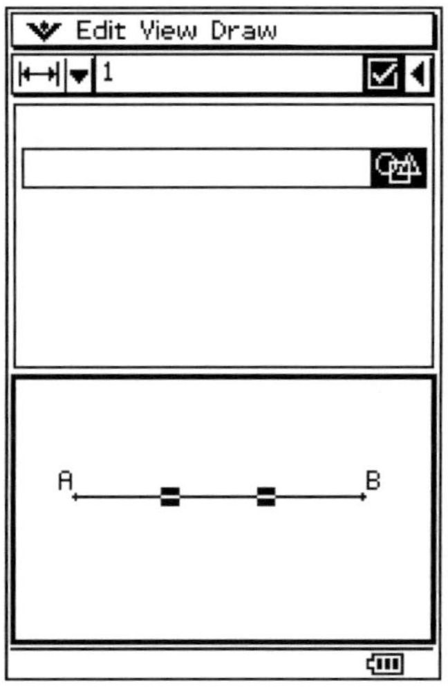

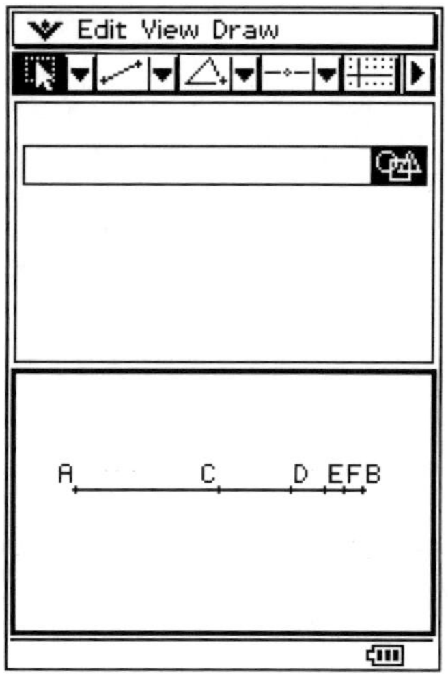

We can see how far the frog has traveled at the end of each jump, simply by selecting point A and the point at the end of the jump and inspecting the measurement box:

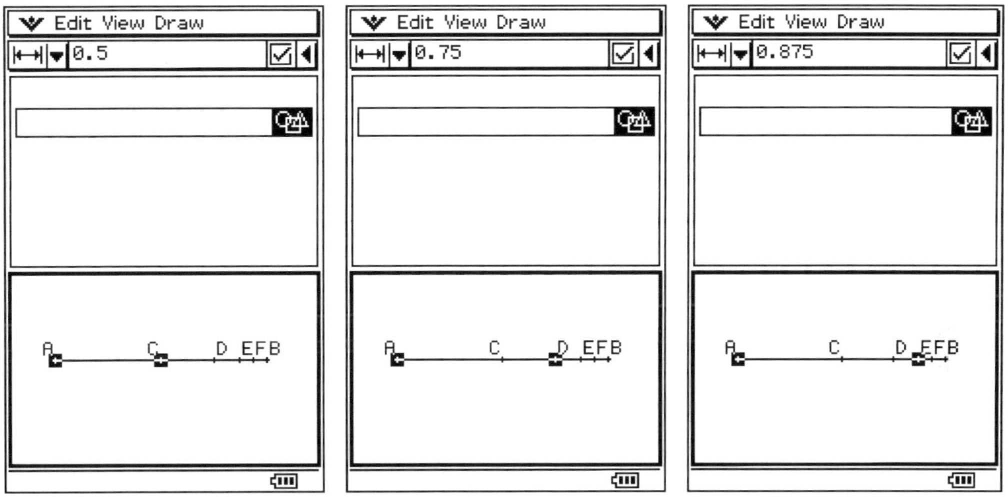

Of course, we can see from the drawing that at each point, the frog can reach the edge of the pond if it makes one more jump of the same size as the previous jump. So the distance traveled is:

1-last jump size

Hence the distances above are: 1-1/2, 1-1/4, 1-1/8 etc.

The distance traveled may also be written out mathematically as a series:

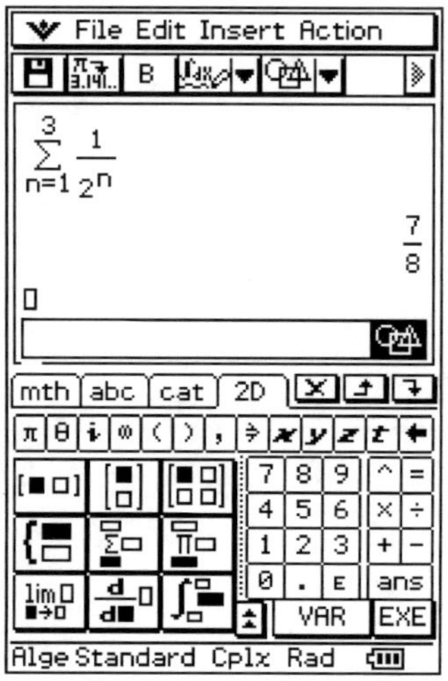

If the frog does 50 jumps, how far will he have gone? Easy just change 3 to 50 in the expression above:

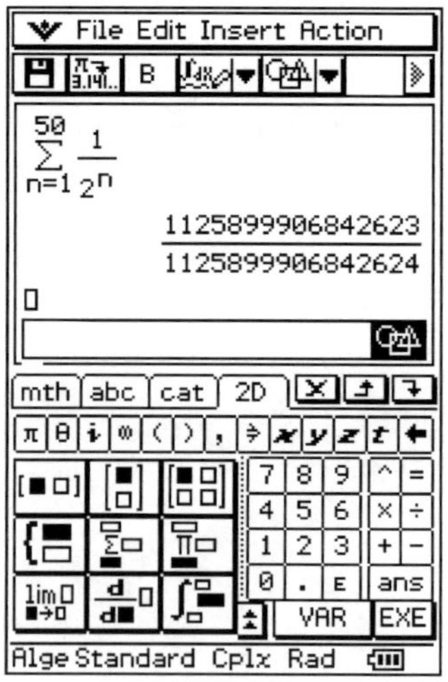

What if he does an infinite number of jumps?

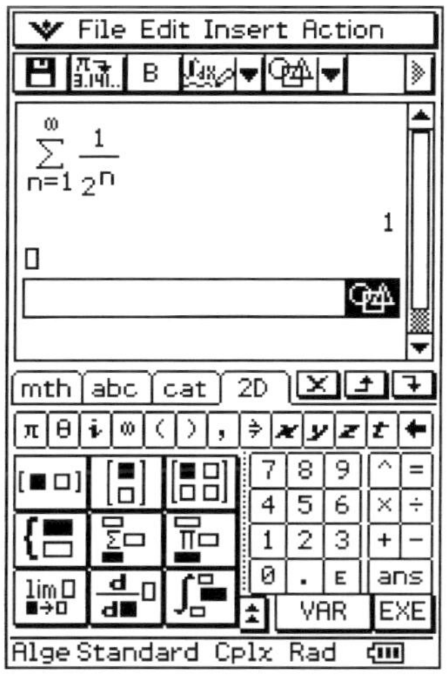

Finally, he reaches the edge of the pond - but not until after an infinite number of jumps!

A Fractal Area

We start with a right angled isosceles triangle of side length 16. The area is, of course, 16*16/2 = 128. We create a smaller triangle, by joining the medians of the larger triangle. As 4 similar small triangles make up the larger triangle, the area of the smaller triangle is ¼ of the larger triangle, which is 32.

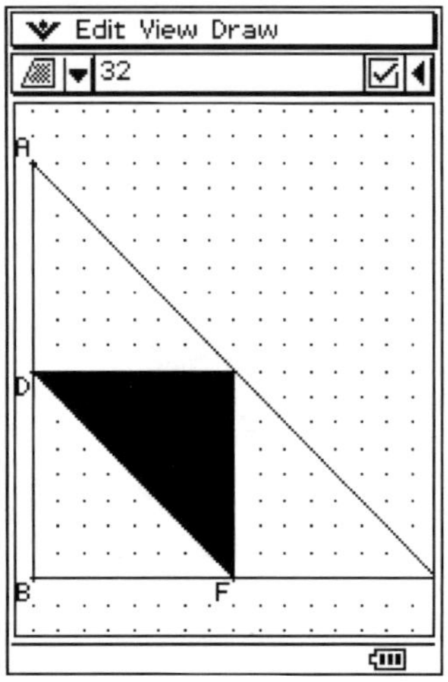

We now take each of the 3 unshaded triangles and shade new triangles defined by joining their medians. (You can hide the point names using **Edit/Select All** then **Edit/Properties/Hide Name**.) Notice that if nothing is selected, the measure box displays the total area of all shaded polygons in the drawing.

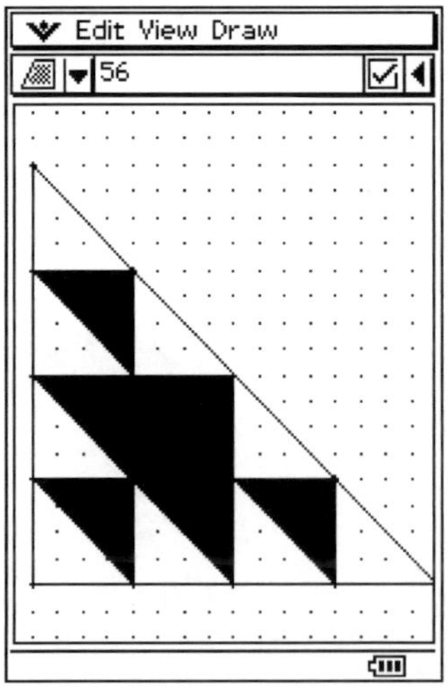

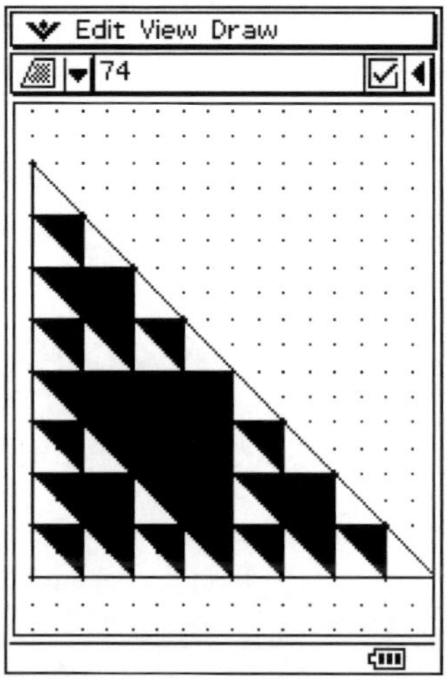

Can we derive a formula for the area?

The first diagram had area 32.

The second diagram had the original 32 plus ¼ of each of 3 unshaded triangles (each of area 32).

32 + 32*3/4 = 56

The third diagram now added 9 new triangles, each of whose areas was 1/16 of the first shaded triangle:

32 + 32*3/4 + 32*9/16 = 74

We could write this as follows:

$$32\cdot(3/4)^0 + 32\cdot(3/4)^1 + 32\cdot(3/4)^2$$

This, we can enter into eActivity:

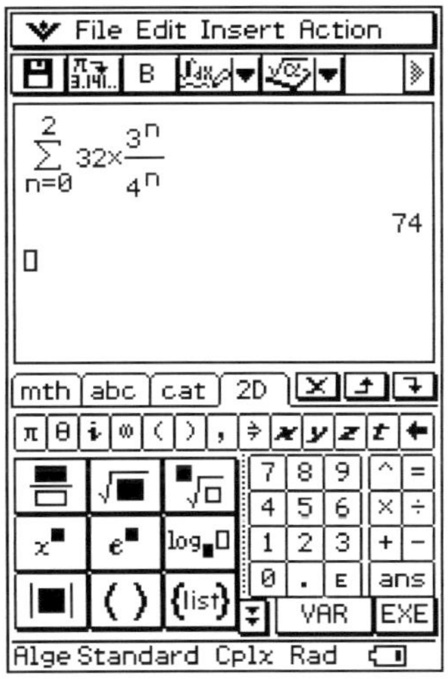

What would the area be at the next step?

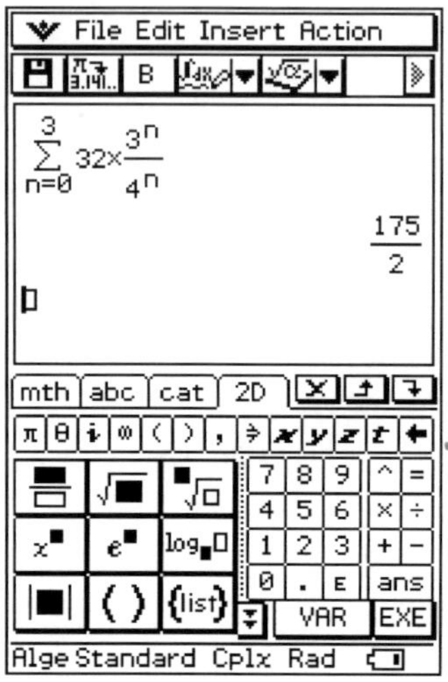

How about after an infinite number of steps?

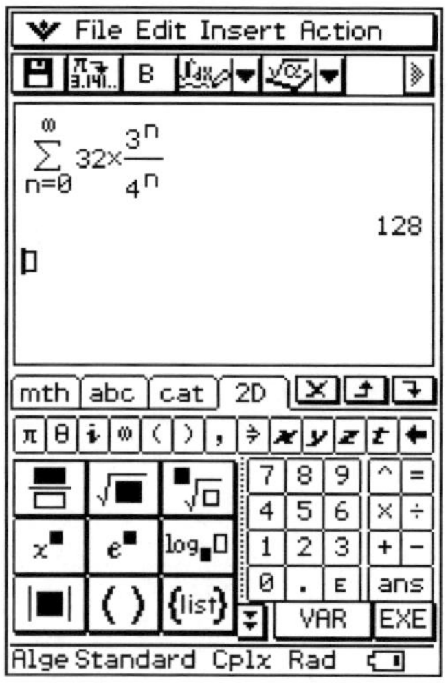

Eventually, we fill up the entire large triangle!

Another Area Problem

A pyramidal building has mosaic patterns on each floor. The patterns on the top three floors look like the following:

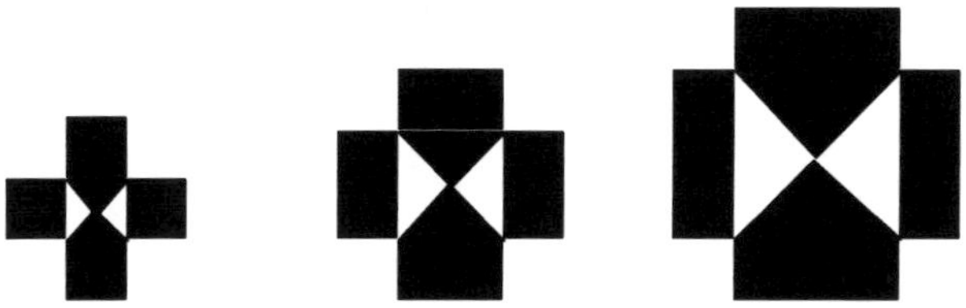

There are fifteen floors. We want to know the total areas.

With the integer grid on, we can quickly draw the figure as four rectangles (it's easiest just to use the freeform polygon tool rather than the rectangle tool). We select and shade these rectangles. Then we can draw the diagonals of the voided square in the middle and create the intersection point.

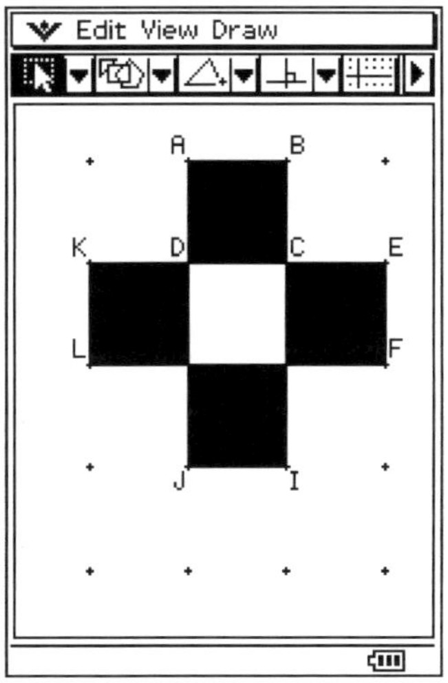

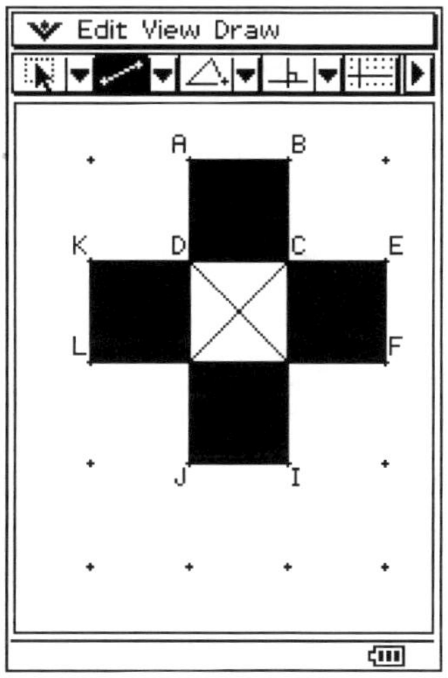

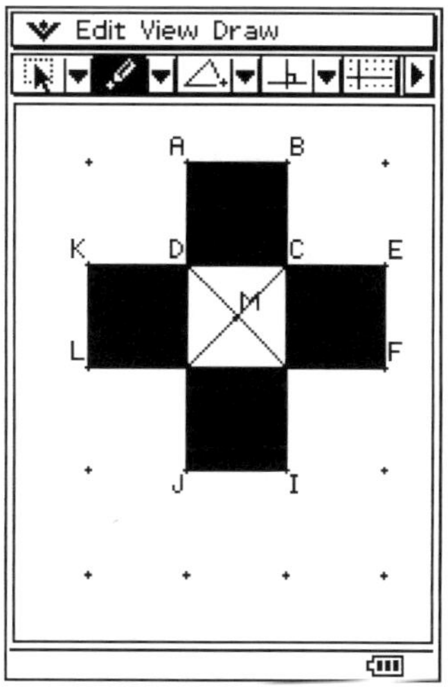

A polygon other than a triangle can only be shaded if it is created with each side a complete line segment. Triangles are different: you can shade a triangle, even if the sides have not been created just by selecting three points. In this case the side DM, for example, does not exist explicitly M is just the intersection point of the two diagonals. However, you can still shade DCM by selecting the three points and clicking **Edit / Shade On** .

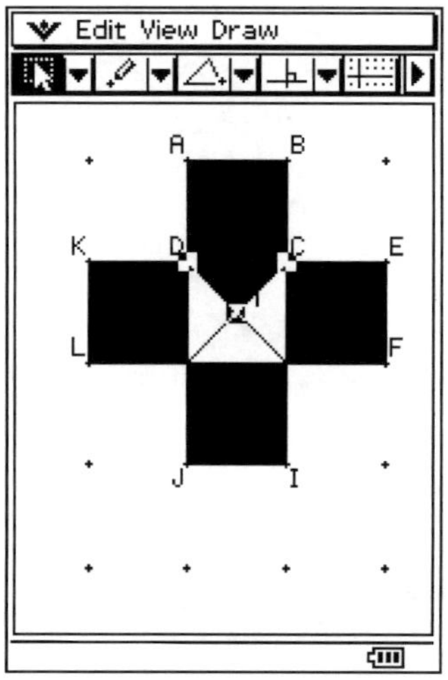

Once you have created the correct shaded area, make sure that nothing is selected. Then the Measure Box contains information about the entire shaded portion of the figure. The two options are perimeter or area..

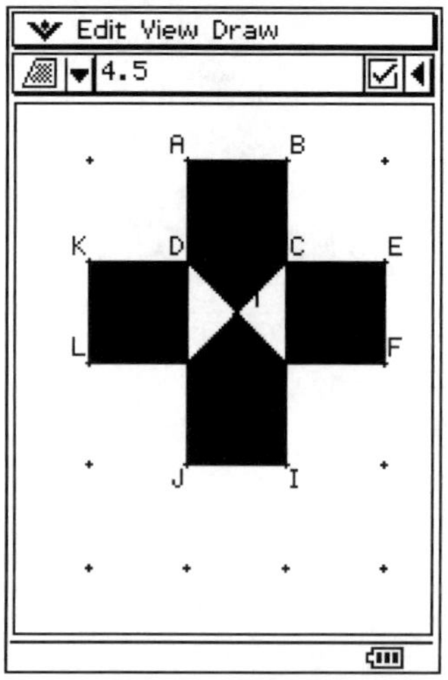

We see above that the area of the first figure is 4.5. We are going to investigate a sequence of area values as we lengthen the sides of the figure. The computer algebra command sequence() takes as its argument a list of numbers and tries to find a formula to fit the sequence. We'll build a call to this function:

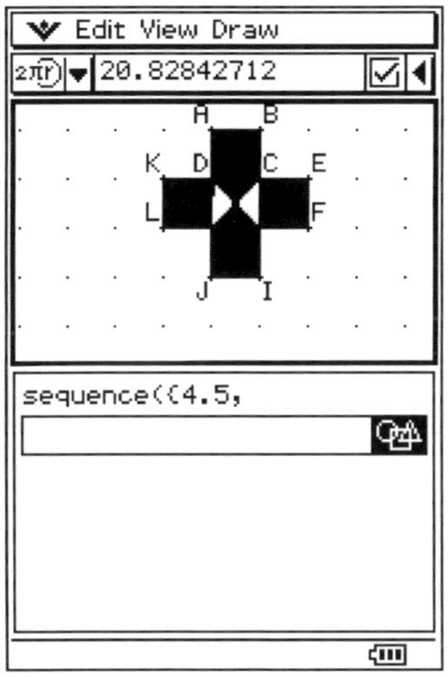

Notice that the syntax for a list is to enclose it in {}.

Let's find the next value for our sequence. We select the points on the right of our figure and drag then 1 unit over.

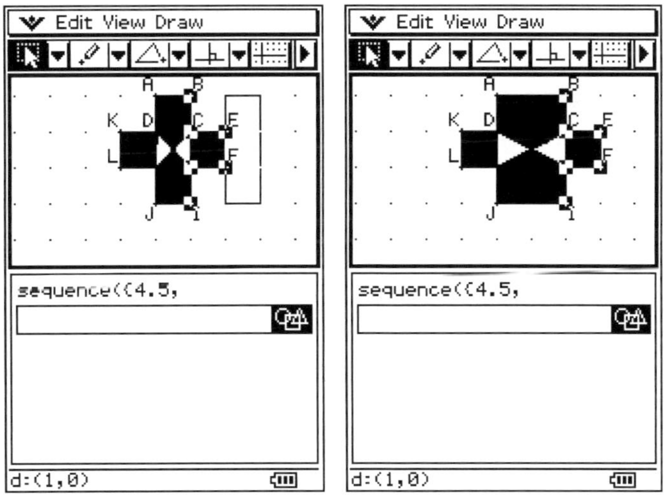

Then we select the points on the bottom of our figure and drag them one unit down.

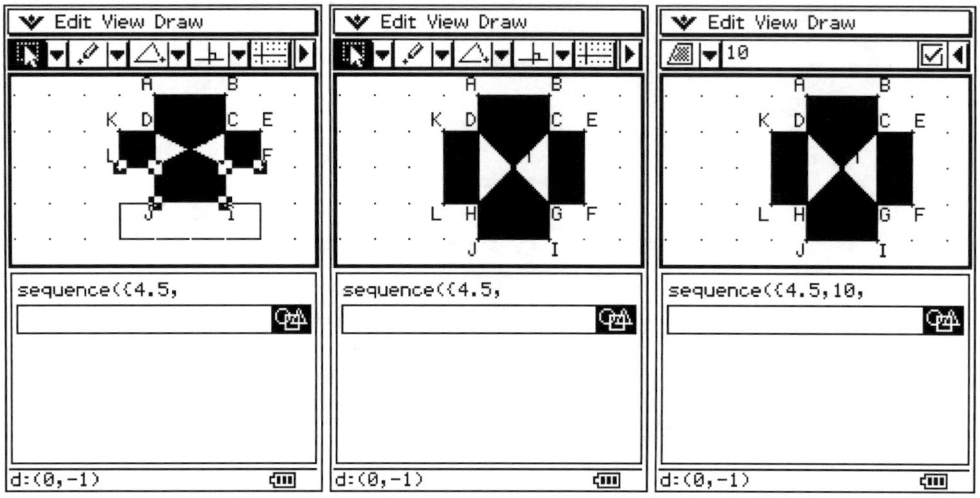

We can now repeat the process to find the next two values:

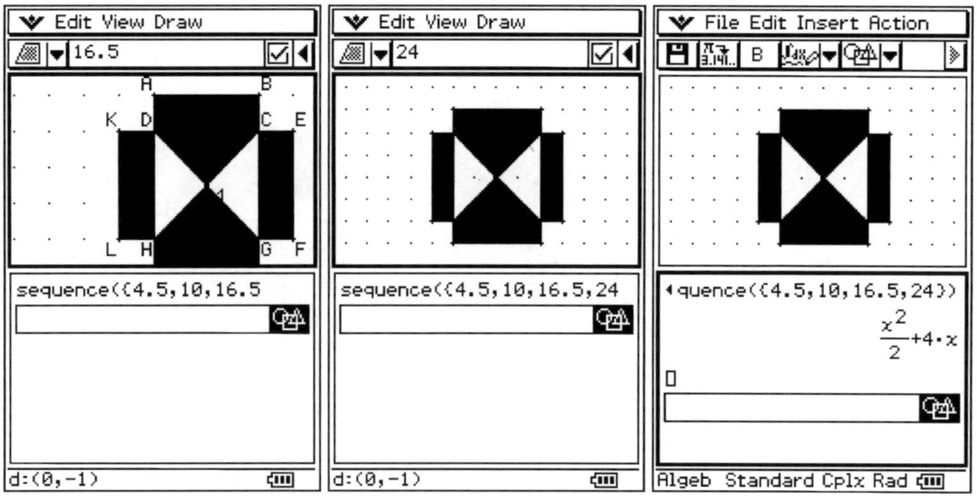

We get an equation for the area on step number x.

We can verify that the equation is correct, for example, by substituting x=4. The vertical line (read as "with") is interpreted by the algebra system as "substitute the equation after the line into the expression before the line.

To build the expression below, you can drag and drop the result from the previous computation into a new math input.

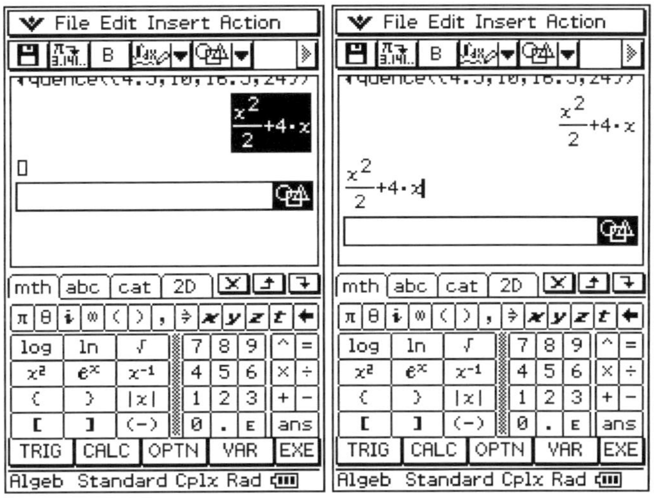

You then have to find the vertical line on the soft keypad - it is buried away in the **Math/Optn** Tab

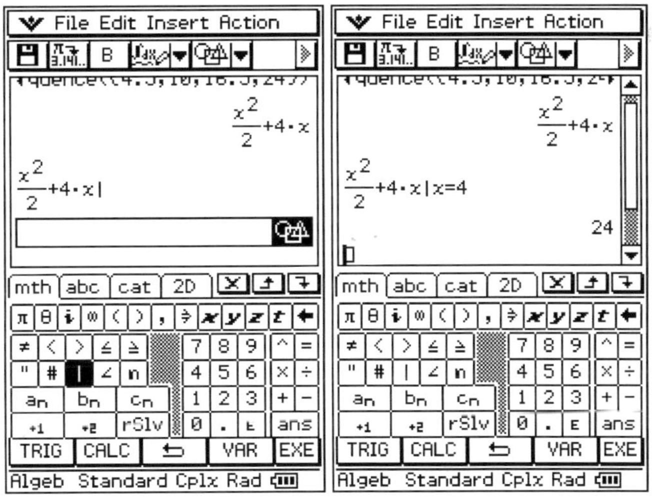

We can find the area of the fifteenth floor, for example by changing x:

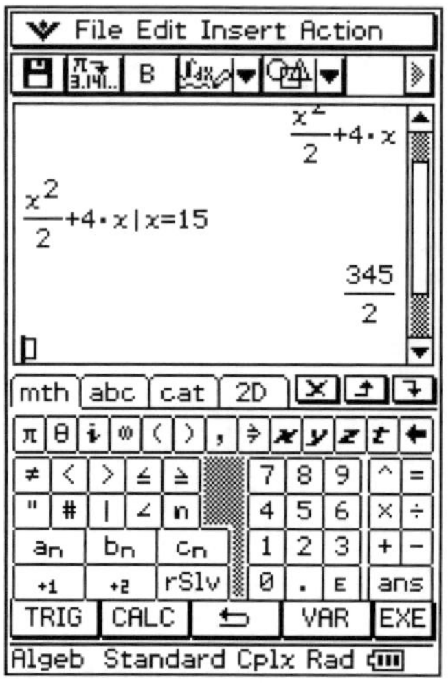

The summation command is on the 2D keyboard. This gives us the total area for all 15 floors.

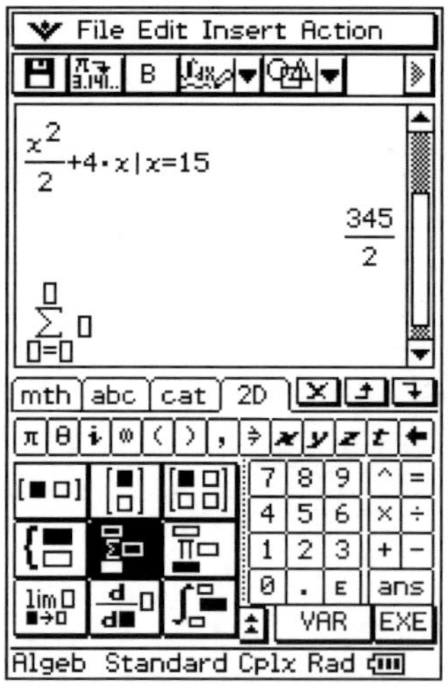

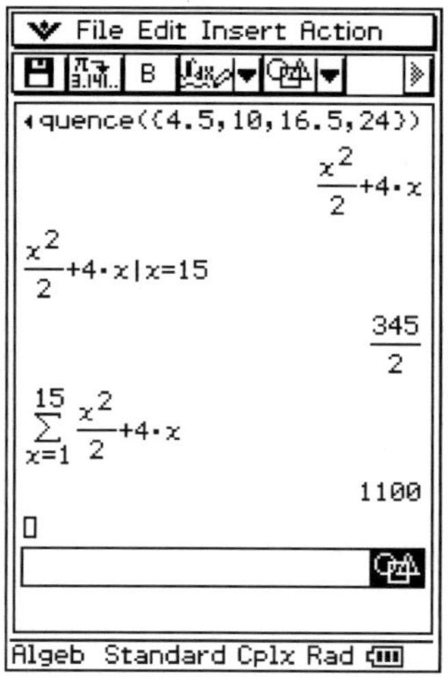

And the answer is --- 1100.

We can also answer a more general question: what is the total area of the top N floors?

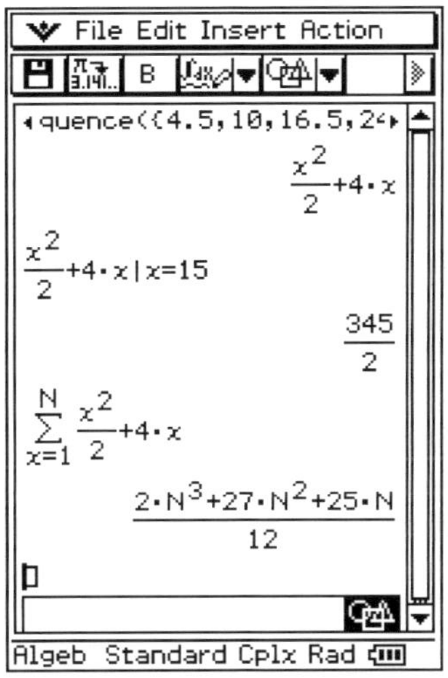

Transformational Geometry

Transformational geometry studies the properties of figures under transformations, such as rotation, translation, reflection, dilation. Transformations are available from the **Draw/Construct** menu or the drop down button:

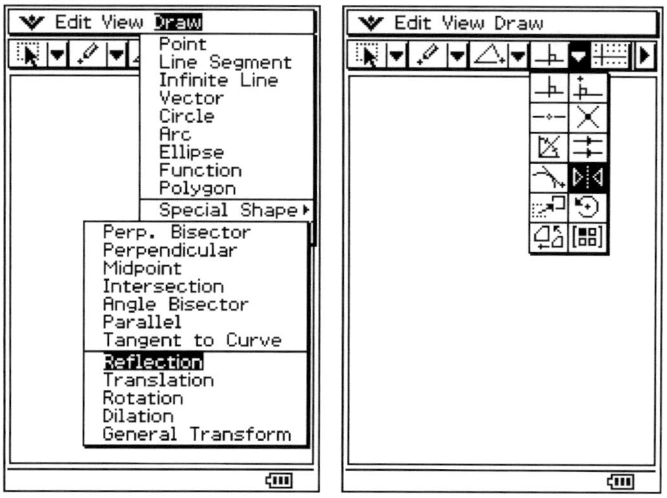

The following are available:

 Reflection

 Translation

 Rotation

 Dilation

 General Transformation (matrix defined).

To apply a transform, you first select the object or objects which are to be transformed, then you click on the appropriate button, or make the corresponding selection from the **Draw/Construct** menu. Depending on which transformation you are applying, you will then be prompted to supply subsidiary information as shown in the table below: If you click on the transformation button without first making a selection, then the entire figure will be transformed.

Transform	**Subsidiary Information**
Reflection	Axis
Translation	Displacement Vector
Rotation	Center and angle
Dilation	Center and scale
General Transform	Rotation / scale matrix plus translation vector

Symmetries of special shapes

Let's start by creating a special shape - a kite - by using ClassPad's transformation tools.

We can create a kite simply by drawing a triangle and reflecting it in one of its sides. To do this in ClassPad, we draw the triangle, then select two of the sides. We then invoke the reflection tool, and select the third side as the axis:

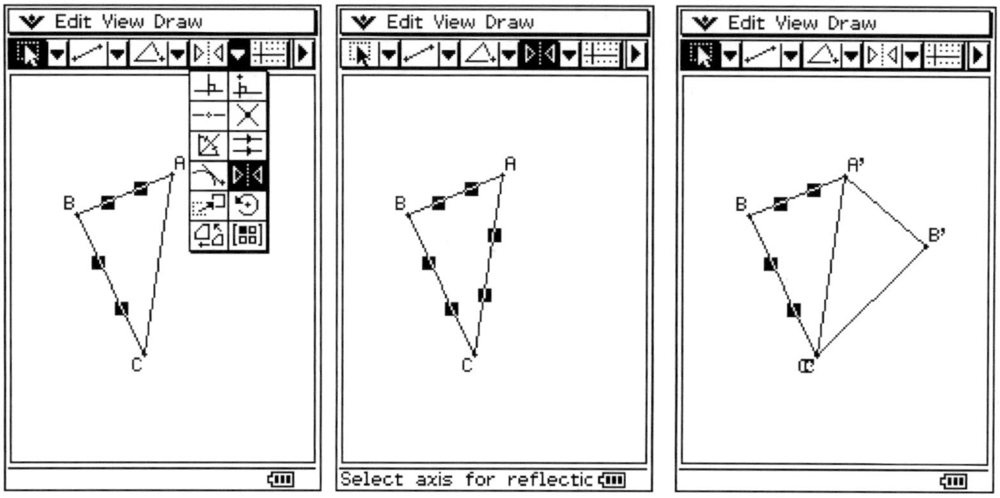

Dragging vertices of the figure, you can see that it retains the kite property.

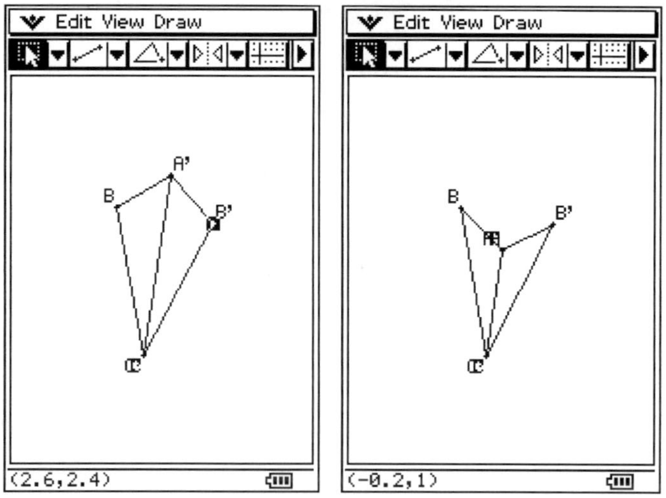

Looking for symmetries

The Special Shapes button or menu in ClassPad lets you quickly create shapes with special properties:

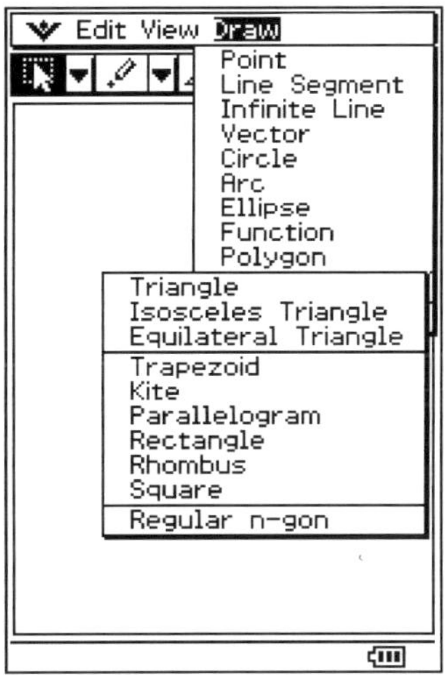

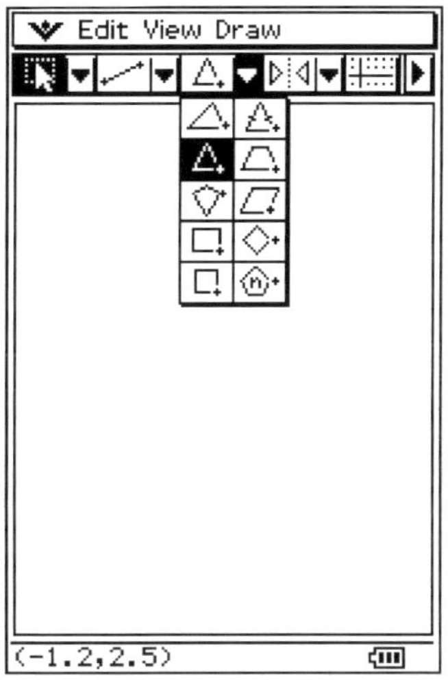

Available shapes are:

 Triangle

 Isosceles Triangle

 Equilateral Triangle

 Trapezoid

 Kite

 Parallelogram

 Rectangle

 Rhombus

 Square

 Regular n-gon

We'll look at the symmetries of a couple of these shapes

Isosceles Triangle

Can we find an axis of symmetry of an isosceles triangle?

We click the isosceles triangle and drag an outline box to create an isosceles triangle:

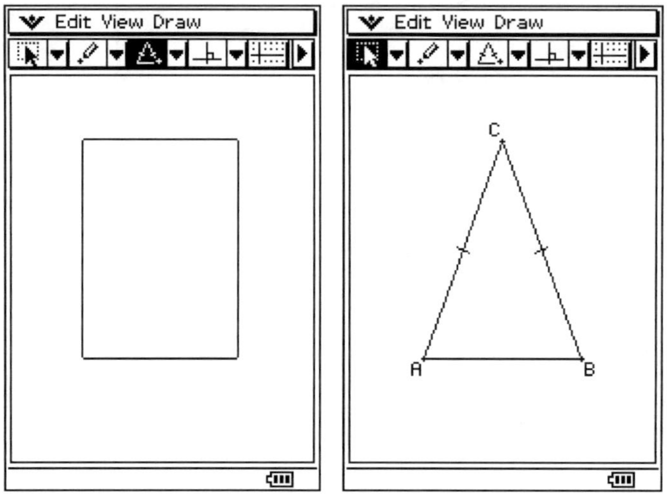

The perpendicular bisector of AC should be an axis of symmetry. We create this by selecting AC and pressing the perpendicular bisector button:

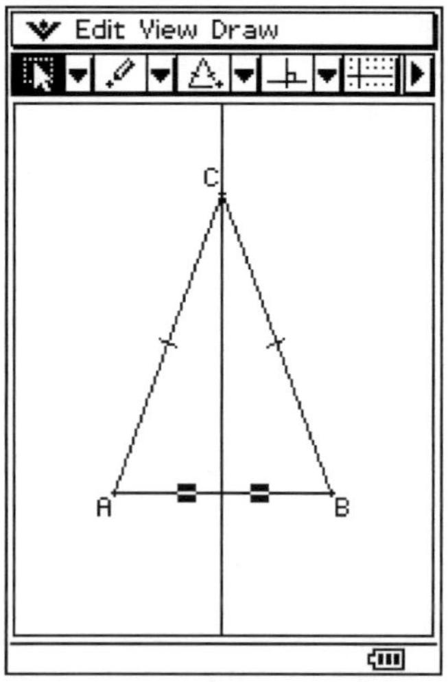

Now lets try reflecting the triangle in this line. Select the triangle, press the Reflection button, then click on the perpendicular bisector:

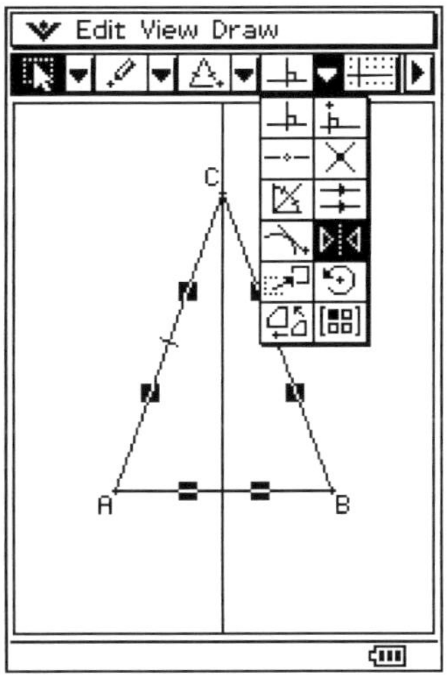

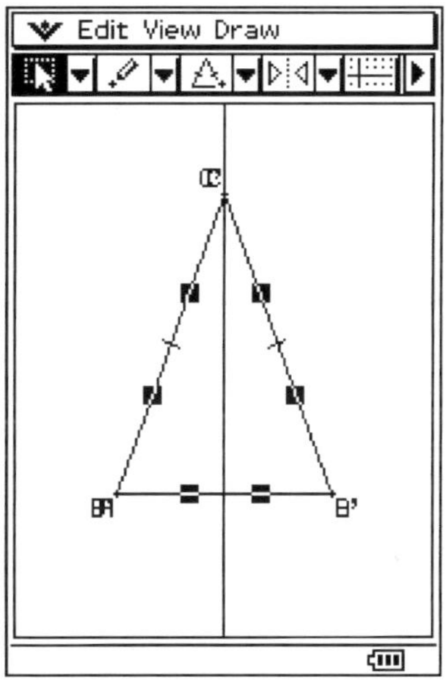

The triangle has convincingly enough reflected into itself.

How about the perpendicular bisectors of the other sides, are they axes of symmetry too?

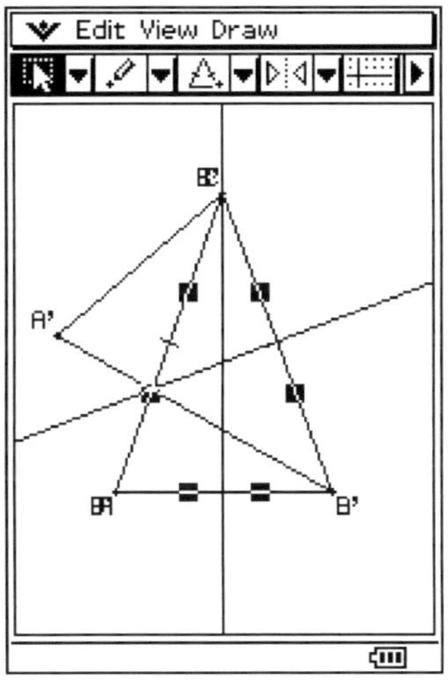

Well, no!

Parallelogram

What kind of symmetry does the parallelogram have?

Using the parallelogram tool you can create a parallelogram in a similar way to an isosceles triangle, by dragging an outline box:

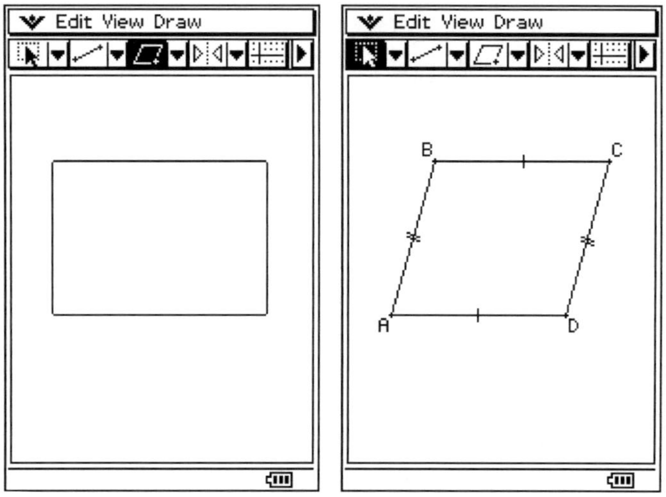

Are the diagonals, perhaps, axes of symmetry? Let's draw in a diagonal and reflect the figure about it to find out:

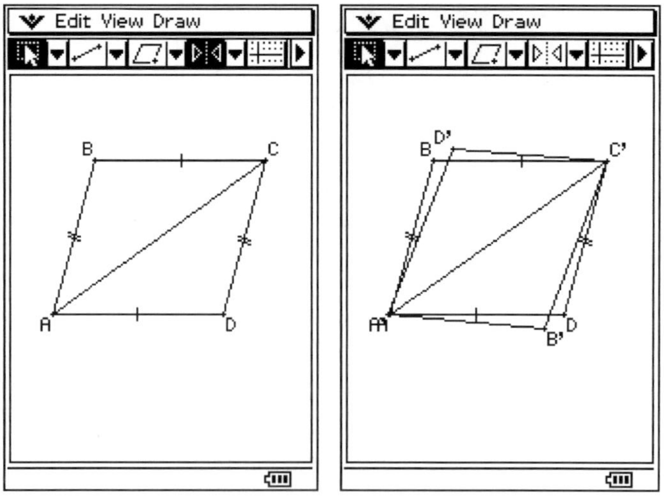

Clearly, not. You can remove the unwanted reflection using the **Edit/Undo** command.

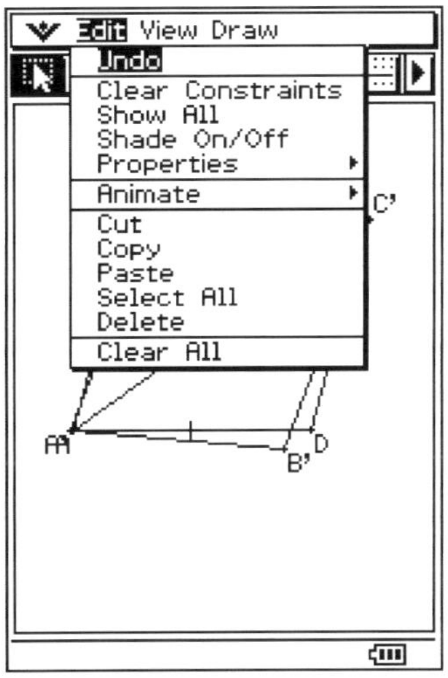

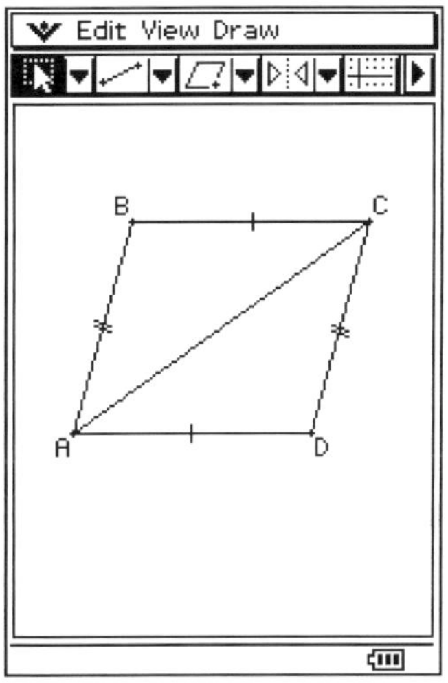

There is a transformation which conserves the parallelogram, however. To define it we need to draw the other diagonal and construct the intersection of the diagonals:

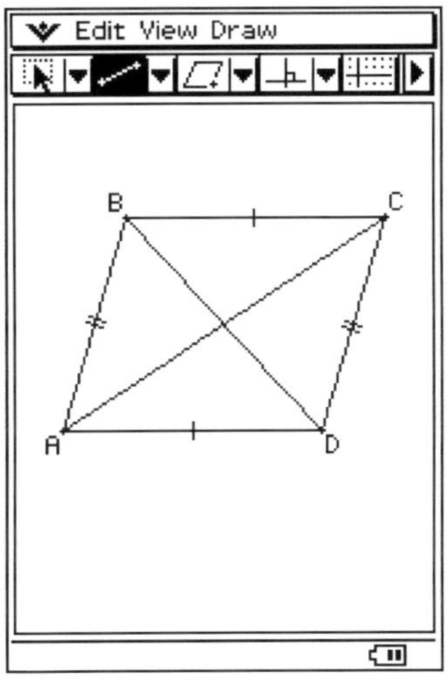

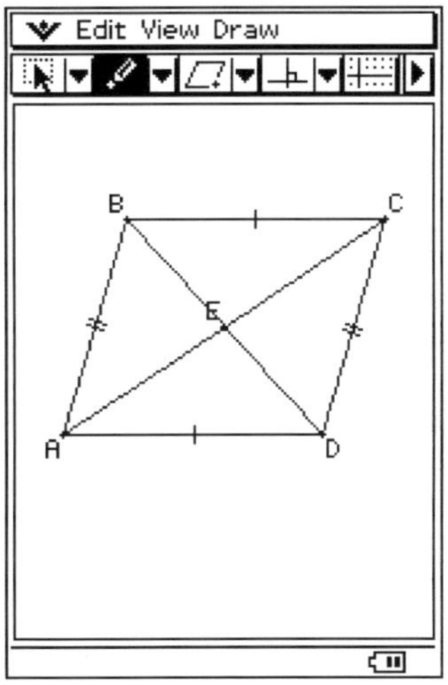

Select the parallelogram and click on the Rotation button. Now specify E, the intersection of the diagonals as the center of the rotation. Make the rotation angle 180 degrees and see the parallelogram transform into itself.

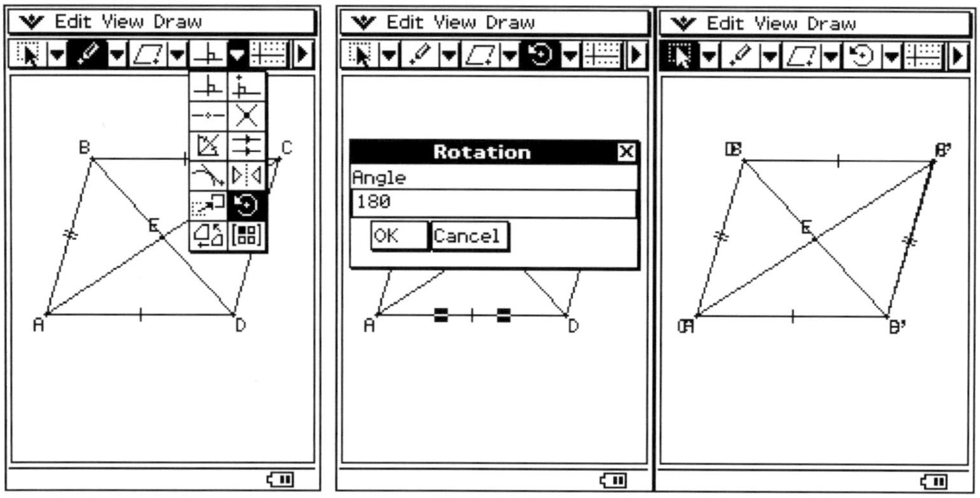

Further Investigation

What symmetries can you find for the other special shapes -

- Equilateral triangle
- Rhombus
- Rectangle
- Square
- Regular Pentagon
- Ellipse

Reflections on Billiards

The game of billiards (at least the American version) requires the player to bounce his cue ball off two sides of the table before hitting one of the other balls. We can use reflections to determine which direction a ball should be shot in order to achieve this.

Let's start with an easier problem, though. Let's work out how to bounce off a single cushion to hit another ball. We wish to find a path from ball F to ball E, bouncing off the cushion AB.

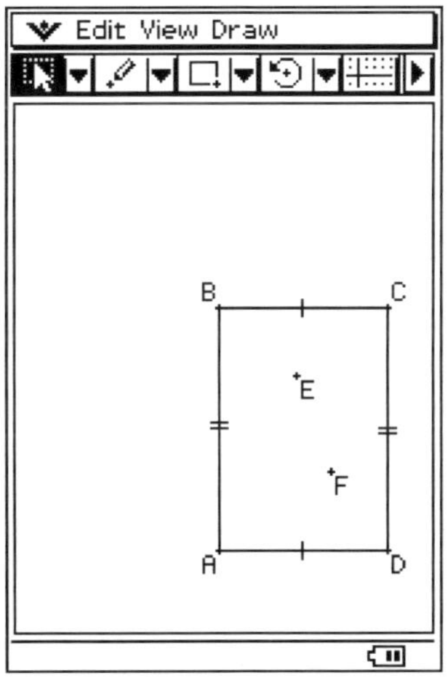

The physics of ball-bouncing guarantee that the ball will leave the cushion in a path which is the reflection of the path which it would have made if it continued through the cushion unimpeded. So to find the correct path we need only reflect our target in AB, join F to the reflected image of E and reflect the resulting line back..

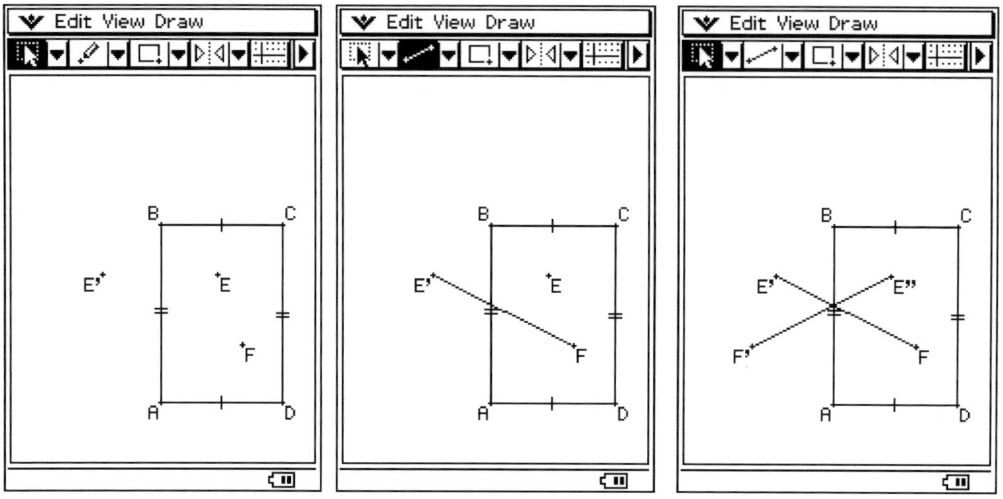

The Two Cushion Problem

Back to the original problem - how to bounce the ball off two cushions and hit the target. A little more reflection (one more to be precise) gives us the answer:

We want F to hit E again, but this time bouncing off cushions AB then BC.

After reflecting E in AB, we then reflect its image E' in BC, then join F to this second image E".

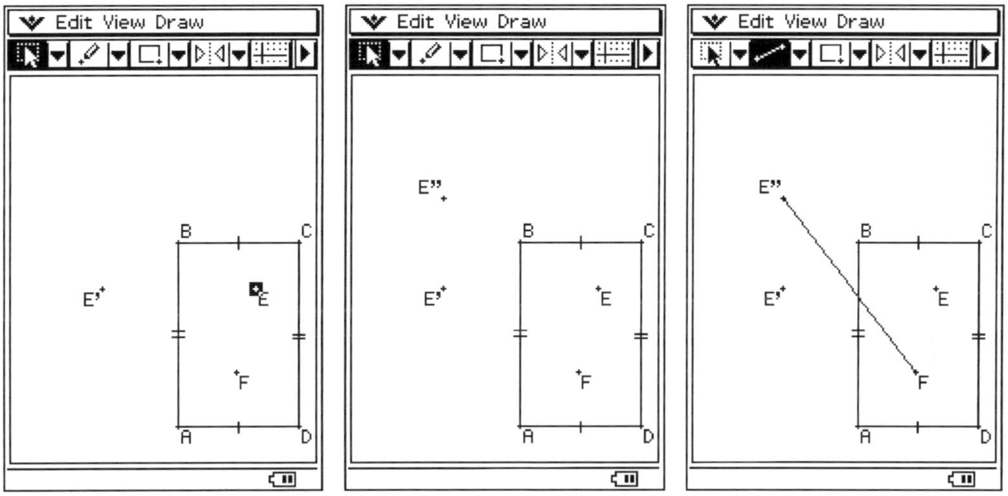

Reflecting the resulting line in first AB then BC yields the desired path.

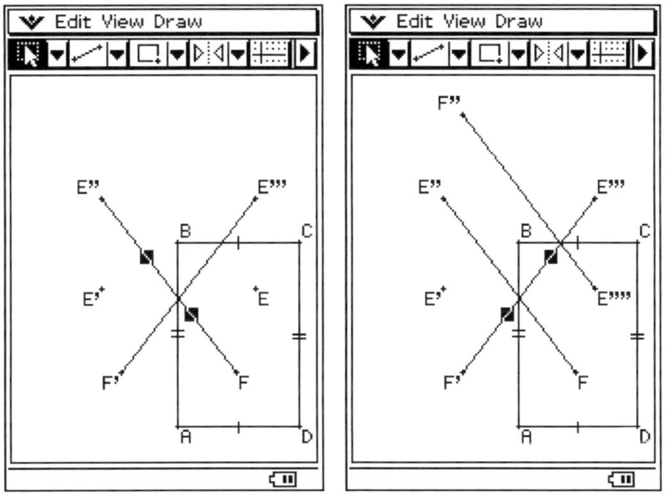

You can drag E and F around to different locations. The configuration maintains a wall to wall path from ball to ball..

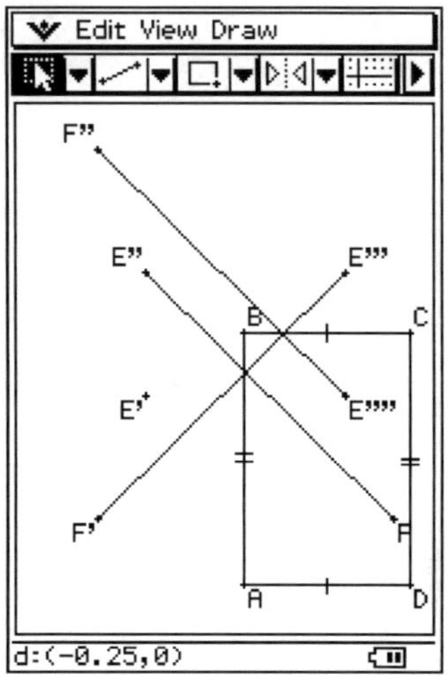

Further Investigation

In the above pictures, the initial path and the final path look to be parallel. Is this guaranteed to be true? Why?

Dilations and similar triangles

Here's a problem:

> Triangle ABC has vertices (-2,-1), (-1,1) and (2,-1). Triangle DEF has vertices (-3,2), (-1,6), (5,2). Show that the two triangles are similar. Find a transformation which takes ABC into DEF.

Locating the end points is easy if you have the integer grid on. One way of showing that the triangles are similar is by showing that they have the same angles. Inspect an angle by selecting two sides and displaying the angle in the measurement box.

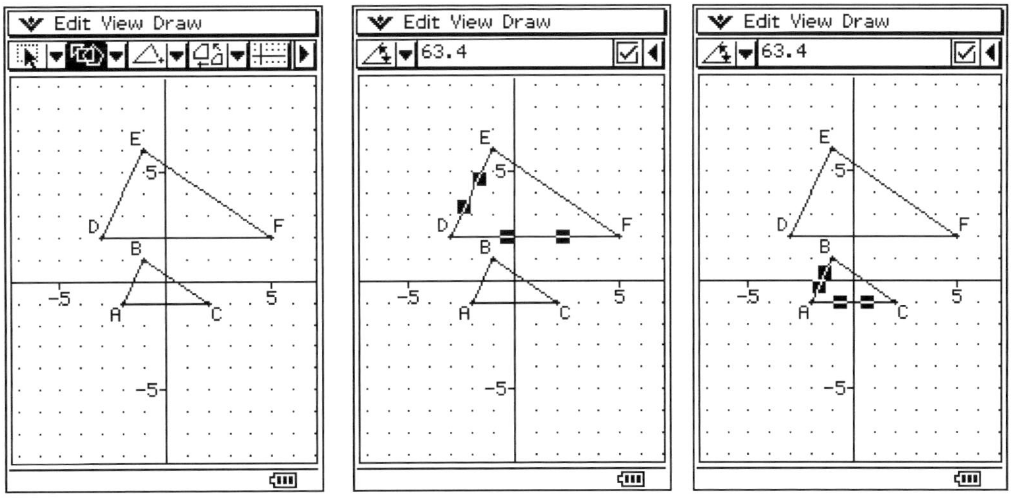

Another way of showing that the triangles are similar is by showing that the ratio of the lengths of corresponding sides is the same. You can inspect the length of a side by selecting it and displaying the length in the measurement box:

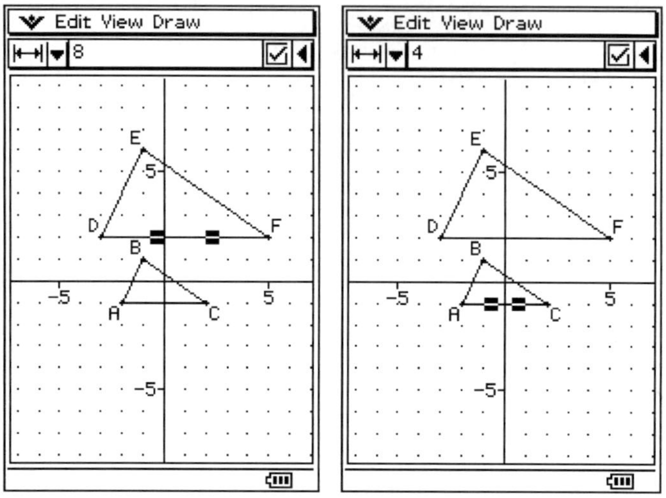

We see that the ratio of side lengths is 2. We can check the other sides to verify that they are in the same ratio.

Now the transformation which takes similar triangles into similar triangles is a dilation. To specify a dilation, we need to know its scale and its center. The scale of the dilation is just the ratio of corresponding sides: in this case 2.

The center of the dilation must lie on the line joining A to D and the line joining B to E and the line joining C to F. We can draw in these lines and verify they meet in a point.:

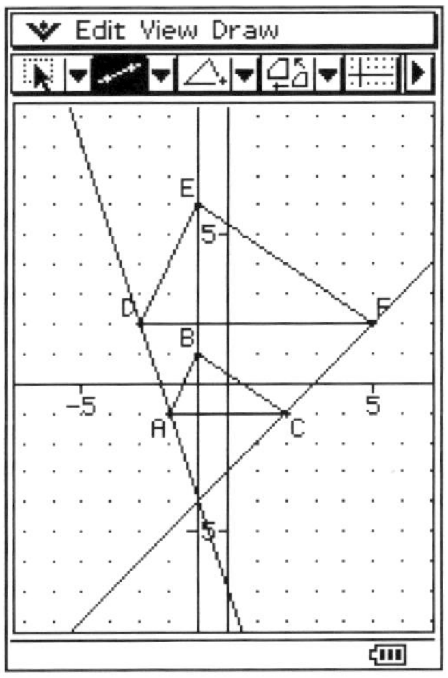

They do indeed join at a point: (-1,-4). So the transformation which takes ABC into DEF is a dilation, center (-1,-4) scale factor 2.

To verify this, let's create such a dilation. First, we clean up the drawing, removing everything but the original triangle ABC. Now we click on the dilation button. ClassPad asks you to specify a center for the dilation. You should click on the point (-1,-4). A dialog box then requests a scale factor for the dilation: you should enter 2.

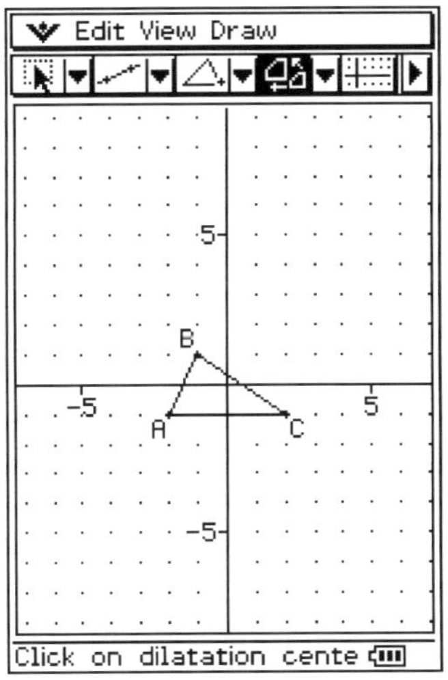

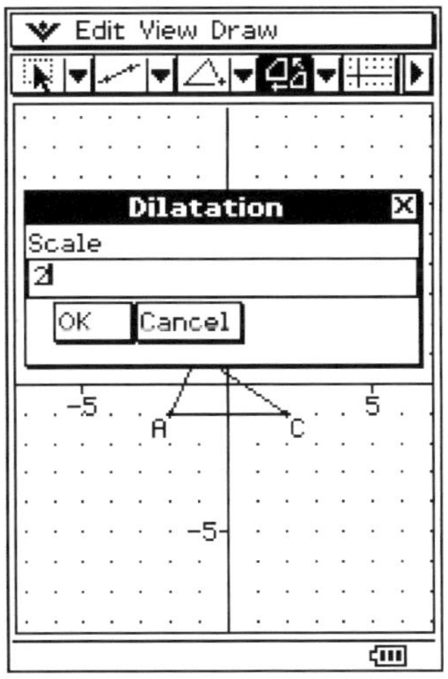

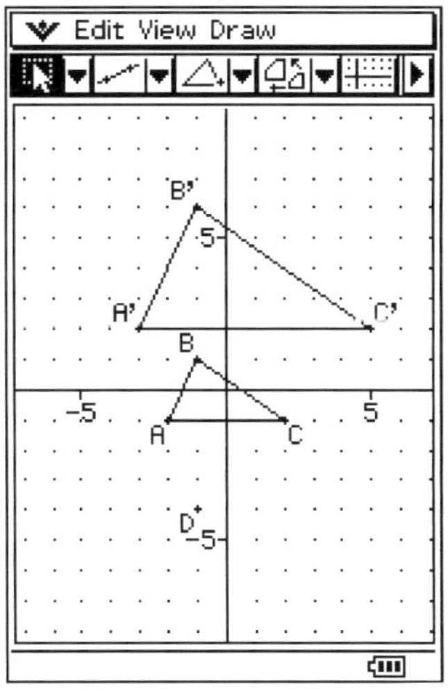

Changing a Dilation

Once you have created a dilation, there are several ways of changing it. First, you can change the center of the dilation simply by dragging it:

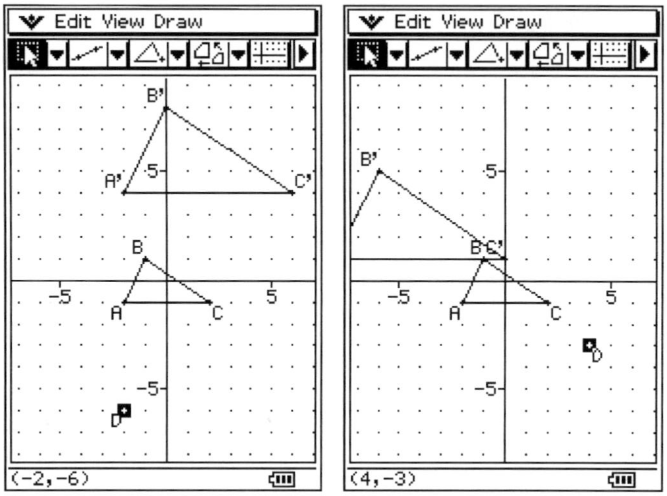

Secondly, you can edit the scale of the transform. To edit the scale, you have to select the transform. You do this by selecting a point and its image. So, for example, select both A and A' and bring up the measurement box. The first option is the distance between the points. The second option is the dilation scale.

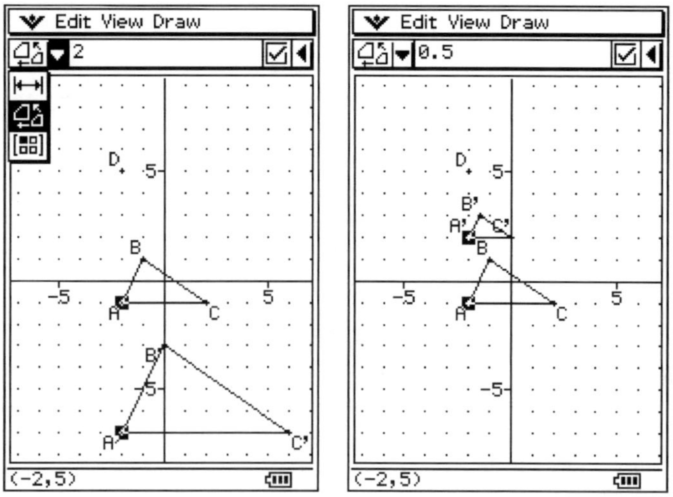

Combining Transformations

You can create new transformations by combining old ones. It turns out however, that these new transformations can be familiar ones in disguise.

Let's have a look at what happens when you do one reflection after another. We'll reflect triangle EFG in line AB, then reflect the image in parallel line CD:

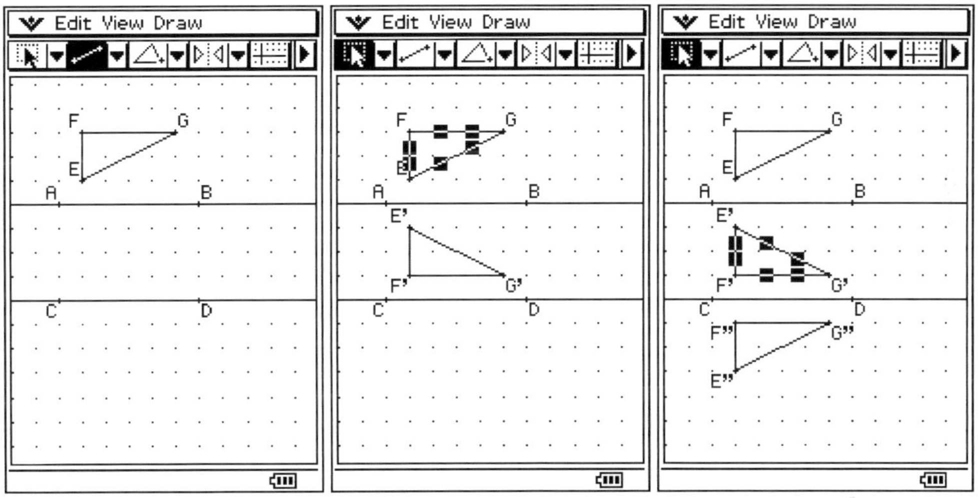

We can see that the result of the combined transformations E”F”G” is simply a translation of the original triangle EFG. We can examine the distance between F and F” and see that it is twice the distance between the parallel lines. By dragging one of the parallel lines we determine that this relationship holds for different distances between the parallel lines.

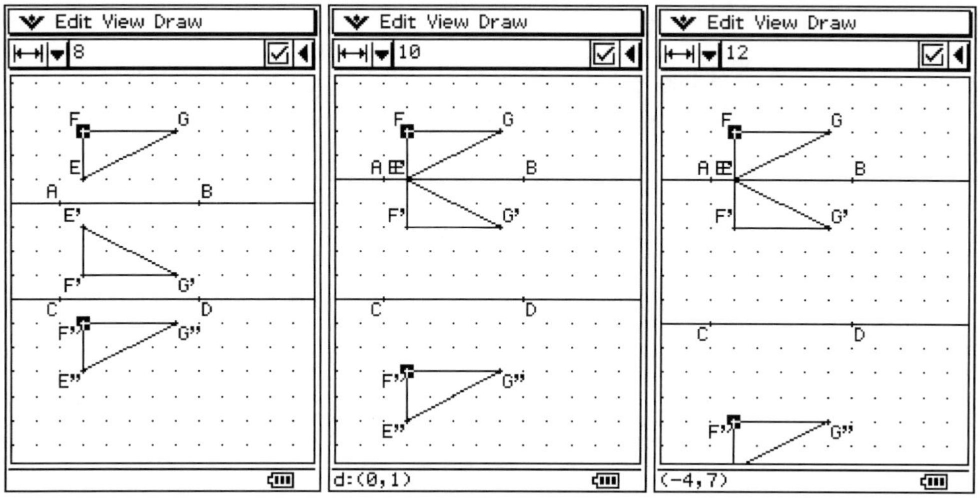

We can drag the lines into different (though still parallel) directions. We observe that there is still a translation, and that the direction is perpendicular to the axes of the reflection (we can check this by drawing a line between E and E”, then examining the angle between this line and the line CD)

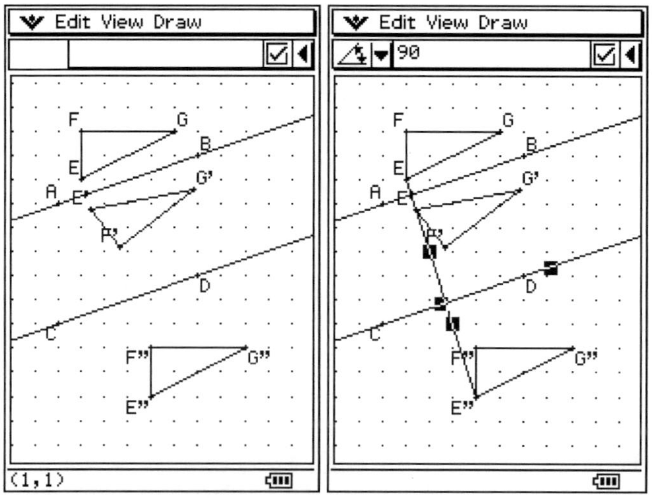

On the slant

Now let's try dragging one of the axes so that it is no longer parallel to the other:

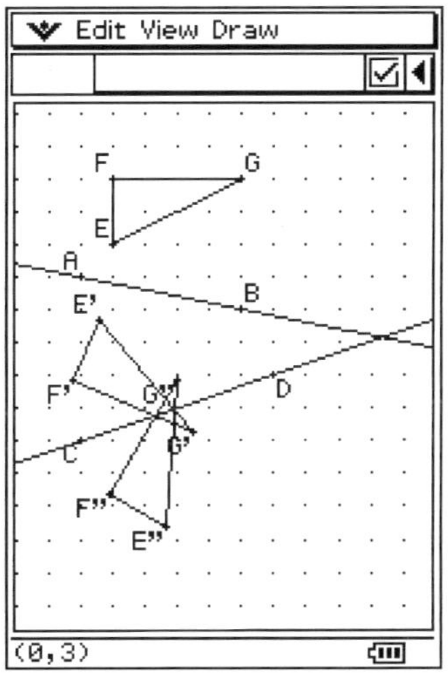

Clearly the combined transformation is no longer a translation. But what is it?

Perhaps we can help our intuition by looking at a few more orientations of the axes:

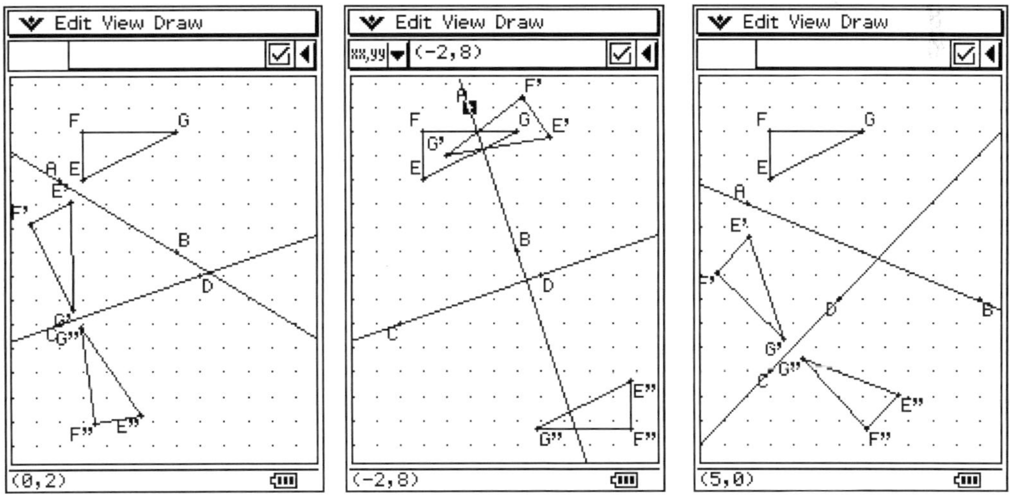

In fact, the combined transformation is a rotation about the point of intersection of the lines. We can verify this by creating the intersection point H and examining the angles EHE" FHF" and GHG":

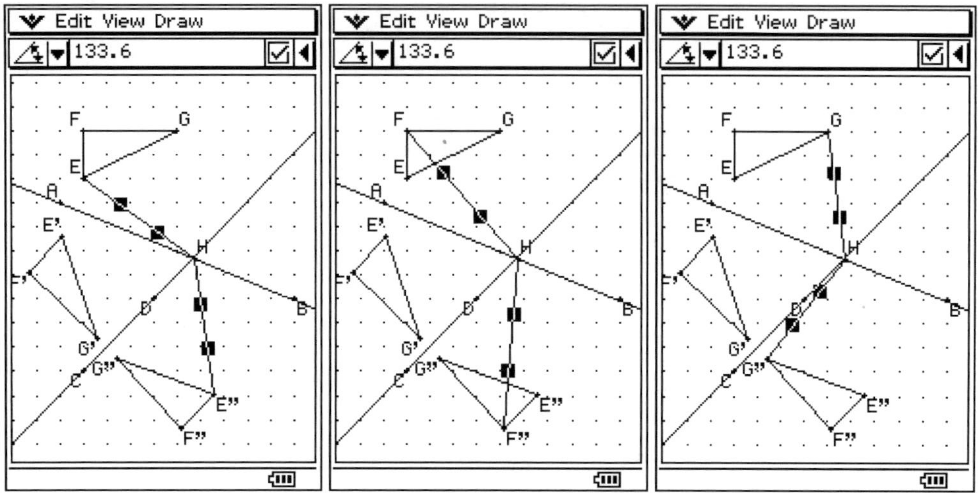

How does this angle relate to the angle between the axes. Let's examine that:

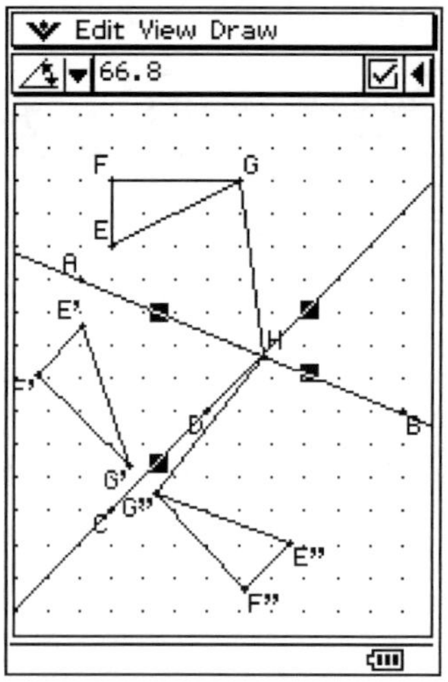

We can see that, in this case at least, the combined reflection is equivalent to a rotation of twice the angle between the lines. Try some more instances to convince yourself that this is generally true.

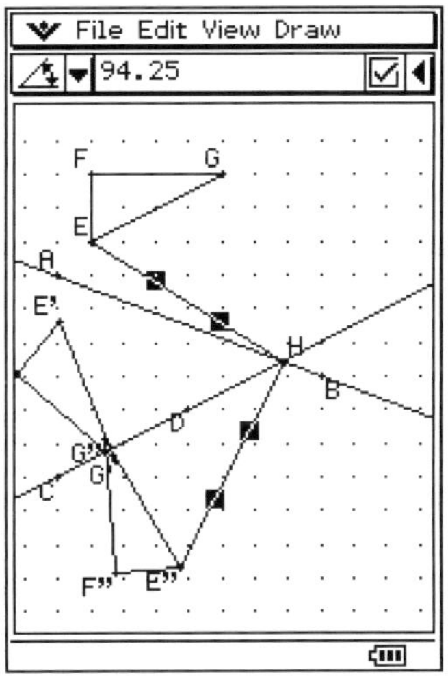

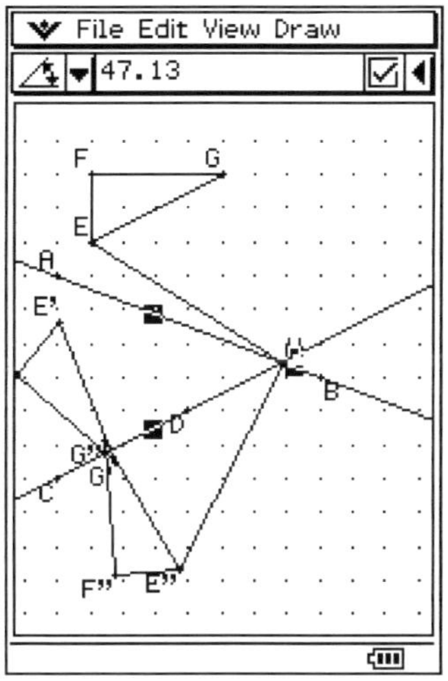

Reflections and their Matrices

Let's reflect the point (2,2) in the line through the origin and the point (1,3):

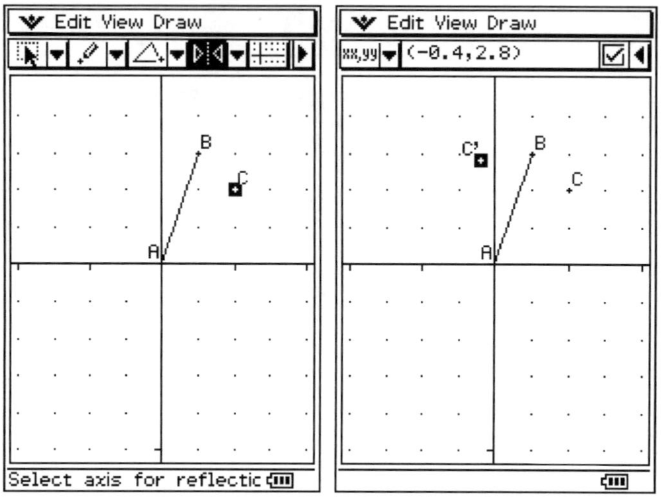

We can look at the matrix form of this transformation by selecting both C and C' and drag them into a calculator strip in eActivity:

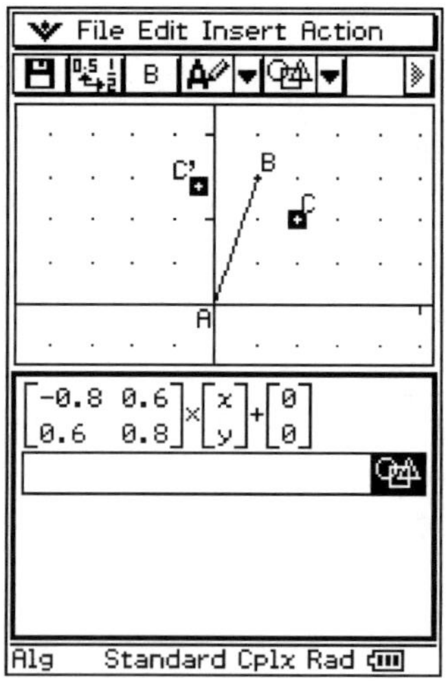

Now evaluating the right hand side of this equation with the coordinates of C: (2,2) substituted for x and y gives us the coordinates of C'.

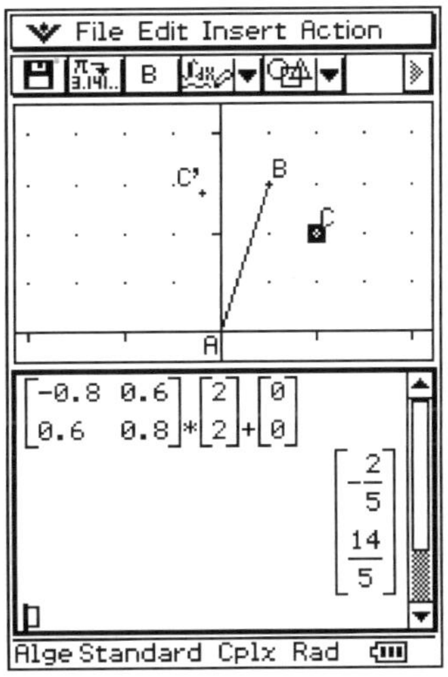

Let's now reflect C' in the line through the origin and the point (-1,2), and copy the equation of this transform into eActivity:

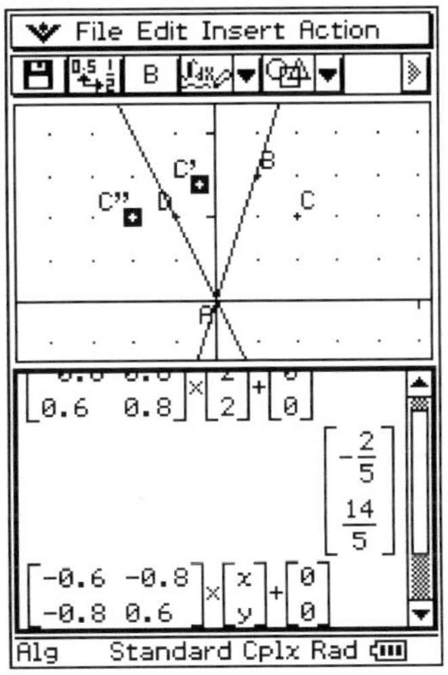

If we now want to find the matrix representation of the transformation which takes C into C", we can do this by multiplying the two matrices (careful of the order, the second transform should be on the left)

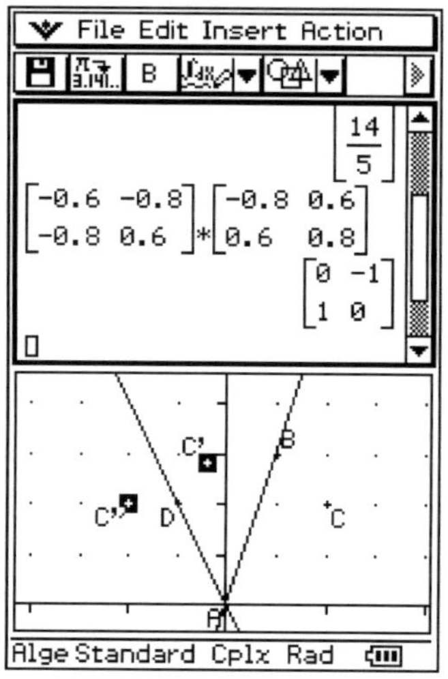

Now to check this against the original geometry, let's create a line segment CE and apply the new transform. To apply a matrix transform, we use the general transform button. First copy the matrix from the eActivity, then you will be able to paste it into the general transform dialog:

Note that to paste into a dialog box, you cannot use Paste from the menu (as there is no menu), however you can use the Paste button from the soft Keyboard.

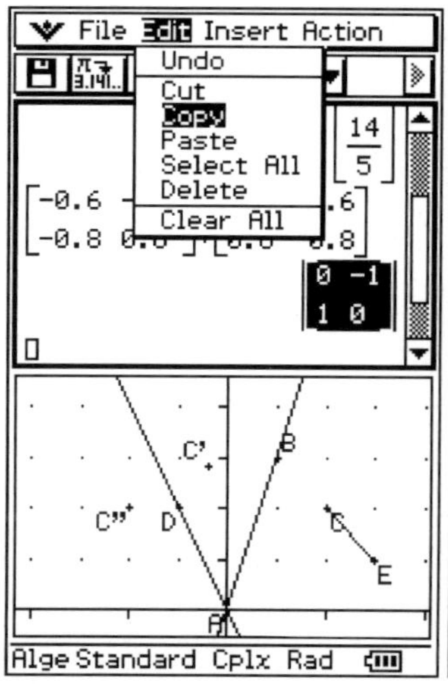

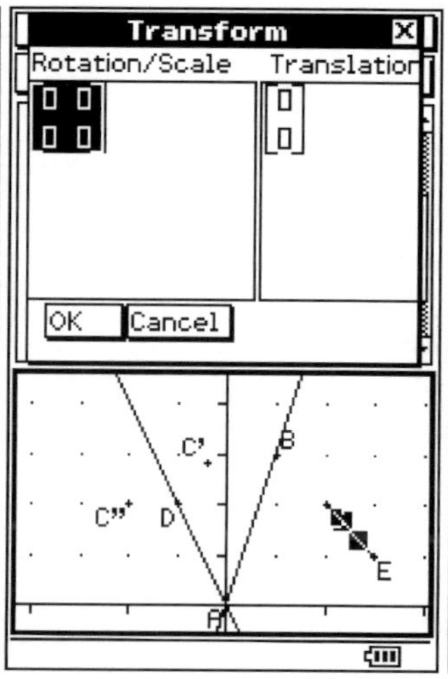

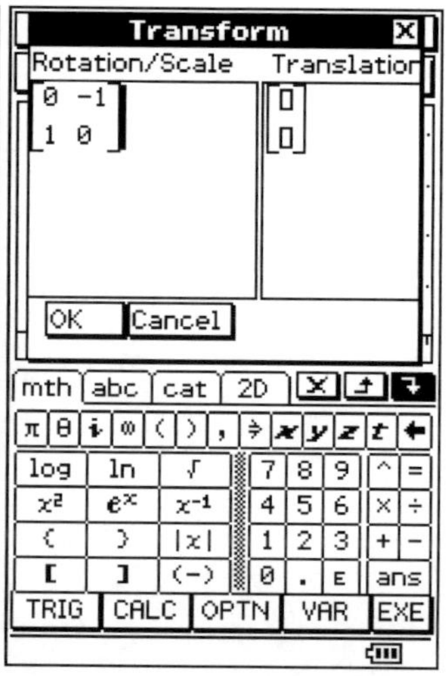

We see that indeed C is transformed into C"

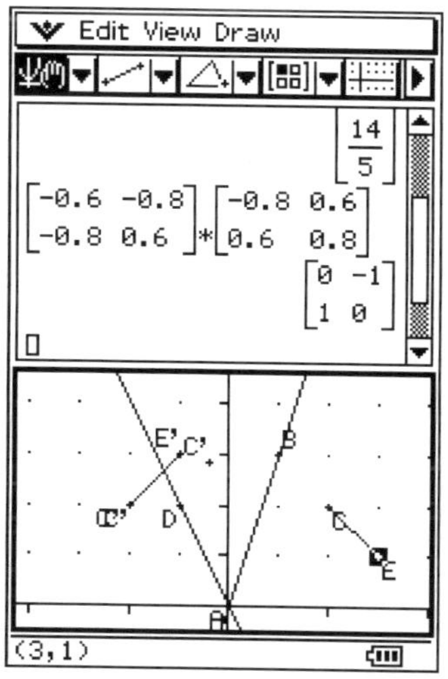

Determinants and Scales

If you drag a polygon into a math entry area, it will appear in the form of a matrix. Each column of the matrix is made up of the x and y coordinates of one vertex of the polygon. Try drawing a quadrilateral and dragging it into eActivity:

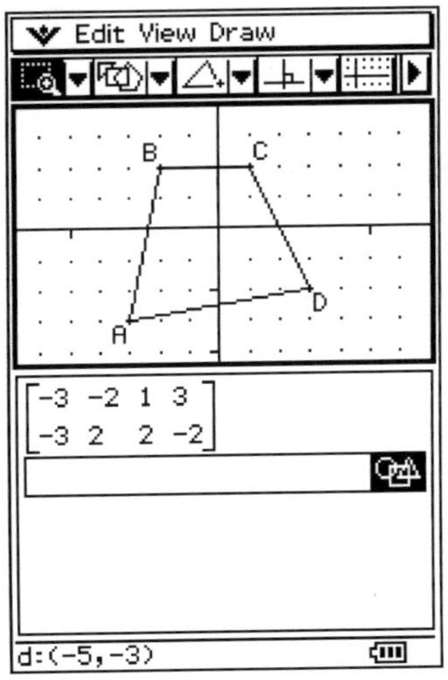

Let's examine the effect of applying the transformation whose matrix is $\begin{bmatrix} 2 & -1 \\ 1 & 1 \end{bmatrix}$

To apply a transform to the polygon, we need simply multiply the polygon's matrix by the matrix for the transform. We can then drag the new matrix and drop it into the geometry to see the transformed polygon:

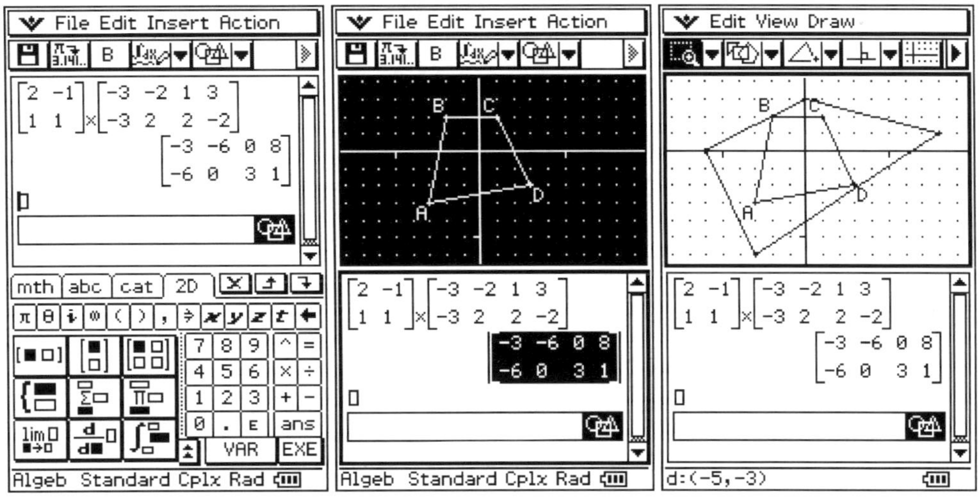

We can examine the area of each polygon by selecting it and looking at the measure box:

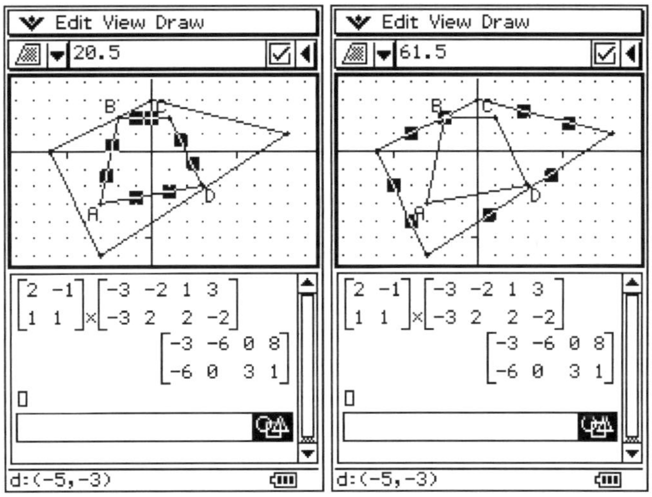

We see that the image is three times the area of the initial polygon. Can we relate this back to the original matrix?

ClassPad's algebra system has a function called det() which computes the determinant of a matrix. Let's apply this to the transformation matrix:

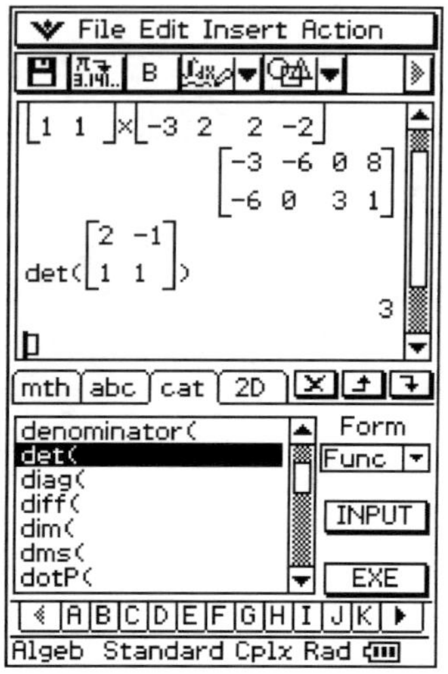

We see that the determinant of the transform is the same as the scaling which the transformation effects on the area of the polygon. Is this generally true? You could try a few more examples to persuade yourself, but in fact the answer is yes.

Of course we remember how to compute the determinant of a 2 by 2 matrix and don't need ClassPad to do it for us, but were we to forget, we could have our friendly (and non-judgmental) ClassPad remind us.

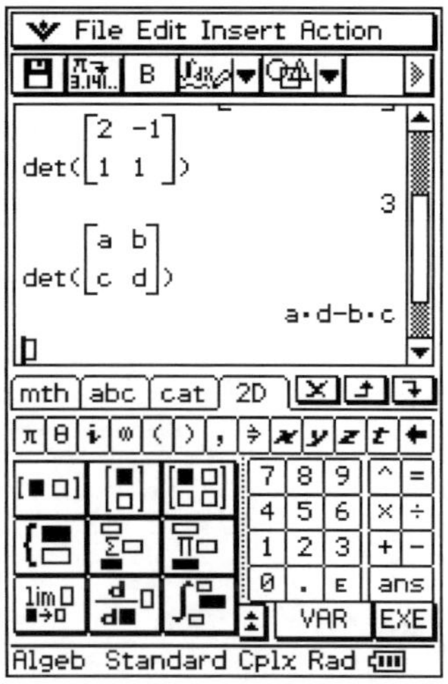

Constraint Based Geometry

> "Take a right angled triangle"
>
> "Let ABC be an isosceles triangle with altitude AD"
>
> "Consider the four common tangents between the circle center at the origin radius 4 and the circle centered at (8,0) radius 2"
>
> "DEF is a triangle with side lengths 5, 12 and 13."

Many geometry problems start with a declaration of the properties of the figure which is to be studied. They don't necessarily tell you how to go about constructing it. Now working out how to construct specific geometrical shapes out of primitive operations is indeed an interesting and sometimes a challenging intellectual activity. But it is not the only intellectual activity connected with geometry.

Constraint based geometry lets you specify a figure by sketching it and imposing constraints on the sketch. It is up to ClassPad to work out how to construct a figure which fits the constraints (or to throw up its arms in despair and decide that it cannot construct it).

Constraints take two forms: structural constraints and measurement constraints.

Structural (qualitative) constraints specify relationships such as incidence and tangency which are on/off in nature (either a line is tangential to a circle or it is not, either a point lies on a line or it does not).

Measurement (quantitative) constraints specify relationships such as distance, angle, slope, direction, which have numeric values.

Both types of constraint are handled in the Measurement Box. The measurement box displays information about the currently selected object or objects, it lies underneath the geometry module's button bar, and is invoked by pressing the tab arrow:

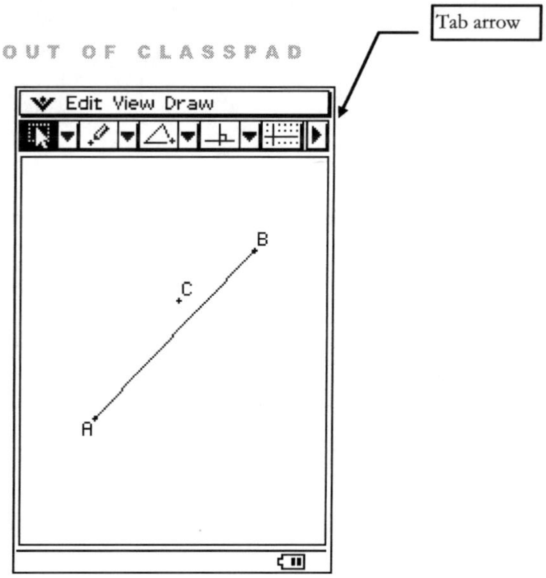

Selecting a point and a line segment displays the perpendicular distance between point and segment in the Measurement Box:

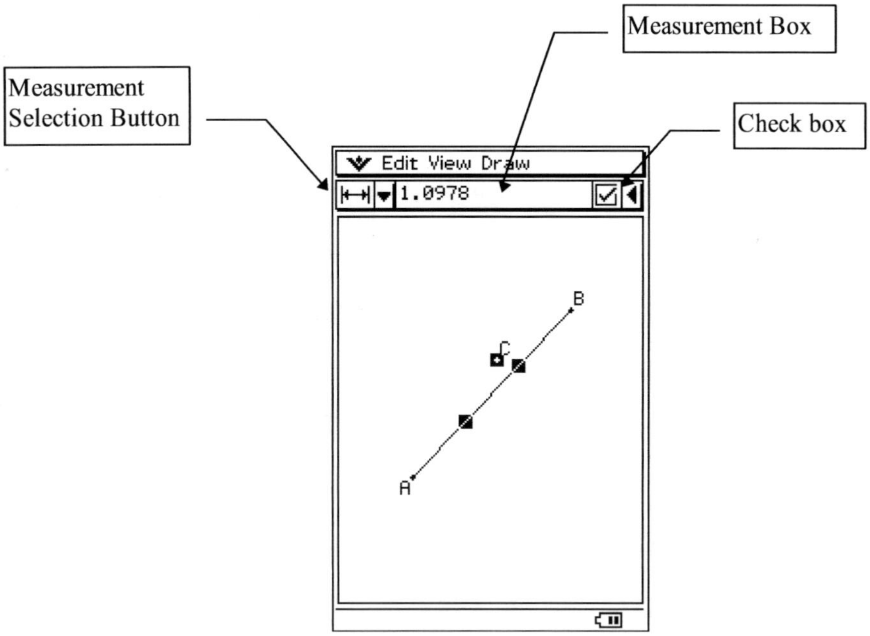

The Measurement Selection Button allows you to select which of the available measurements is to be displayed in the Measurement Box. If our selection consists of a point and a line, then the available measurements are:

- perpendicular distance
- incidence

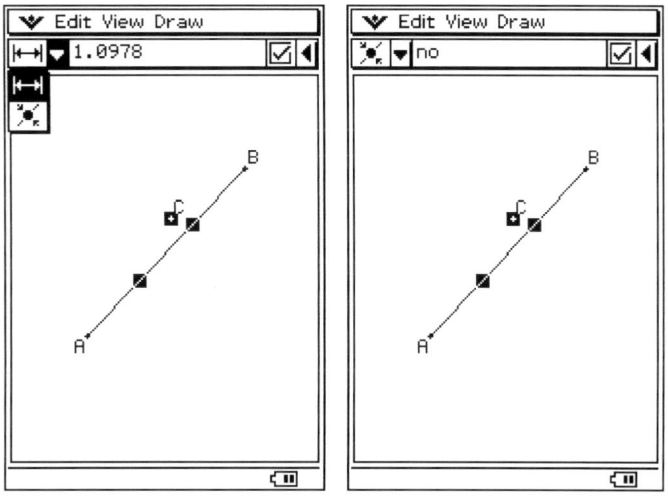

Perpendicular distance is a quantitative property, and its numeric value is displayed. Incidence, on the other hand, is qualitative. The measurement box displays **yes** or **no** depending on whether the property holds or does not.

So far, we have looked at properties, but not set any constraints. The button we use to set a constraint is the **Check Button** to the right of the Measurement Box.

If you click on this button when a structural constraint such as incidence is displayed, ClassPad will attempt to satisfy the constraint:

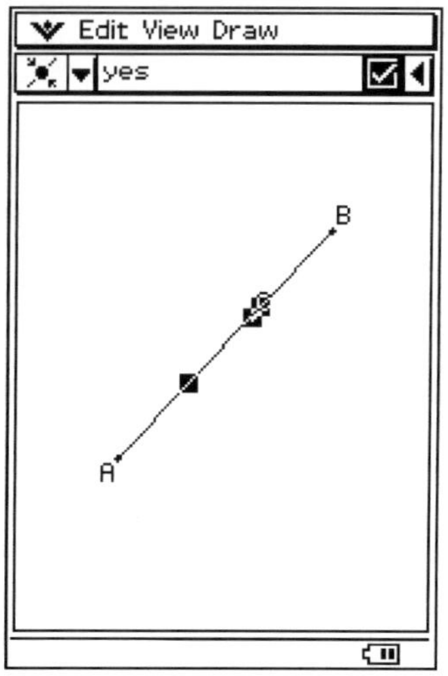

Now if we select point B and point C, we see displayed the distance between the points. We can edit the distance and press the Check button. ClassPad will attempt to position C at the new distance from B (while continuing to maintain the previously established incidence relationship)

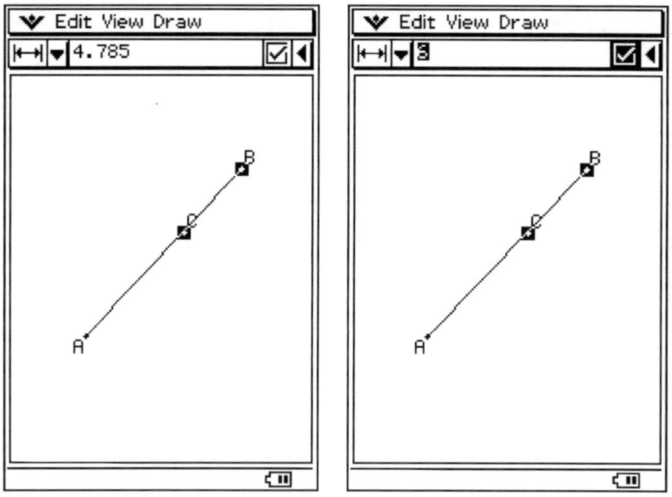

Notice, however, the drawing did not change - this is because ClassPad actually scales the whole drawing based on the first distance entered! At first sight this may seem a strange thing to do - but actually it is handy. If you want to draw a 3,4,5 triangle for example, you can go ahead and sketch it without worrying what size your sketch is, and ClassPad will scale automatically as soon as you enter the first distance.

Back to our point and line pair. You can now drag points A and B around the screen, and the two relationships we have established for C will be maintained:

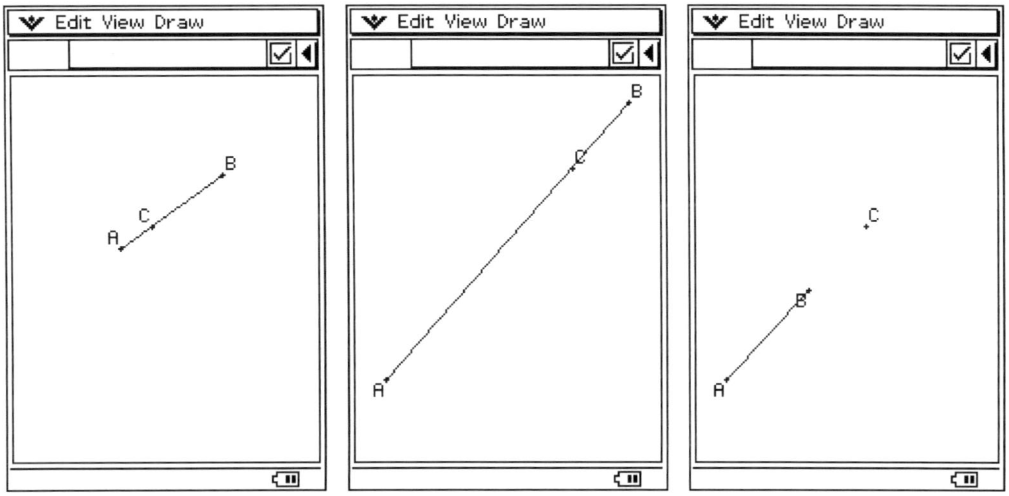

The last picture illustrates a couple of features of the system:

1. Incidence between a point and a line segment is treated as being incidence between the point and the infinitely extended line segment.

2. There can be more than one possible solution to a particular set of constraints.

3. You can always use the Undo - but remember it is only one level deep so do it immediately!

Notice that the constraints are not directional: not only can you drag A and B but you can also drag C:

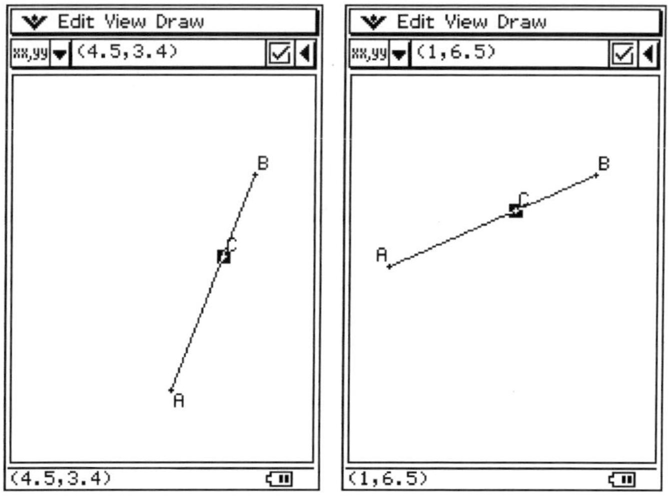

You can remove a constraint simply by clicking on the check box again to unshade it. The constraint applies if the check box is shaded and does not apply if the check box is unshaded:

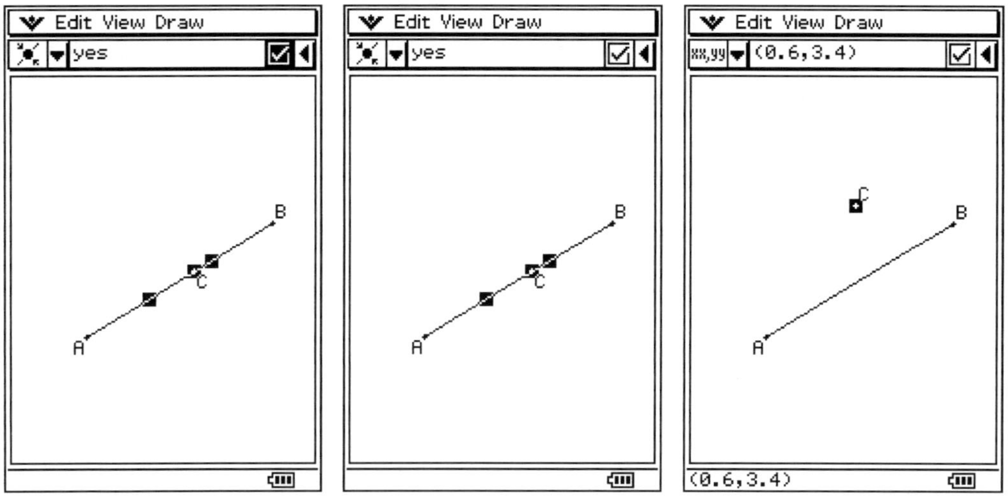

Triangles and Constraints

How big is that tree? Measure it without climbing to the top? Sure, bring out the protractor and string and pace out a line from the base of the tree. Measure the angle of a sightline to the top of the tree.

Now its a simple matter of trigonometry to work out the height.

Alternatively, we can sketch the triangle into ClassPad and use constraints to specify the measurements. Let's say we took 50 paces, and found the angle of elevation to be 58 degrees:

First sketch the triangle, no need to be precise - ClassPad will take care of that when you specify the appropriate measurements:

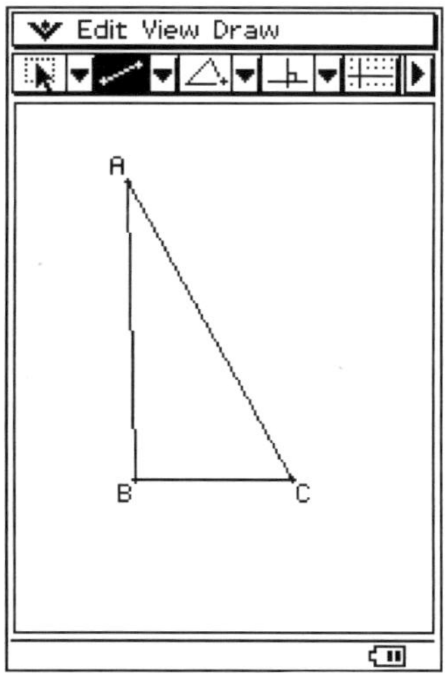

Now you need to enter these measurements.

First select the segments AB and BC, and set the angle between them to be 90 degrees.

Then select the segment BC and set its length to be 50.

Then select the segments AC and BC and set the angle between them to be 58 degrees.

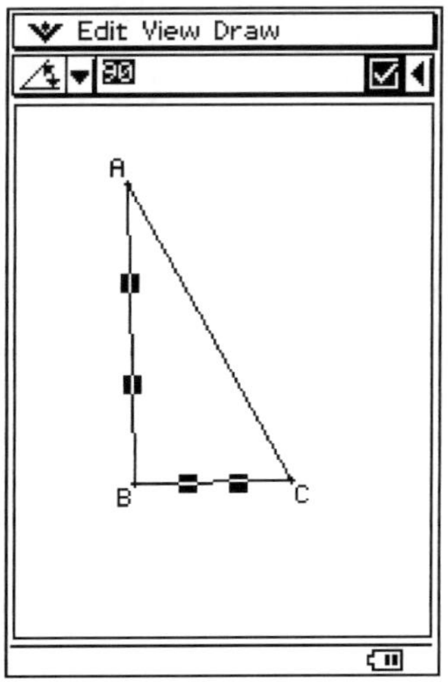

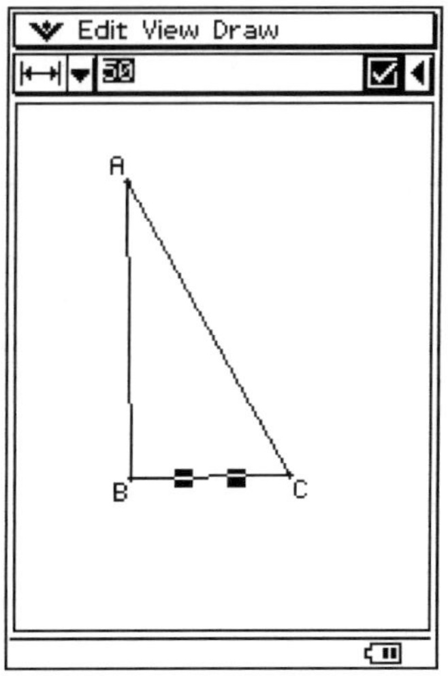

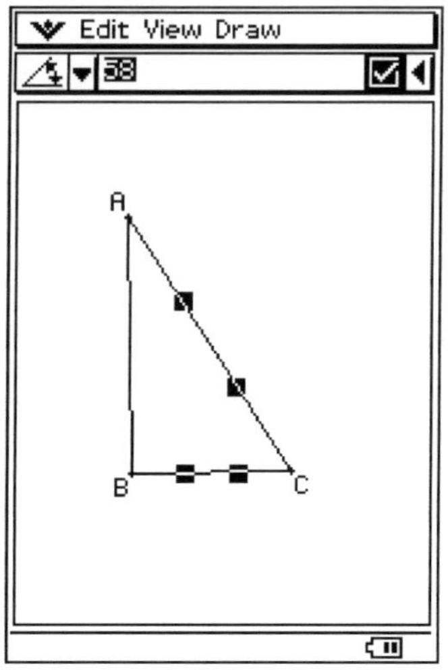

The height of the tree is the height of AB (in paces):

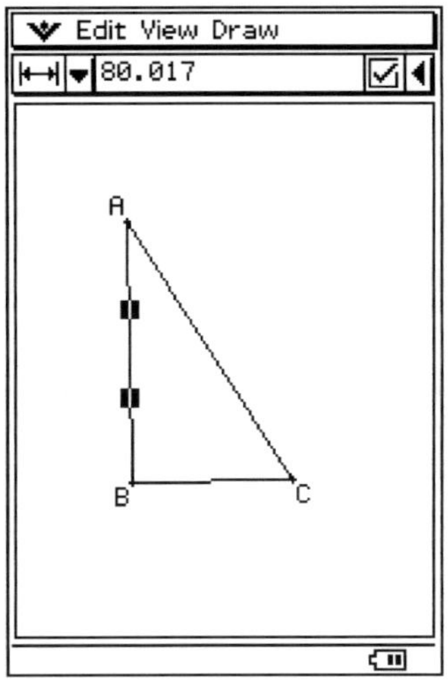

Triangulation

So, we've traded in our protractor for a compass, and are out in a field trying to position our tree and make a map by triangulation. We've paced out our base and are using our compass to measure angles to the tree from each end of the base.

Let's say our baseline is 100 meters long and runs due east - west. The tree is measured to be at compass direction 147 degrees from one end of the baseline and 226 degrees from the other.

With a single line selected, one of the measurement options is the direction. To set AB to run East West, we can set its direction to 0 (that is aligned with the x-axis). We are working with the ethnocentric assumption that north is up.

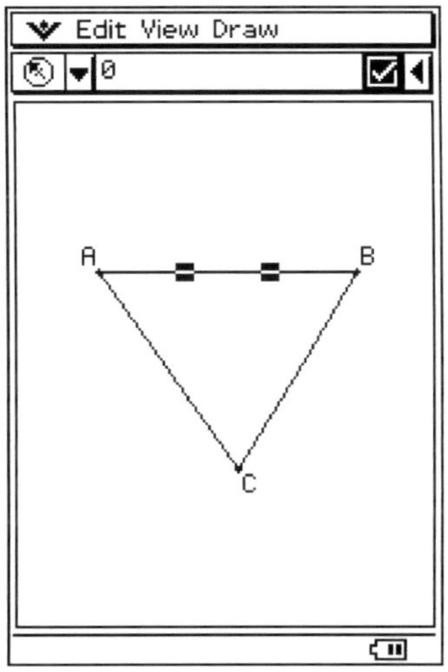

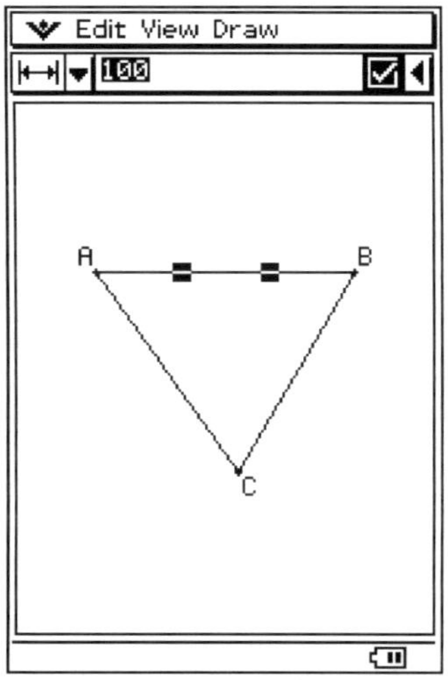

Next we want to set the angles BAC and ABC, and we have a problem. Geographers measure angles clockwise from north. So do compasses. Mathematicians, whose compasses don't measure angles at all, start at east and work counter-clockwise. Fortunately ClassPad will evaluate an expression in the measurement box:

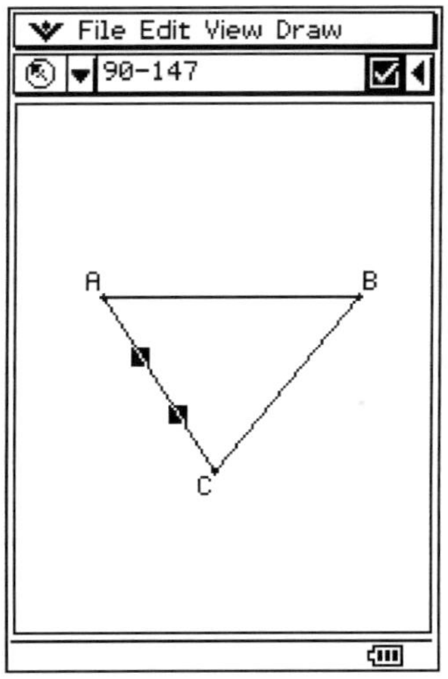

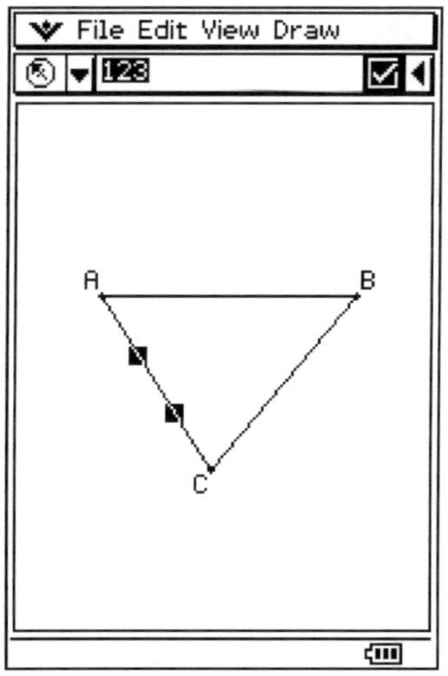

Notice the angle displayed in the measurement box can be the reciprocal direction.

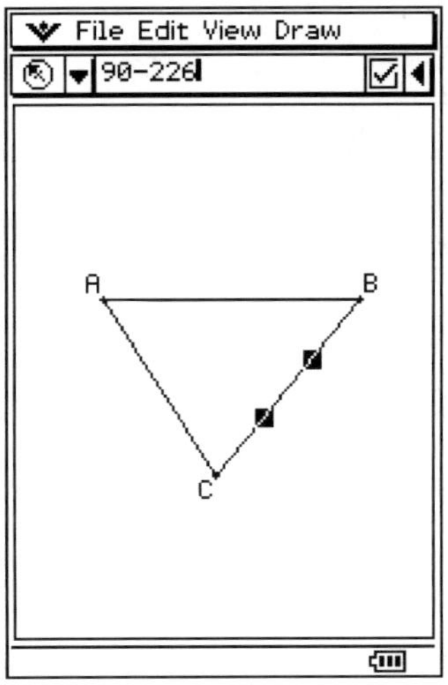

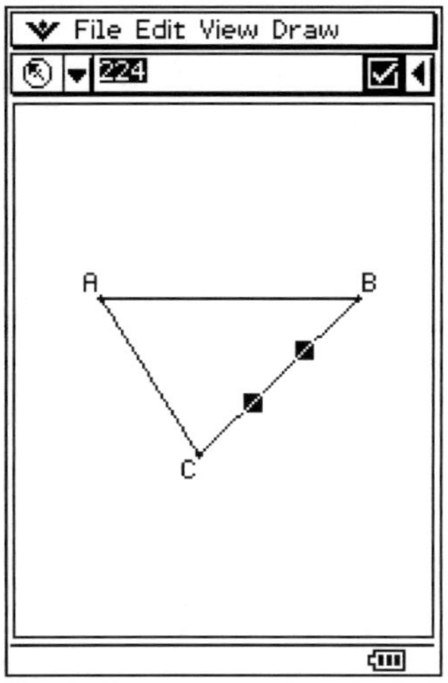

With C selected, its coordinates are displayed in the measurement box.

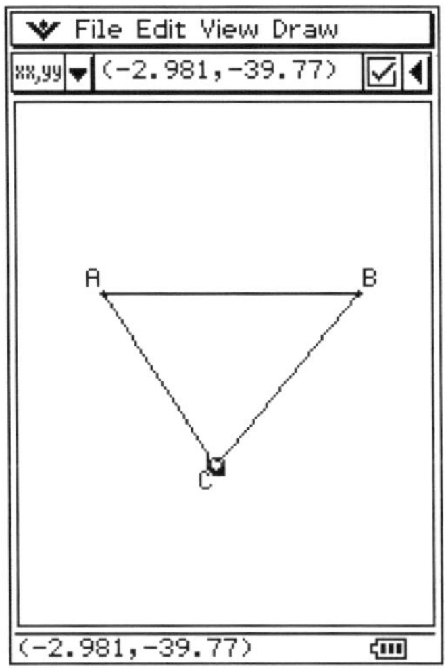

Coordinates relative to what, you might ask. Relative to the origin, of course, but as we positioned AB without reference to any origin, these coordinates are in fact entirely arbitrary (i.e. useless).

We can remedy this by putting A at the origin: Select A, change its coordinates and press the check button. Then select C and you will see its coordinates relative to A.

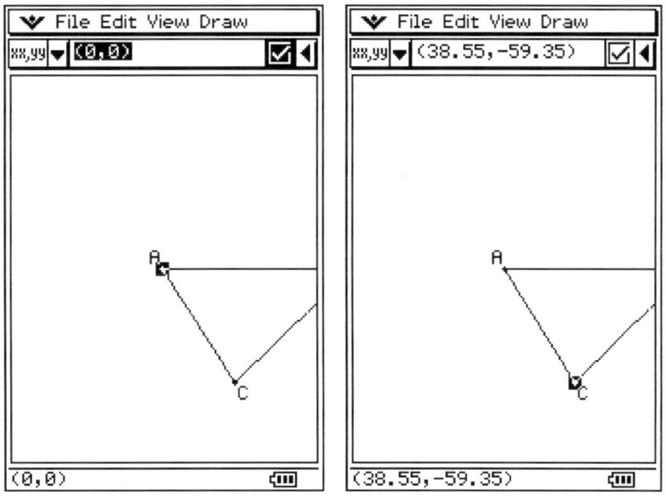

Pythagoras

What is the most popular triangle in all of geometry? If you go by the frequency it occurs in exam problems, it has to be the triangle whose sides are 3,4, 5.

What is the radius of its incircle?

We follow the usual constraint routine to sketch the triangle and set the lengths of its sides:

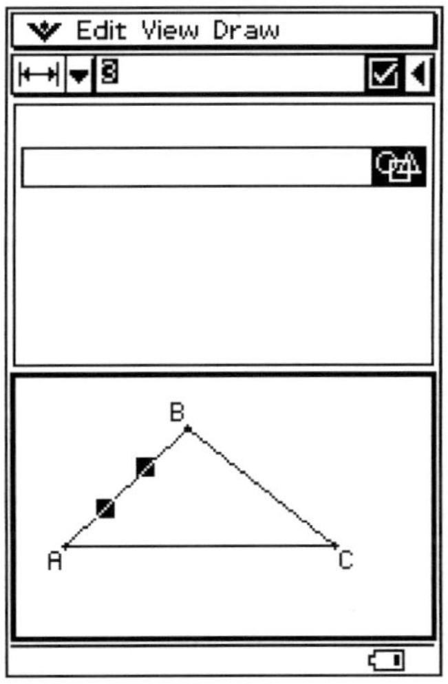

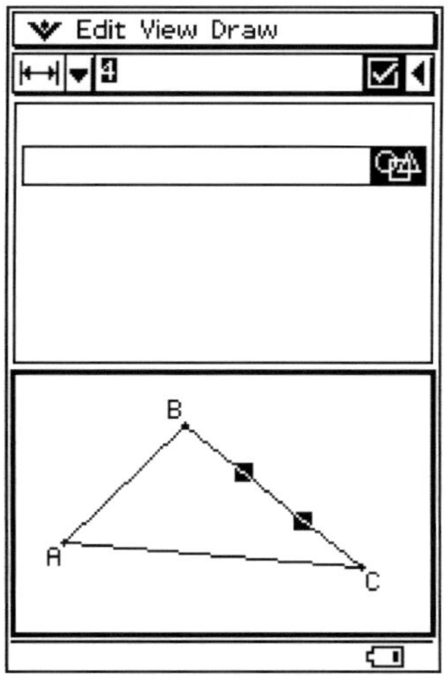

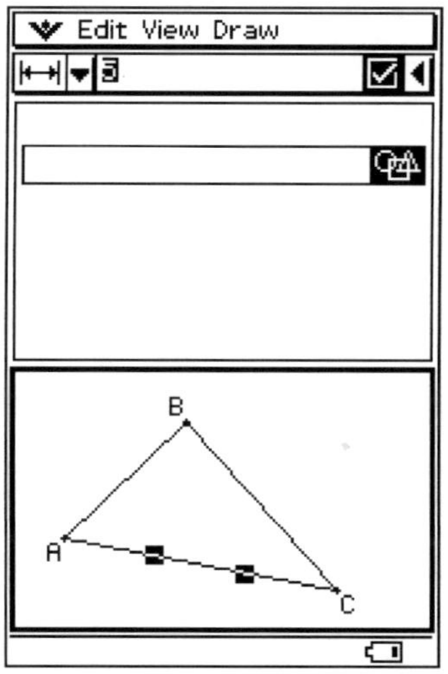

We can check up on Pythagoras (or, strictly speaking, his converse) by selecting sides AB and BC and inspecting their angle:

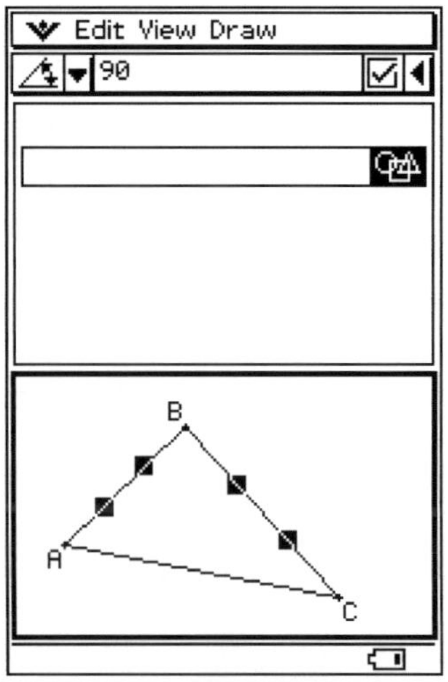

Our mission, however, lay with the incircle. The incircle is the circle which is tangent to the three sides of the triangle. We create this simply by sketching it, then setting tangency constraints between the circle and each side of the triangle:

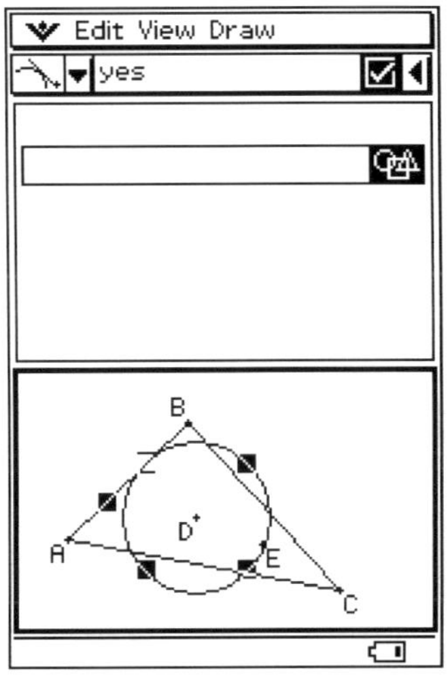

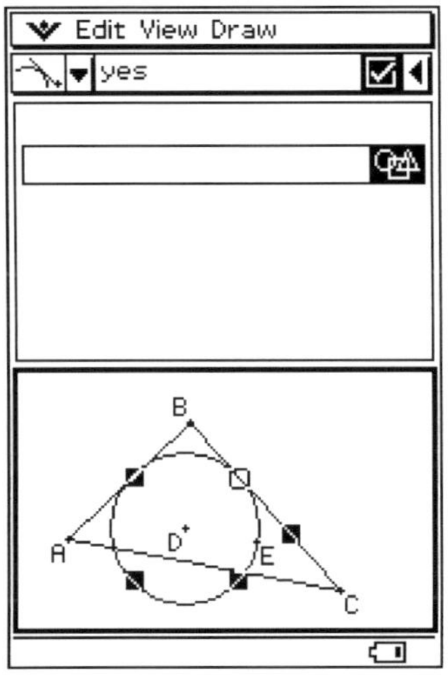

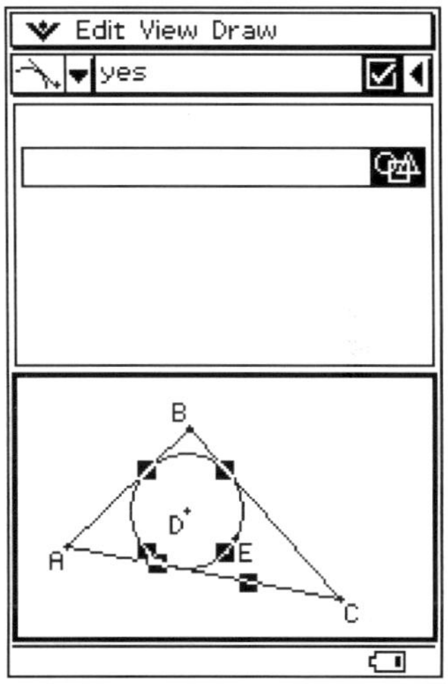

To find its radius, of course, we select the circle and the appropriate drop down in the Measure Selection button:

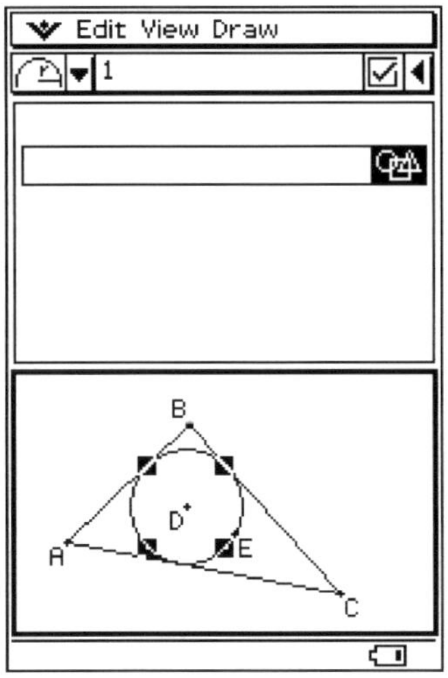

The incircle for this simplest of Pythagorean triangles has unit radius.

Which immediately starts us thinking in terms of a triangle whose sides are 5, 12 and 13:

We already have a side of length 5, so let's make the side whose length is 3 into a side of length 12:

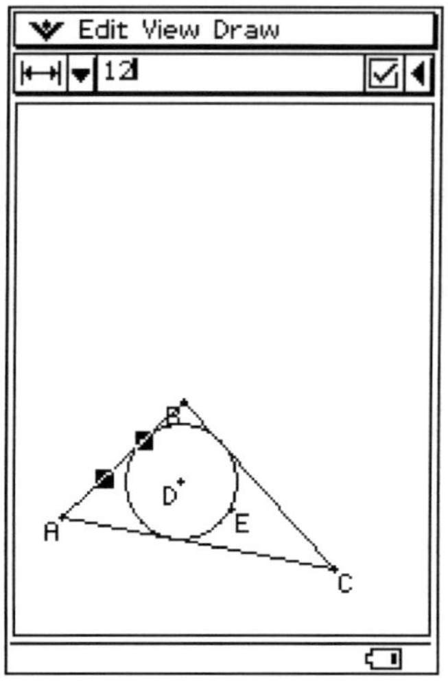

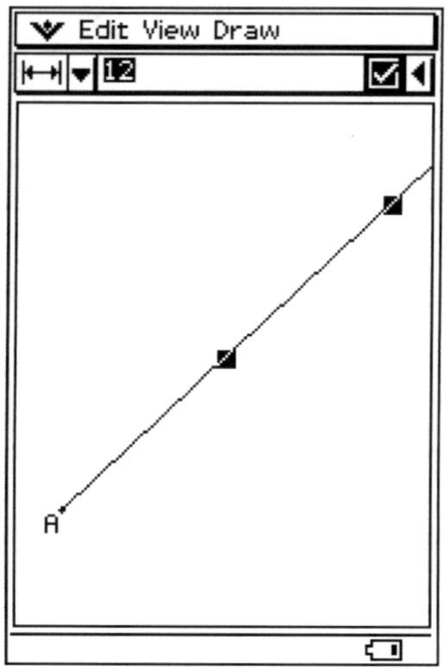

The first tap on the check box unlocks the old length. The second tap sets the length to 12. Which is all very well, but what happened to the rest of the triangle? The problem is, of course, that the third side is still locked to length 4; so we have specified a 4,5,12 triangle, which violates the triangle inequality. Faced with situation where the particular values of the constraints make the drawing impossible, ClassPad draws what it can and leaves out things which it is unable to make conform to the specified constraints.

Now is the time to **Undo** that last action:

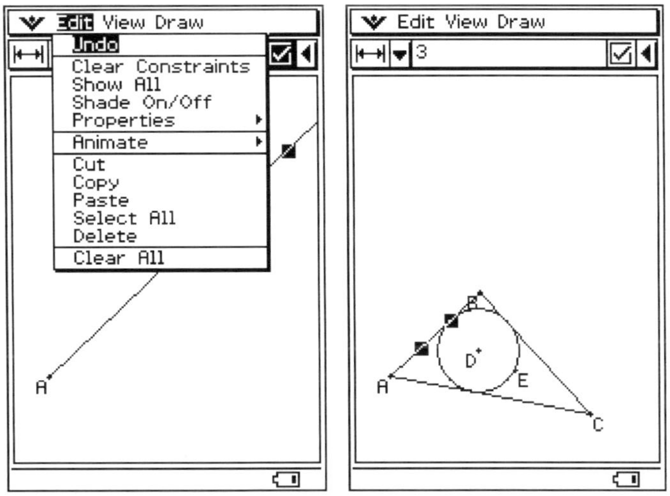

Before setting this side's length, we should unlock the length of BC, by highlighting it and unselecting the Check box:

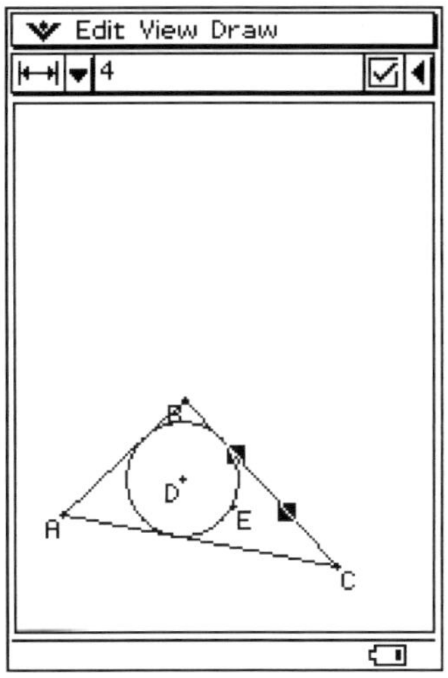

Now we can safely set the length of AB, and then of BC:

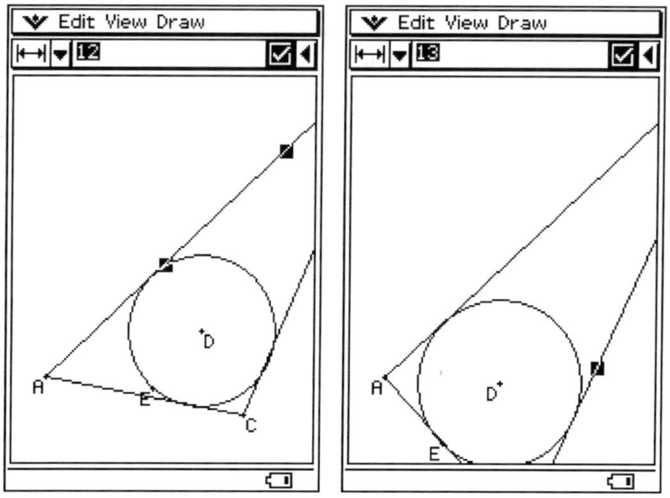

Again, we can check up on old Pythagoras, and look at that radius:

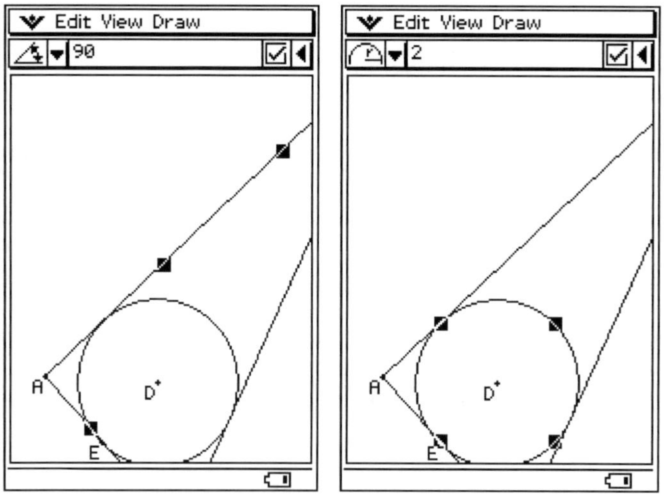

Is there a pattern here? If so, why?

Dot Products and Cross Products

To project a vector onto a line, we can take the dot product of the vector with a unit vector in the direction of the line. Let's illustrate this in ClassPad: First we draw the line and vector. Then we add a second vector which we will align with AB, by constraining the angle between the vector and the line to be 0. We constrain the length of this second vector to be 1:

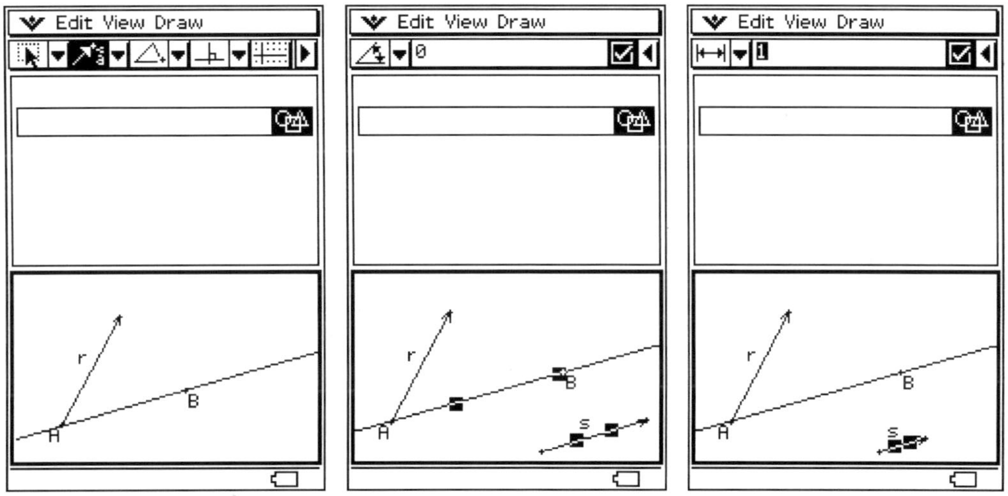

Now we can find the projection of r onto AB by dropping a line from the tip of the vector to the line AB, then constraining it to be perpendicular: If the perpendicular joins AB at C, then the distance AC is the projection of our vector on AB.

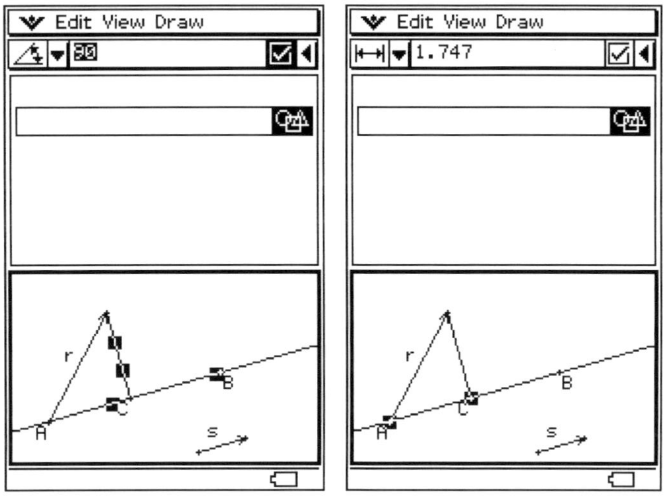

Alternatively, we can compute the dot product of r and s in the eActivity window. We can simply drag the vector from the geometry window into the dot product function, found on the **cat** tab on the soft keyboard, and input into the eActivity window:

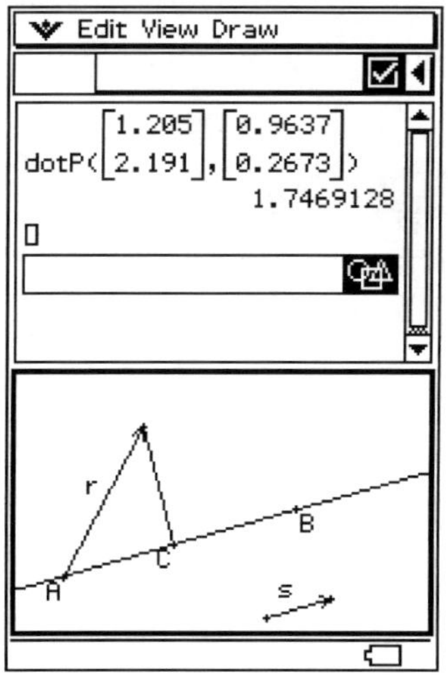

Cross Product

Let's say we are asked to interpret the cross product of the two vectors $\begin{bmatrix}2\\3\end{bmatrix}$, and $\begin{bmatrix}4\\-1\end{bmatrix}$ geometrically:

First we can do the cross product in eActivity:

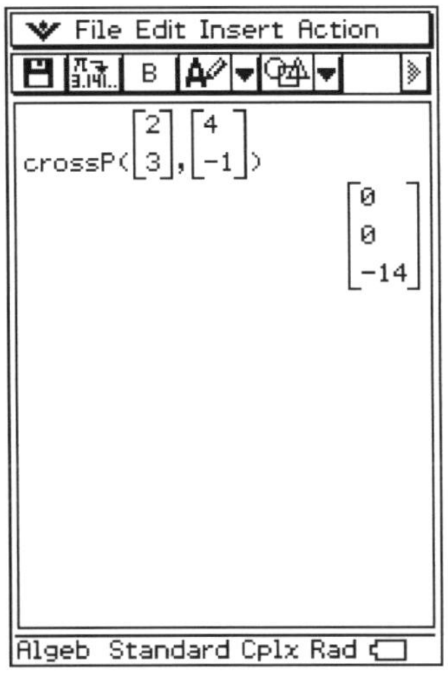

Now, let's drag these vectors into a geometry window. We get two points, whose position vectors are $\begin{bmatrix}2\\3\end{bmatrix}$, and $\begin{bmatrix}4\\-1\end{bmatrix}$

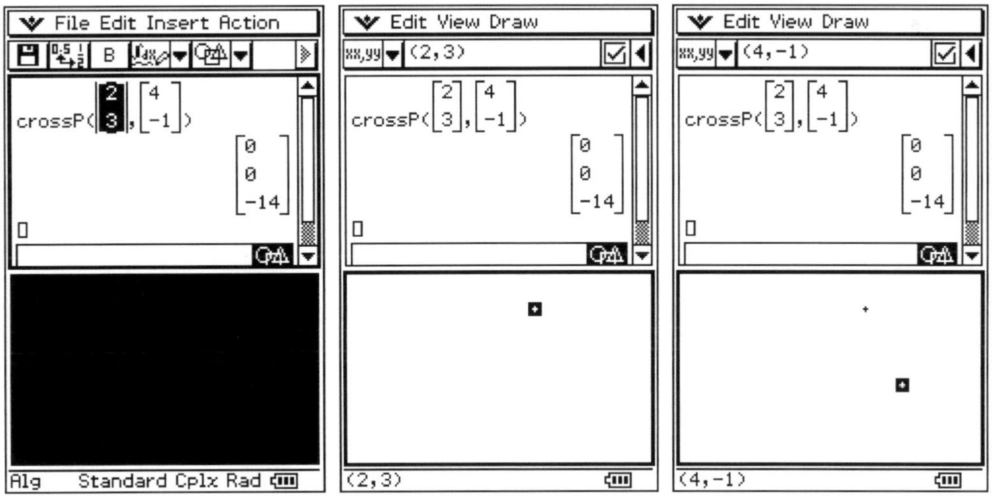

To create vectors, we need to join these up to the origin. To create the origin, we need to draw a point and constrain its location.

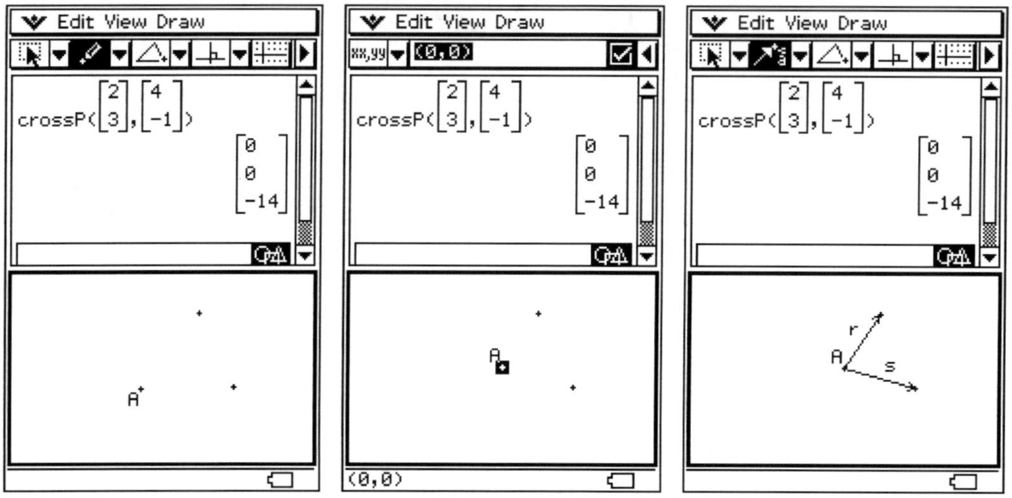

To interpret the cross product, we wish to create the parallelogram formed by the vectors. We do this by adding two sides to make a quadrilateral, then constraining each new side to be parallel to the opposite vector. We create a parallel constraint by specifying a 0 angle.

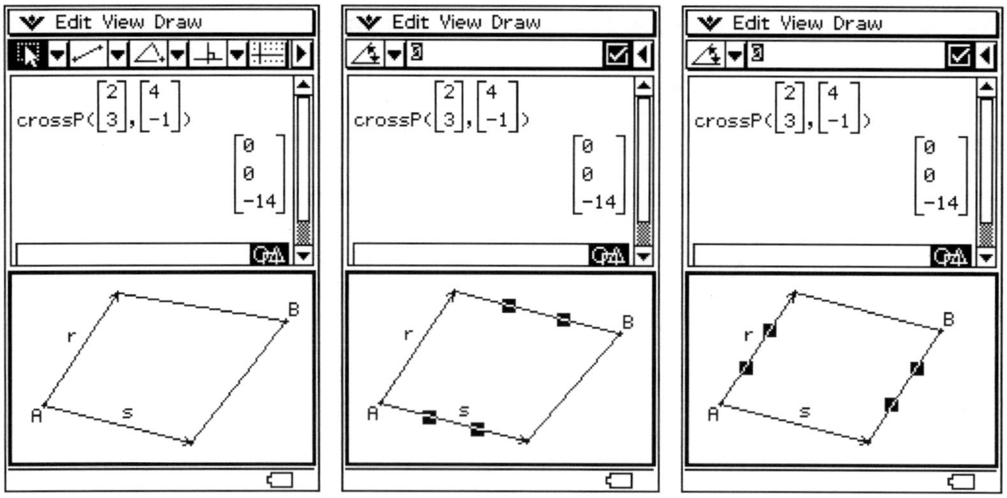

The area of the parallelogram can be evaluated by selecting its four sides:

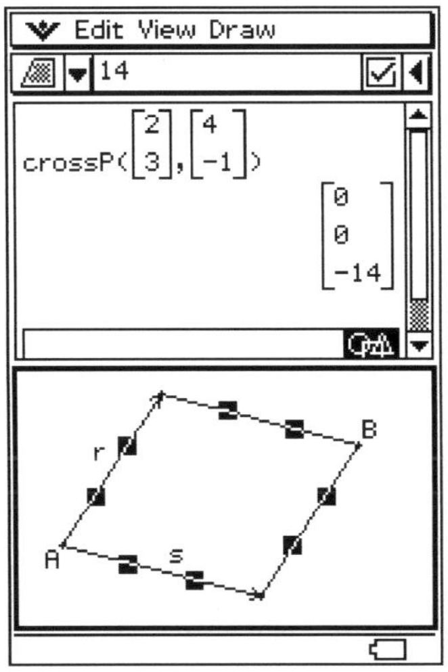

Mechanisms

Mechanisms are pieces of geometry which move in interesting ways. Typical geometry moves in a boring (or sometimes interesting but meaningless) way, when you drag it. With the right constraints set, however, dragging the geometry will mimic the motion of a real-life mechanism.

4 Bar Linkage

The standard non-trivial mechanism is a four bar linkage. Aptly named, as it is drawn with three bars:

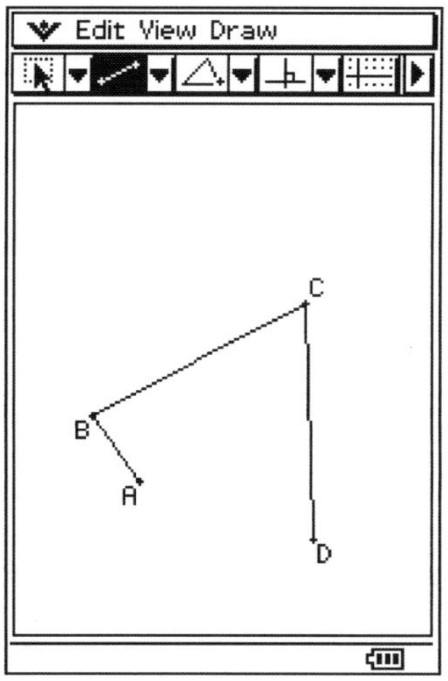

(An imaginary 4th bar is assumed between A and D).

Now if we drag points on the drawing without constraints being set, it certainly does not behave like a mechanism – the "bars" stretch or shrink, which would not be possible in real life, unless they were made of elastic.

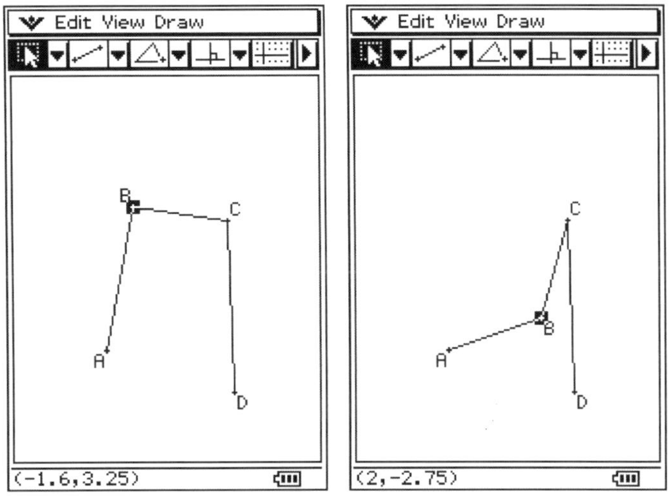

In order to have our drawing behave like the mechanism, we need to constrain the lengths of its sides. We also need to constrain the location of points A and D, where the mechanism is attached to a rigid base:

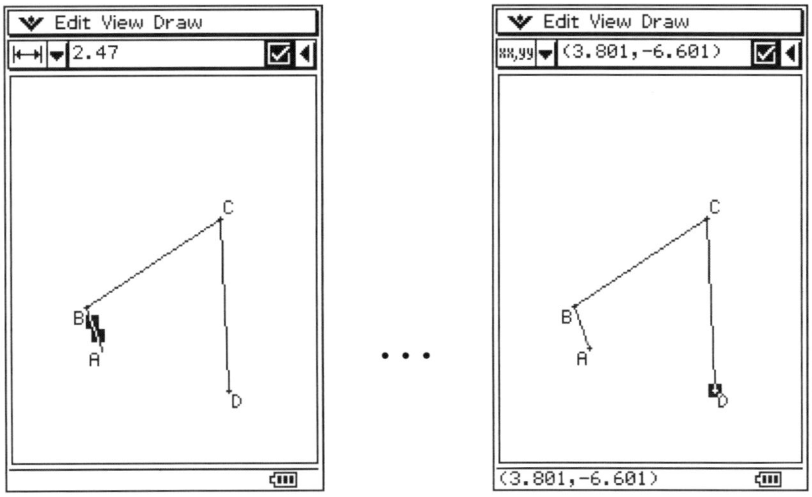

Now if we drag point B, we see the drawing moves in such a way that the lengths |AB|, |BC|, |CD| are conserved, and A and D stay fixed. This is the (interesting) mechanism motion.

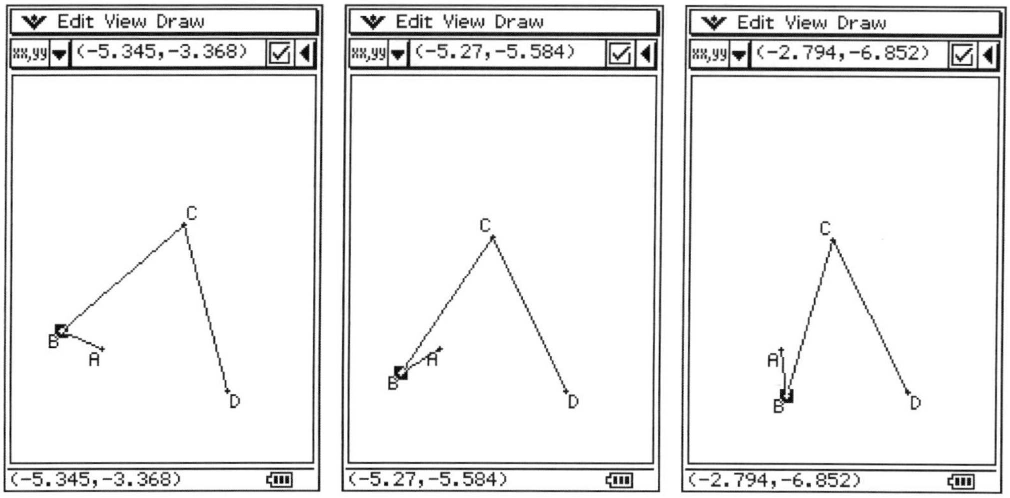

Four bar linkages often have an end-effecter rigidly attached to BC, but at some distance from the centerline. We can add this to our model by drawing a triangle, and constraining the lengths of its sides:

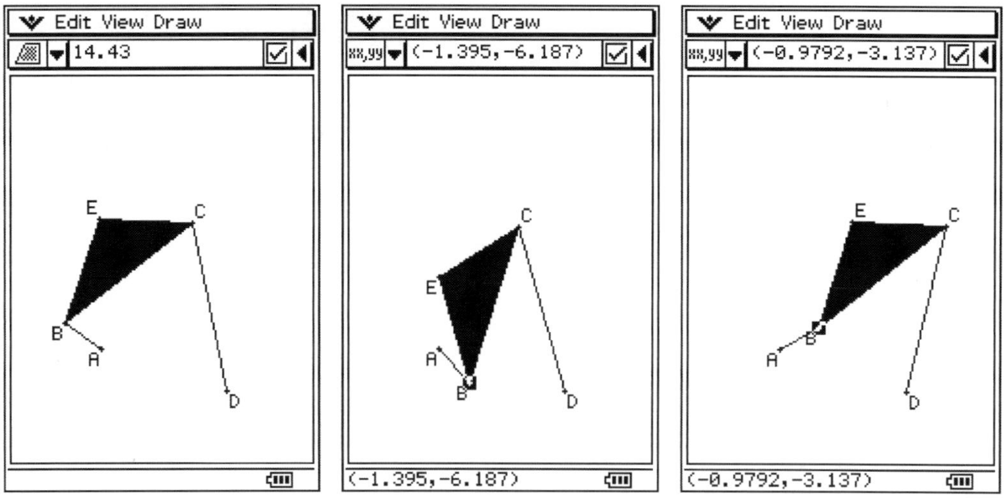

Paucellier's Linkage

Coincidentally, the Paucellier linkage was invented by a French Naval Lieutenant by the name of Paucellier.

It was the first mechanism employing only rotational joints to convert rotary motion to true, rather than approximate straight line motion. (The crank/piston arrangement found in most engines is an example of a mechanism which converts rotary motion of the crank to linear motion of the piston, however it involves a sliding joint at the piston.)

The mechanism remained a curiosity rather than a particularly useful one, probably because it only worked for a portion of a complete revolution.

Notwithstanding this, that great center for useless curiosities – the Houses of Parliament in London apparently employs Paucellier's linkage in its radiator ducts.

Here's what it looks like:

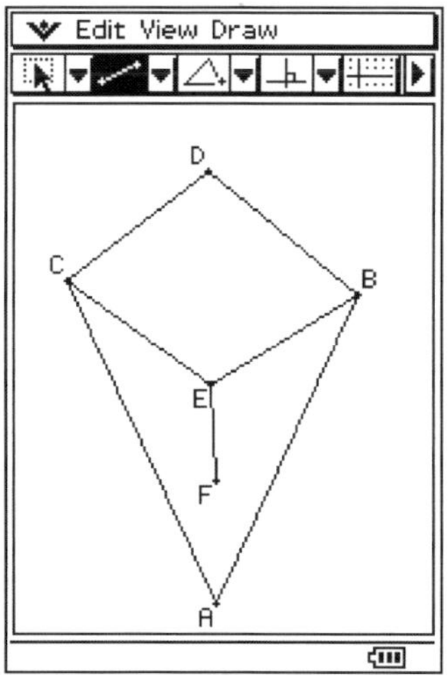

The lines of the figure will all have their lengths constrained. In addition, the location of points A and F will be constrained.

We will need to constrain the lengths in such a way that the following hold:

$$|AB| = |AC|$$

$$|CE| = |EB| = |BD| = |CD|$$

|EF| = |FA|

The crank is the line EF. The output motion is acquired at point D.

For convenience in analyzing the output, we'll put A and F both on the y-axis:

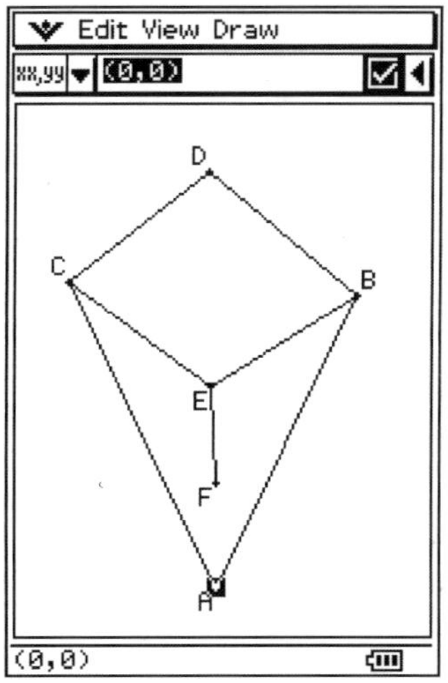

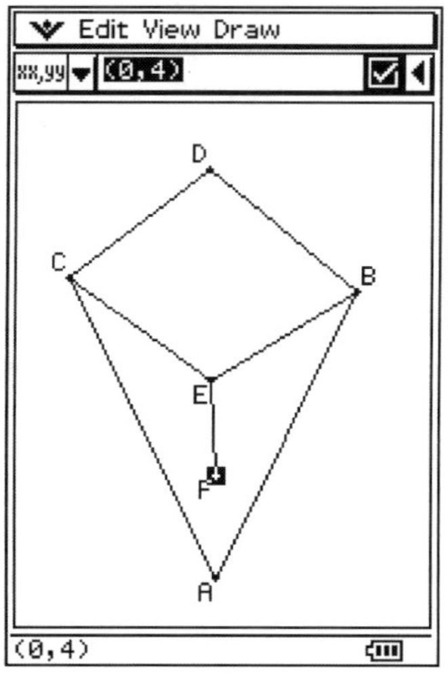

As AF is 4, we need to set EF to 4 also:

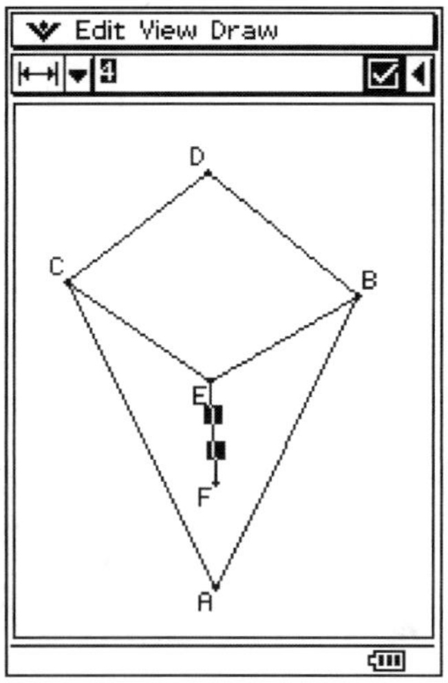

Set AB and AC to the same value: 14 in this case, and the four smaller links to their value: 8

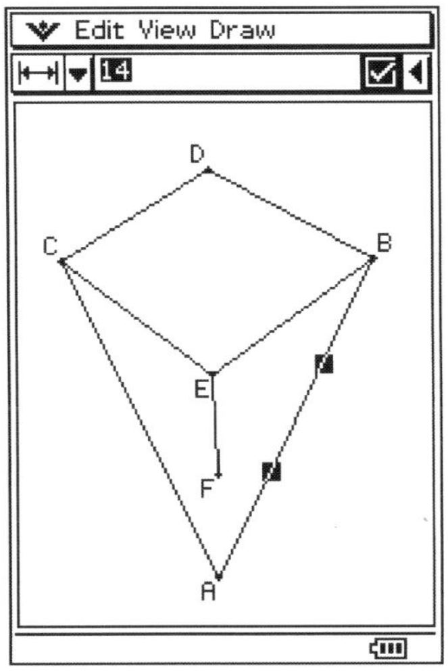

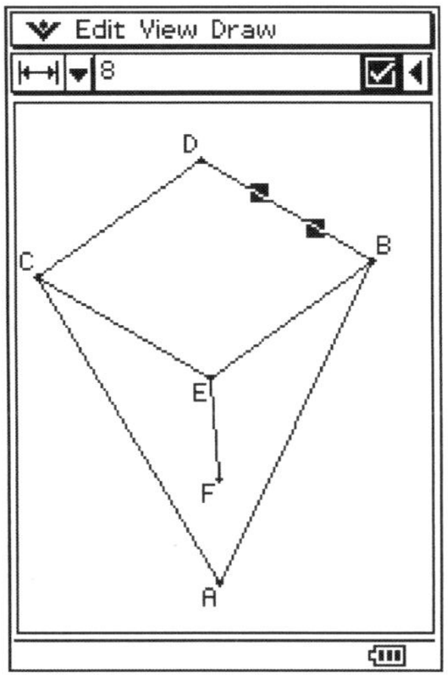

With all the links constrained, we can move the mechanism (if you break it, don't worry, use the undo key, or read on…)

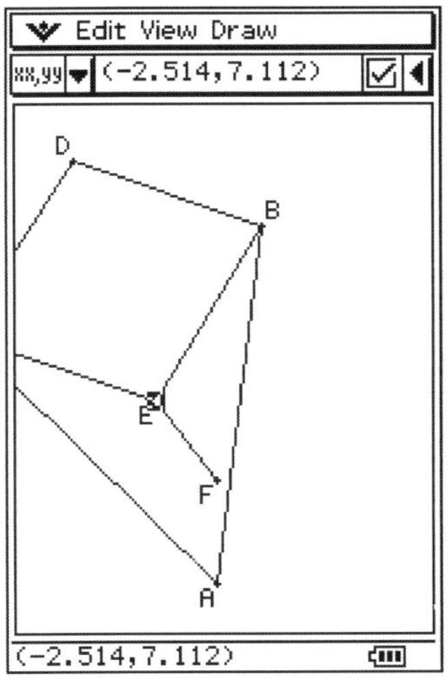

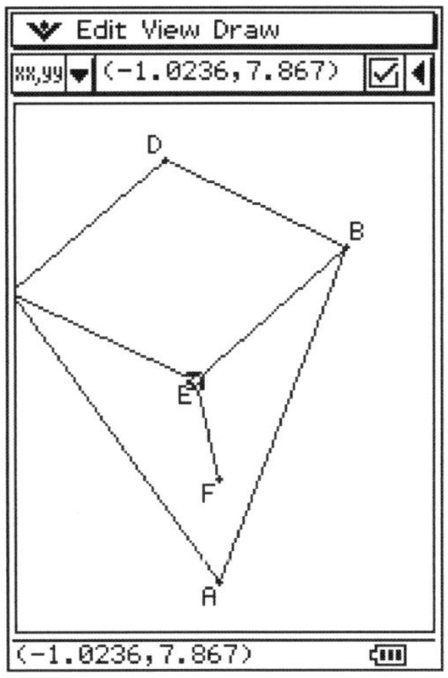

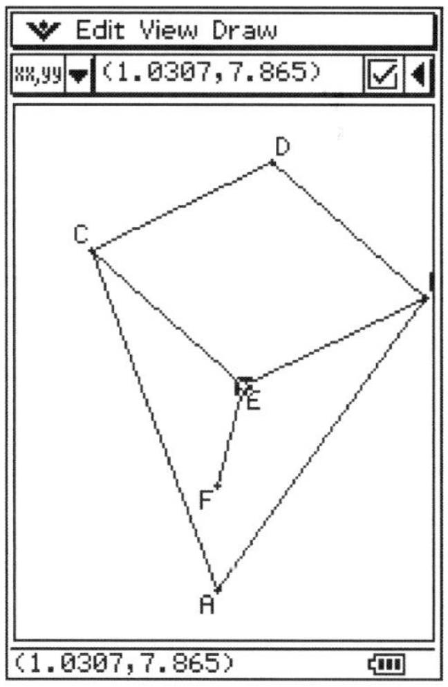

Did D actually move in a straight line? To find out, let's drag D into a geometry link in eActivity:

First create a Geometry Link from the eActivity menu:

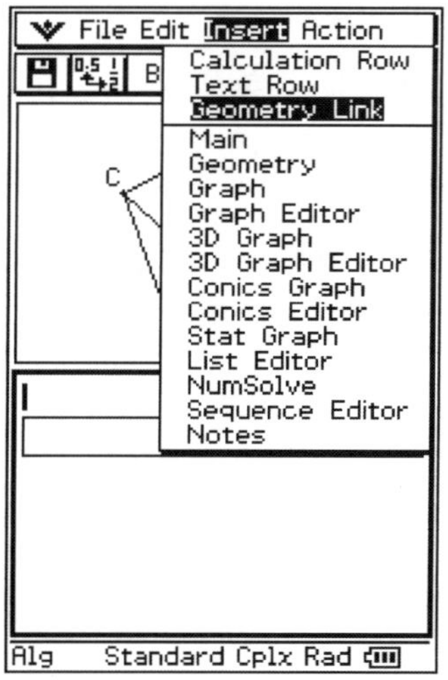

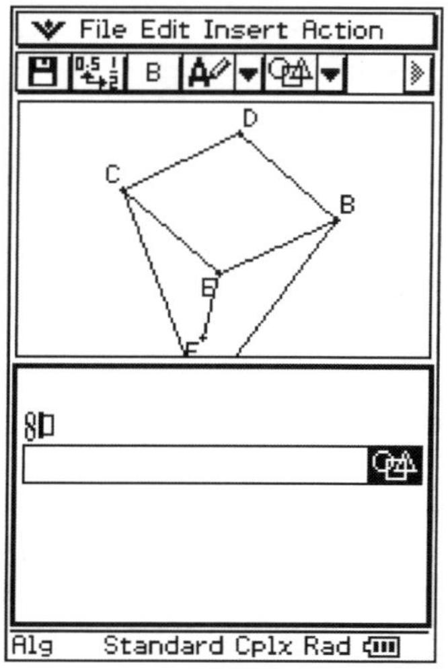

Now drag D and drop it into the link:

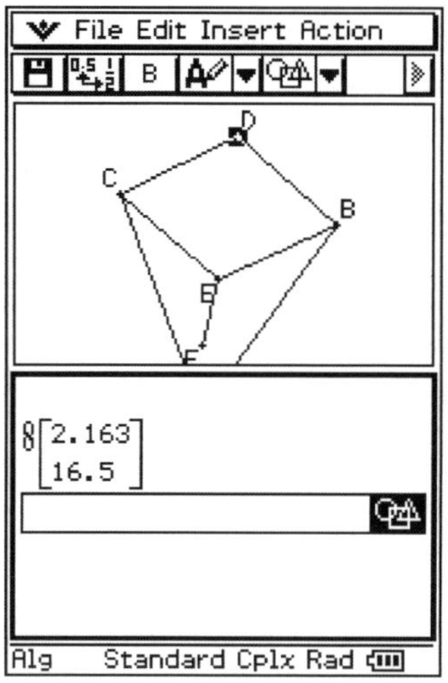

We can now drag point E through its range of motion and watch the coordinates update in the geometry link field:

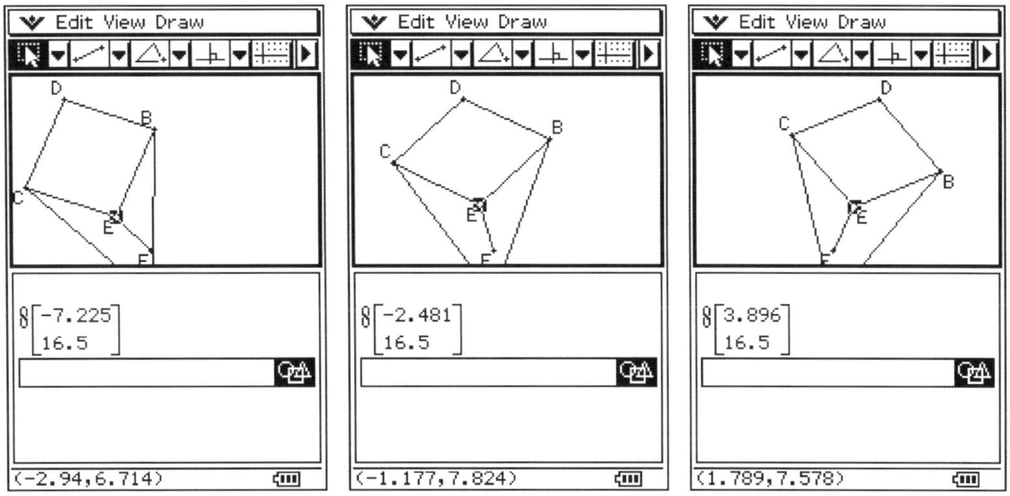

We see that the y coordinate stays constant, and hence traces a straight line.

You'll notice that we were very careful in only dragging E through a small arc of its possible motion. What happens if we drag it farther?

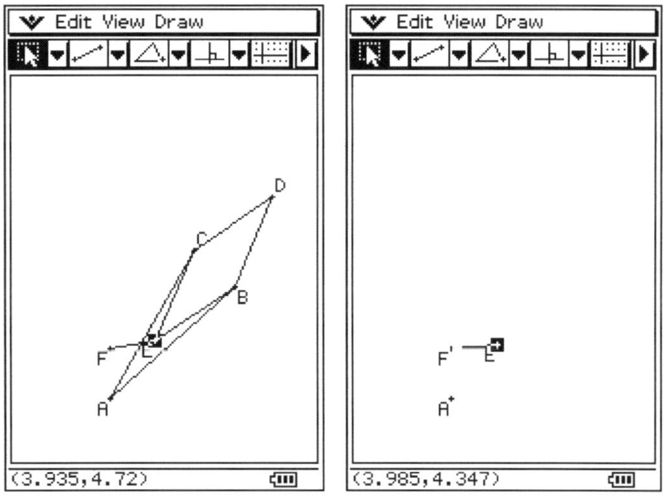

If E goes too far round, then it is impossible to find locations for C and B which match the constraints. These elements and any others which depend on them will disappear from the screen. (Don't worry, you can get them back either with Undo, or by dragging E back.)

Another interesting situation can be created if you drag E too far in one jump. Specifically if you drag it outside of the triangle ABC:

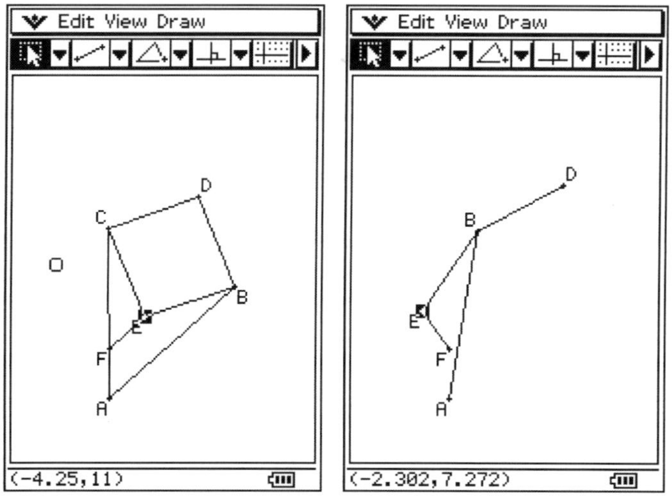

What has happened here is that we have jerked the mechanism into another configuration: one in which B and C are superimposed. You can Undo this, or drag B back out beyond E:

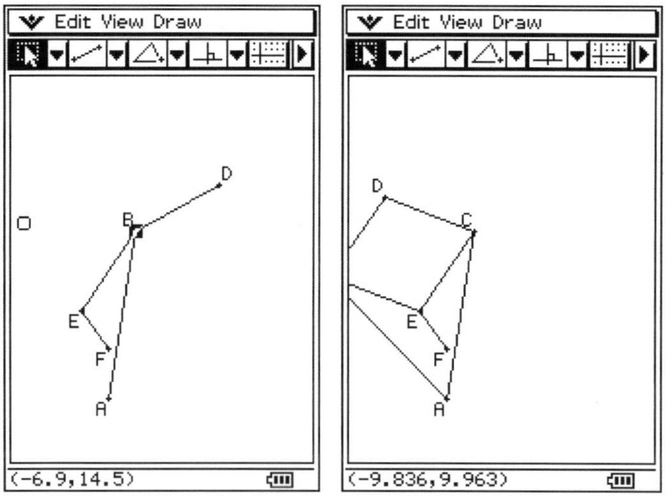

Napoleon's Theorem

Napoleon's theorem offers a tour de force for constraint geometry. The theorem states that for any arbitrary triangle, if you construct an equilateral triangle on each edge, and join the centers of the incircles of these triangles, then the resulting triangle is equilateral. The theorem is named for, and supposedly discovered by, Napoleon Bonaparte, himself no stranger to tours de force.

Take a triangle and draw a triangle on each of its sides

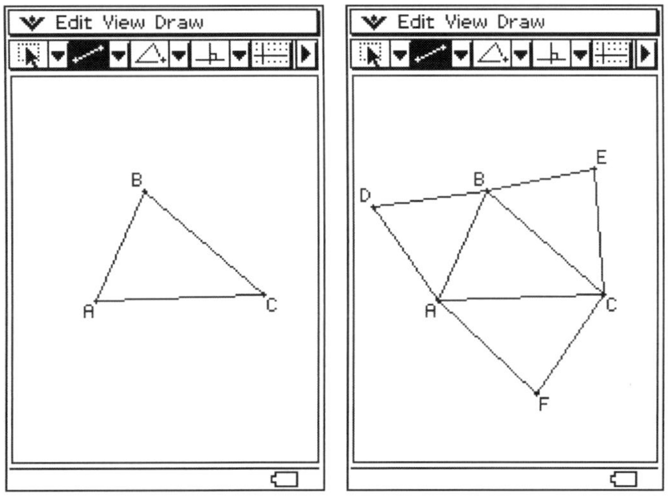

We can make the subtended triangles equilateral simply by specifying congruence constraints. Start by making AC congruent to AF and AC congruent to FC… (Congruence is the second option on the Measurement Selection drop down button when two segments are selected.)

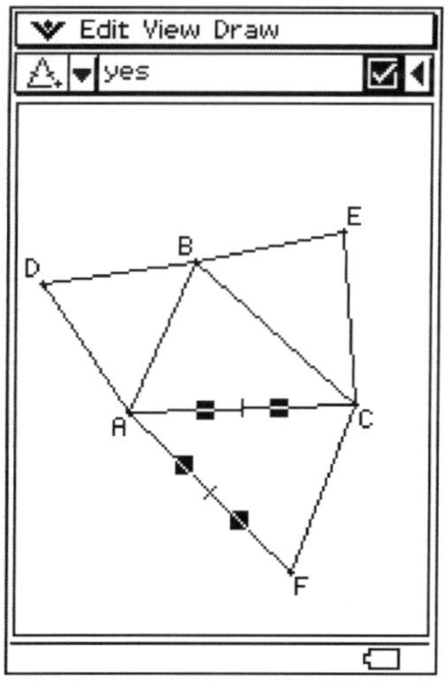

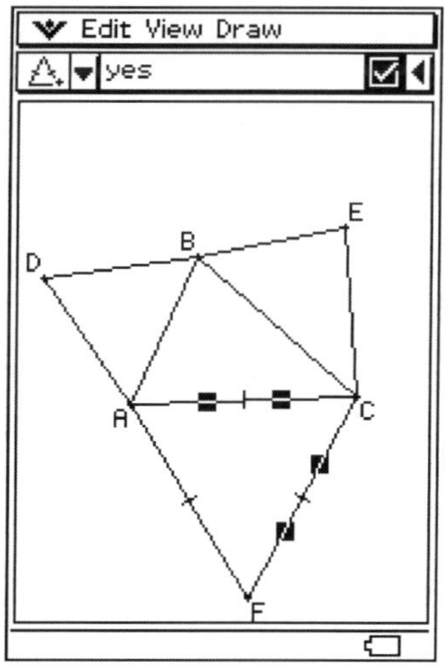

We can follow the same procedure for the other two triangles:

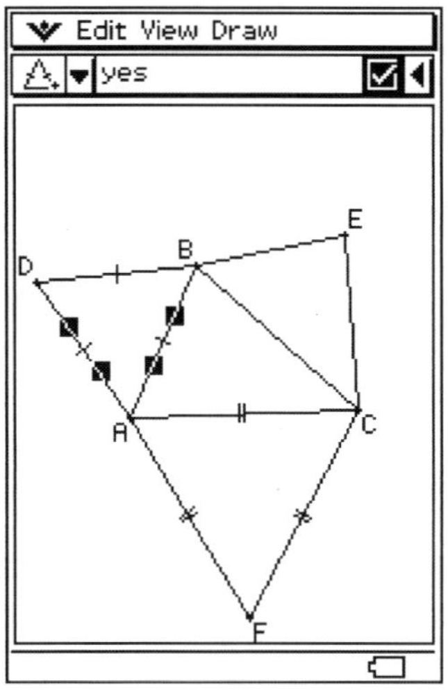

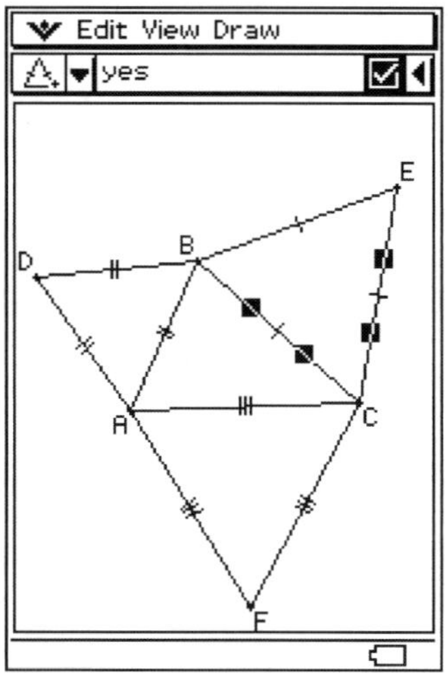

We have created the required equilateral triangles. Now for the incircles. We sketch the circles then set tangency constraints - three for each.

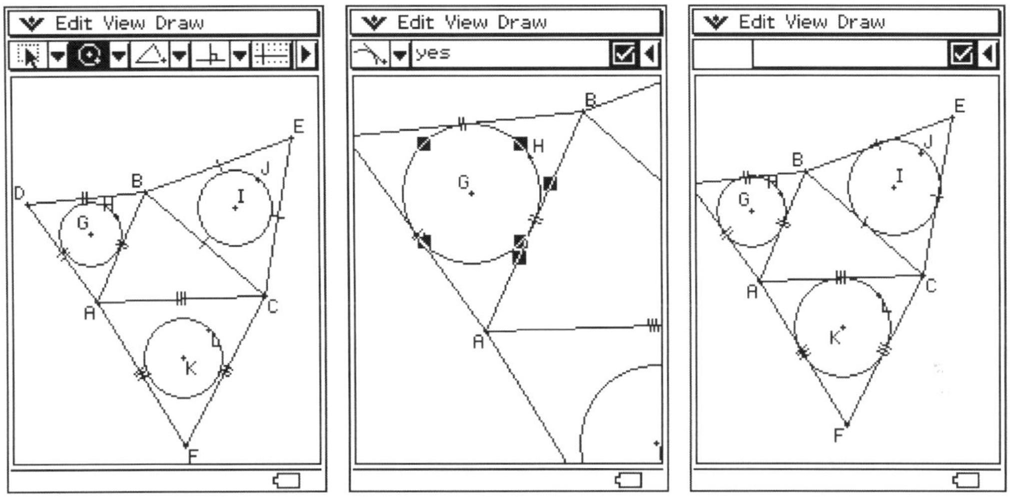

Now let's join the centers of these circles. Shading the resulting triangle makes it stand out from the cat's cradle of lines and circles. If you're not convinced that it is indeed equilateral - and why should you be, Napoleon was more famous for geopolitics than geometry - inspect its side lengths

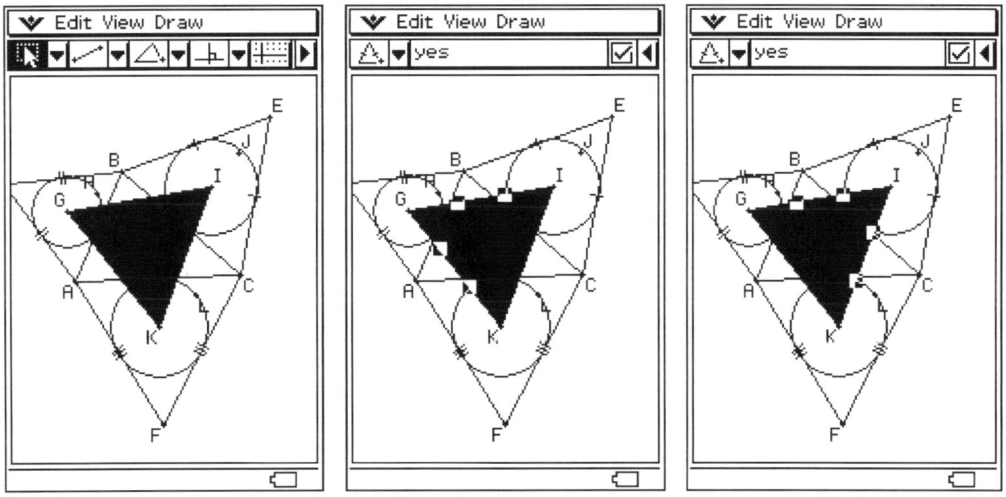

Dragging the original triangle shows that the derived one still remains equilateral, obedient as a Grenadier of the Old Guard to his Emperor.

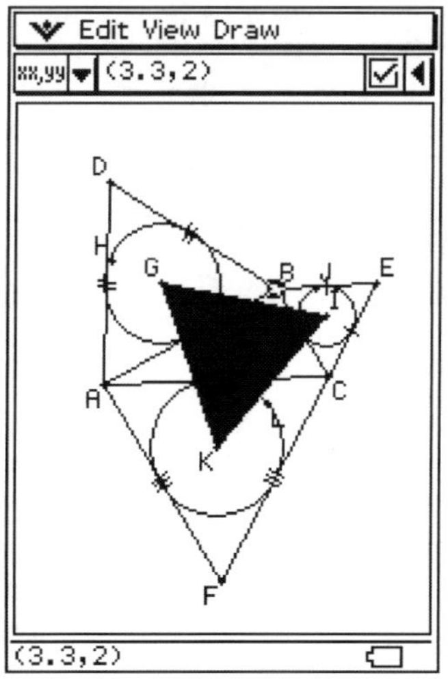

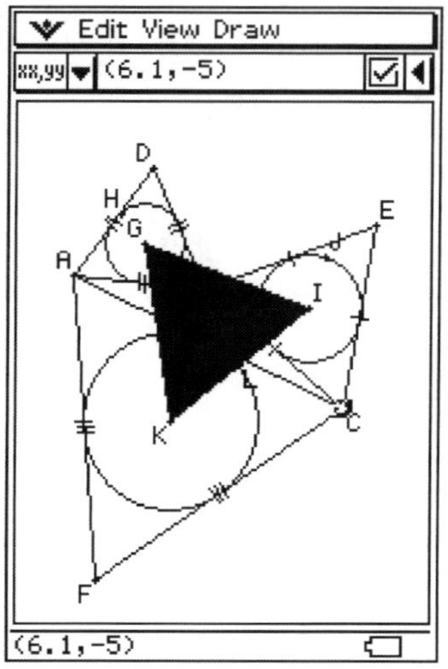

Functions

In addition to ClassPad's fully featured, traditional calculator graphing capability, you can display functions in ClassPad's geometry module. This allows you to explore more fully the interplay between functions and other geometrical entities: lines, circles, points, etc.

To add a function to ClassPad's geometry module, you simply click on the function button in the Draw drop down (or select **Draw/Function** from the menu). Then fill in the function definition dialog.

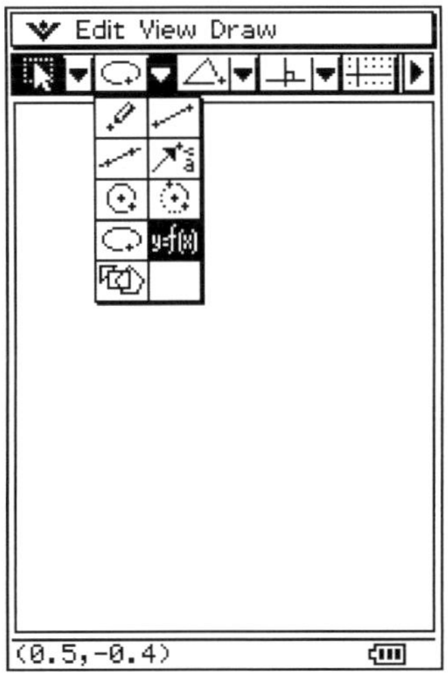

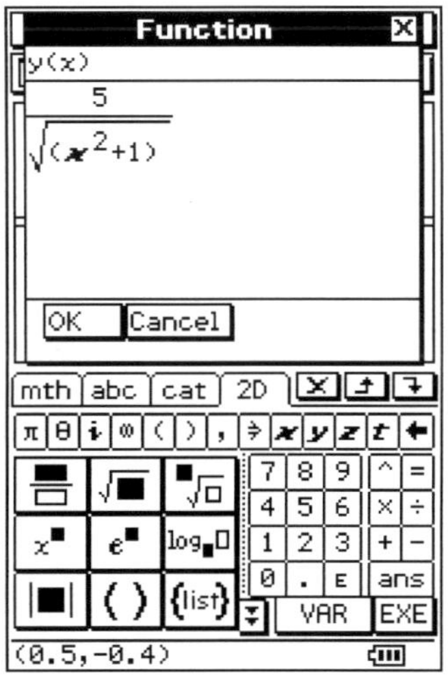

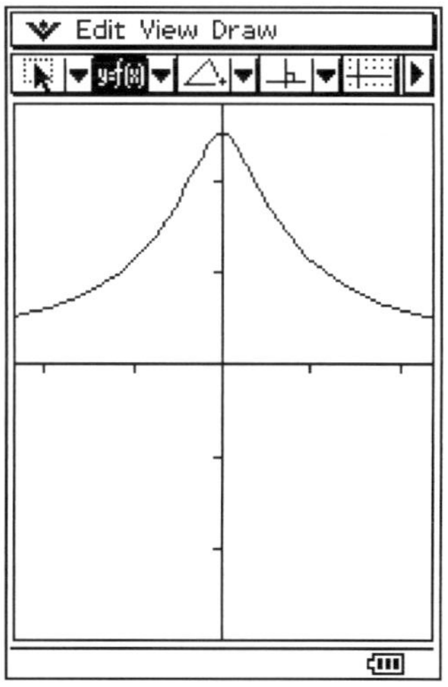

Alternatively, functions can be dragged and dropped into the geometry module from eActivity, or elsewhere, but more of that later.

First, let's look at quadratics.

Quadratic Functions

We'll do some exploration of different quadratic functions. Let's start with the simplest. Turn on the axes and the Integer grid and create the function y=x^2

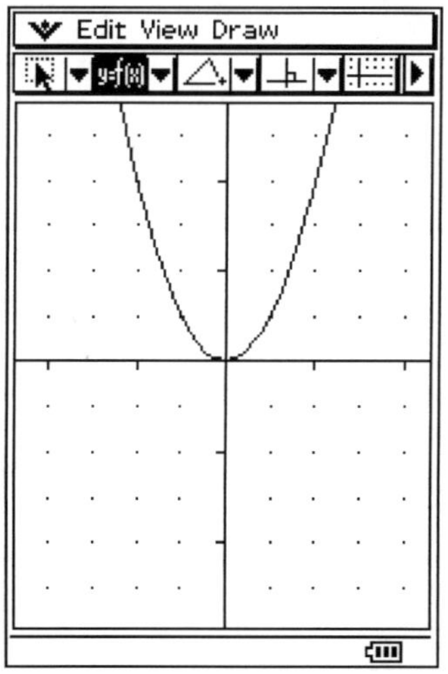

Selecting the function and displaying the measure box shows you the equation and allows you to alter it:

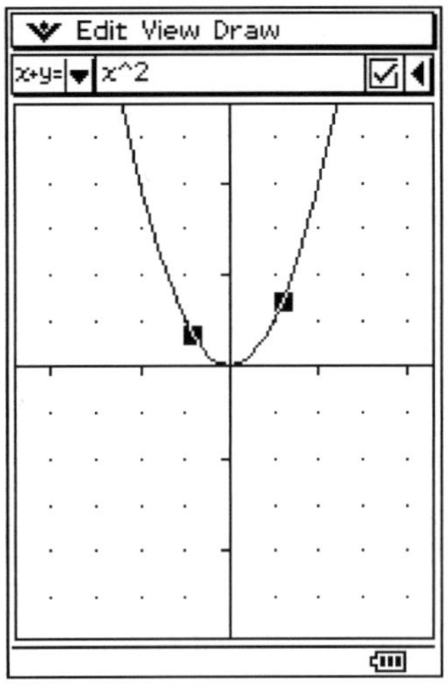

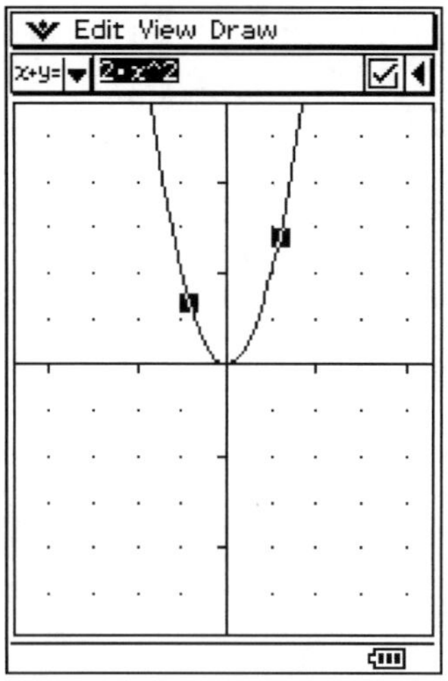

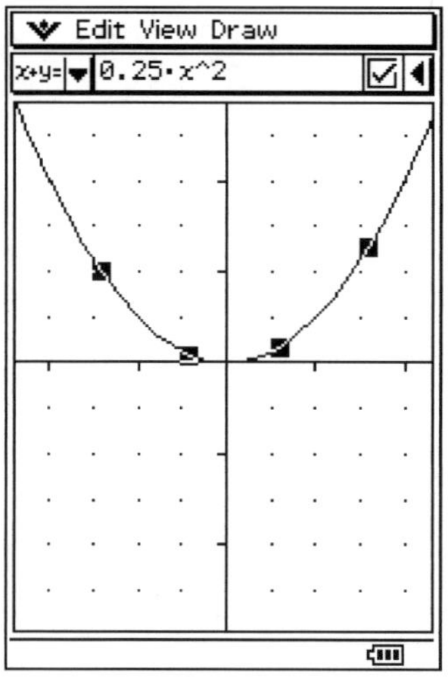

You can also drag the function and observe how its equation changes:

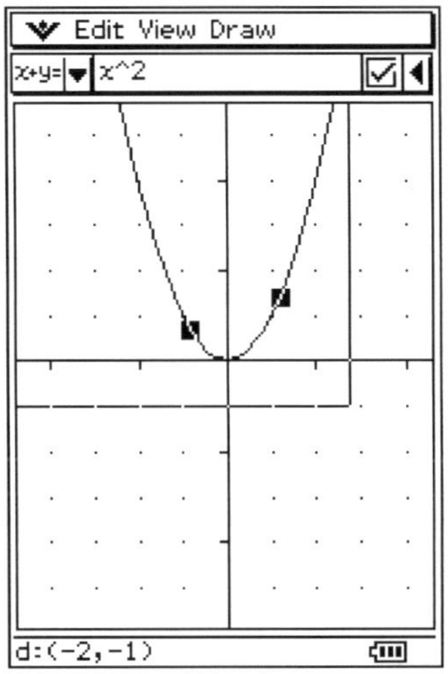

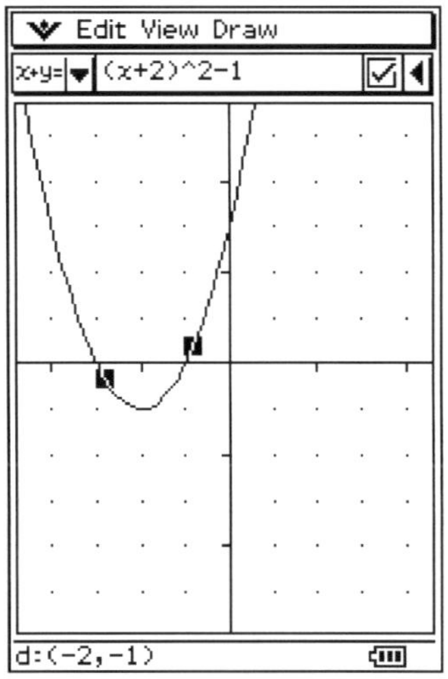

The minimum of $y=x^2$ is at $x = 0$, and has value 0. We have shifted the function down 1 and to the left 2. The minimum of the new function is therefore at $x = -2$ and $y = -1$.

The y intercept of the new function is easily computed: we just substitute $x = 0$ into the equation to see the result $y = (0+2)^2-1 = 3$.

The x intercept involves solving

$$(x+2)^2-1 = 0$$

We can drag the function into eActivity and solve there:

The function is expressed in the form $(x+2)^2-1 = 0$.

We can pull the -1 over to the other side of the equation to give:

$$(x+2)^2 = 1$$

solving this for (x+2) gives:

$$x+2 = 1 \text{ or } -1$$

hence $x = -1$ or -3.

We can verify this by inspection of the graph.

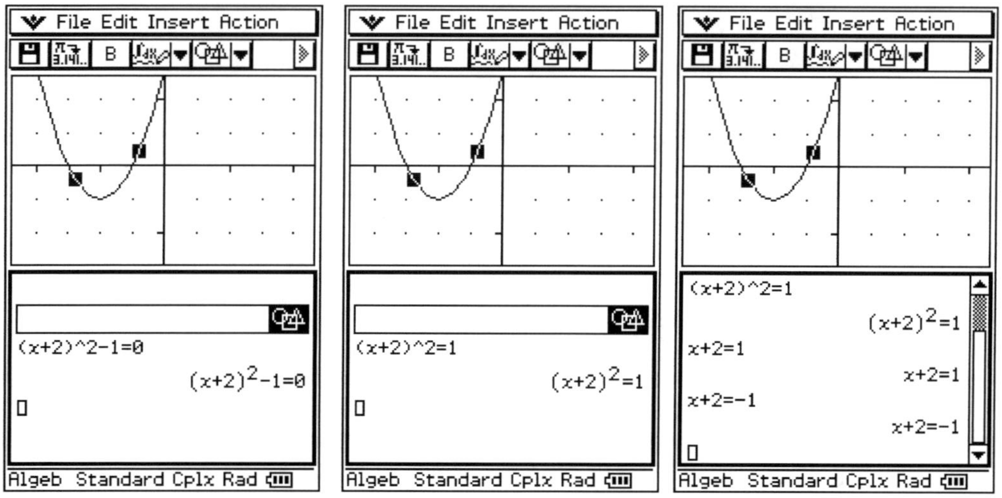

Try dragging the function to another location:

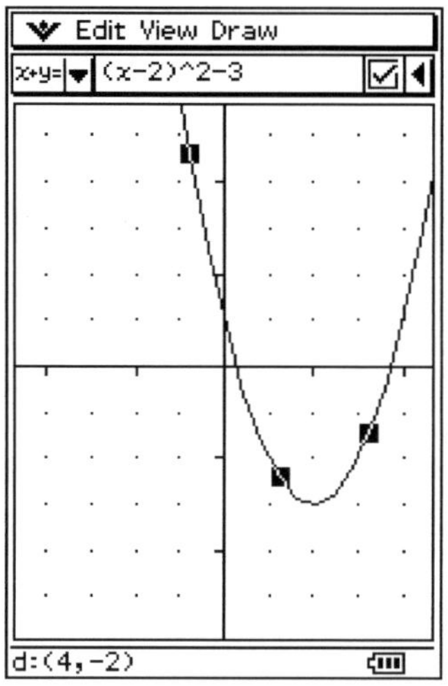

We can see by inspection of the form of the equation that the minimum is at the point (2,-3) and the y intercept is at (0,1).

Again, by inspection we can see that the x intercepts will be located where (x-2)^2 = 3,

hence x-2 = ±√3

and x = 2±√3.

We could jump straight to this solution by solving the equation directly in eActivity: faster but affording less insight.

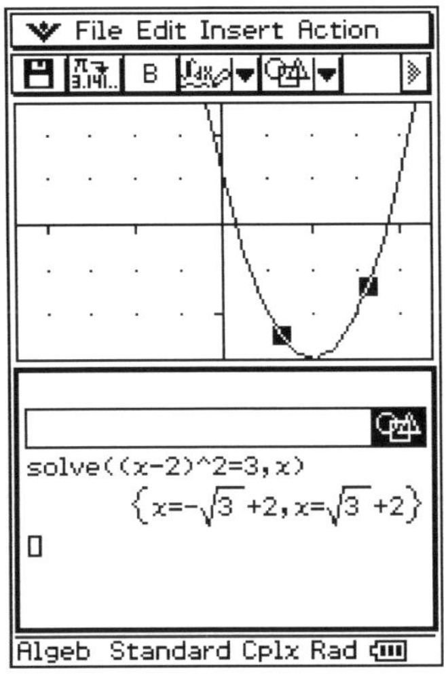

Completing the Square

Analysis of the functions above was greatly aided by their particularly convenient format. What if this format is not available?

Let's analyze the function y=x^2+6*x+5

Enter this into eActivity, then drag and drop into geometry to display the function:

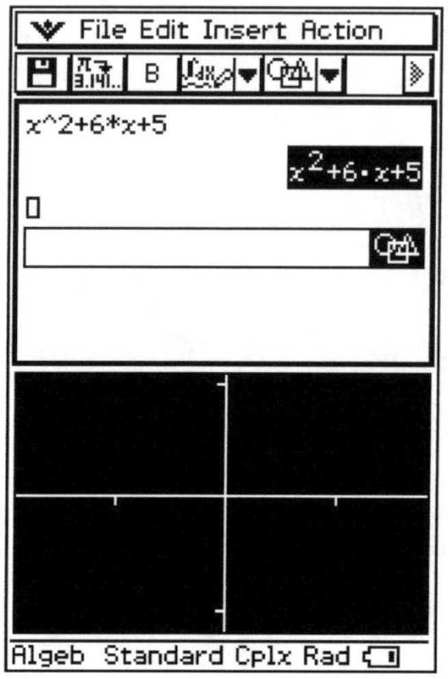

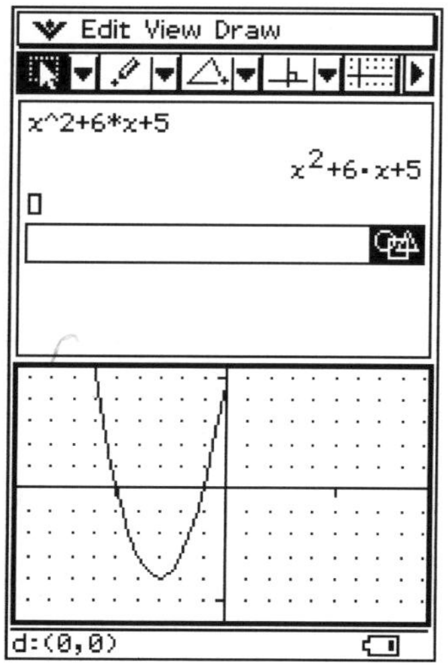

Inspection of the graph suggests that the minimum of the function is at the point (-3,-4). This would suggest the equation:

y=(x+3)^2-4

We can use the expand() function in eActivity to verify this hunch:

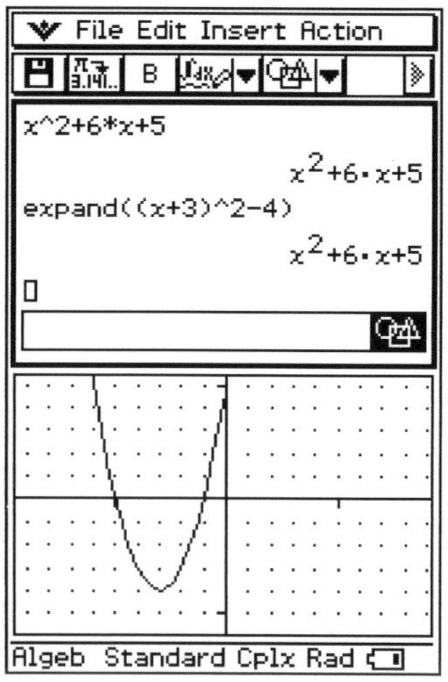

So if we want to find our x-intercepts, we need to solve:

(x+3)^2=4

hence x+3 = 2 or -2

hence x = -1 or -5.

Which coincides with the impression we get from the graph, and which can be verified by a quick call to eActivity's solve() command.

Leading Coefficients

Let's say we wish to solve the equation

2*x^2+6*x+3=0

Let's start off by graphing

y=2*x^2+6*x+3

(Just drag the expression into the geometry window)

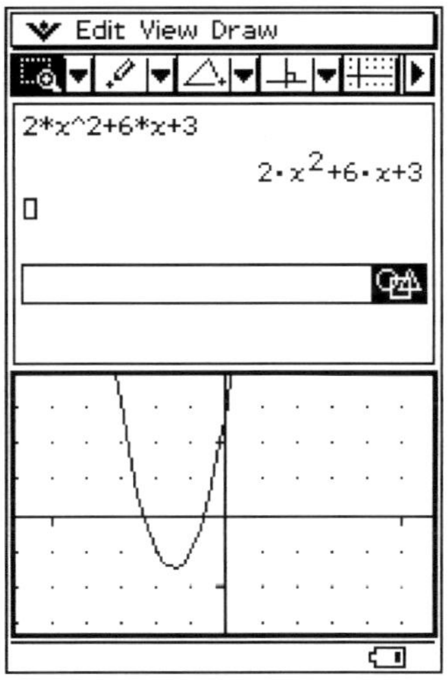

The solutions of the equation are the x intercepts of this function. It would be nice to use the technique of the previous example to solve this equation. There is one distinct difference with this problem, however: the coefficient of x^2 is not 1, hence completing the square is a little harder.

What if we divide the function through by 2 and drag it into the geometry window:

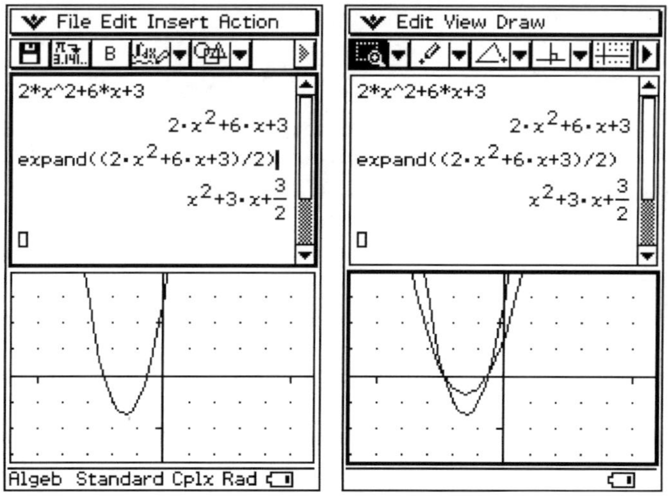

We see that the function is different, but has the same x-intercepts. Zoom in on them if you are not convinced!

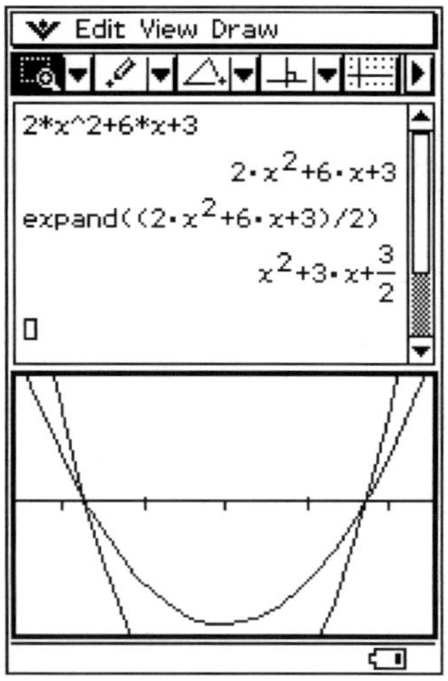

Hence if we want to find the x-intercepts, we can work just as well with the expression

x^2+3*x+3/2

It is easy to see that this can be massaged into the form:

(x+3/2)^2 + b

for some constant b which we need to calculate:

let's first expand (x+3/2)^2:

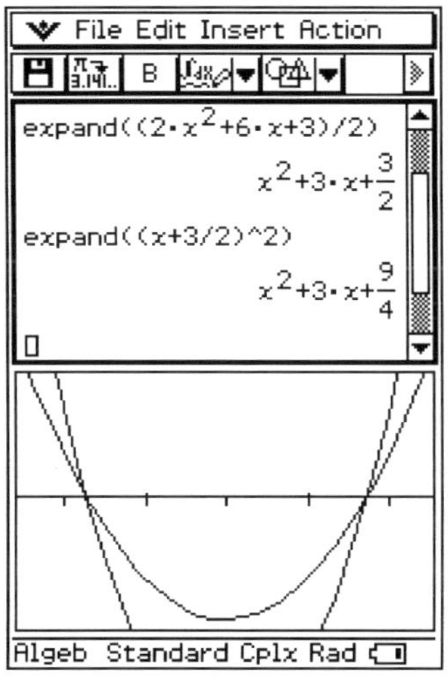

Clearly we need to add on 3/2 - 9/4:

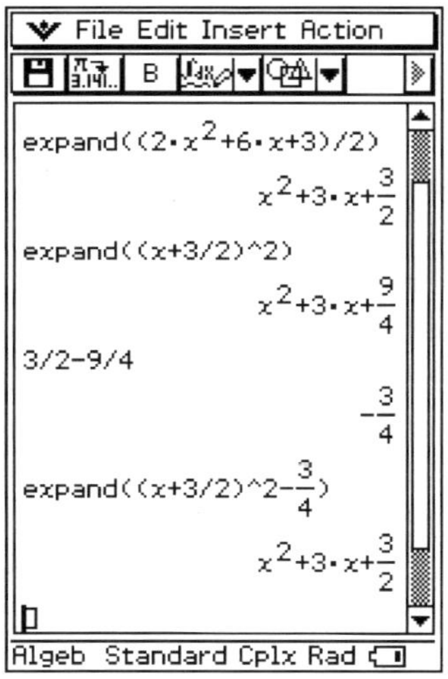

Now if we solve the equation for x, we can compare the solution to this expression and observe the family resemblance between equation and solution:

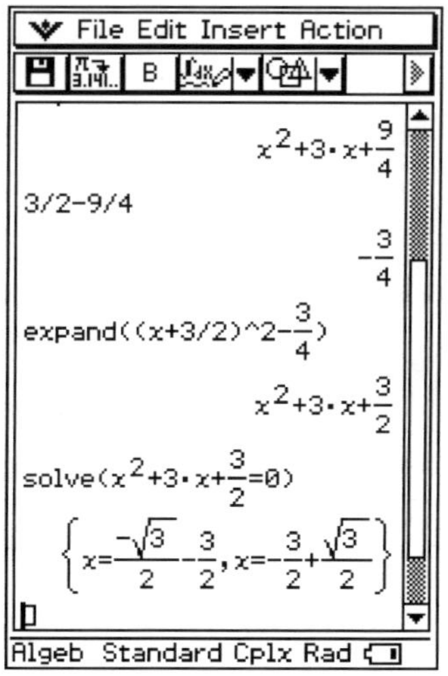

Quadratic Formula

The approach taken in the above example can be generalized to come up with a formula for the general quadratic equation:

a*x^2+b*x+c=0

First we divide through by a:

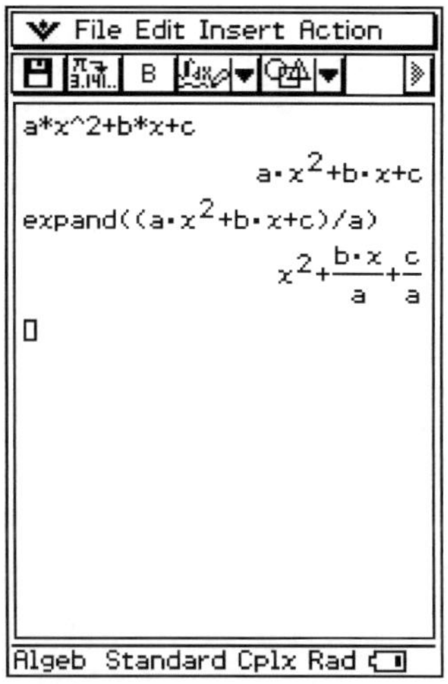

Then we divide the coefficient of x by 2 to come up with the completed square:

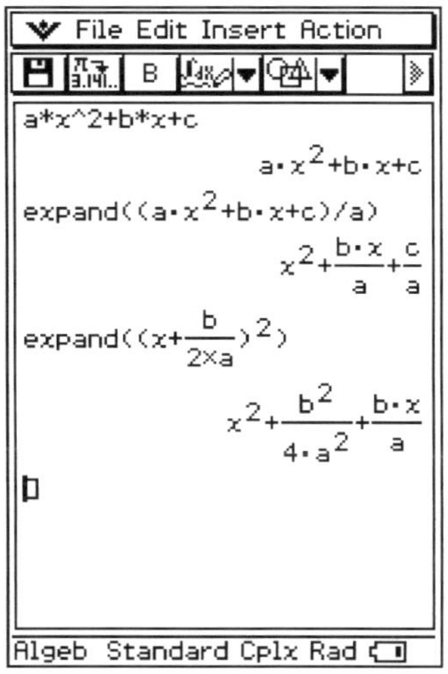

Now we determine the constant term in the completed square:

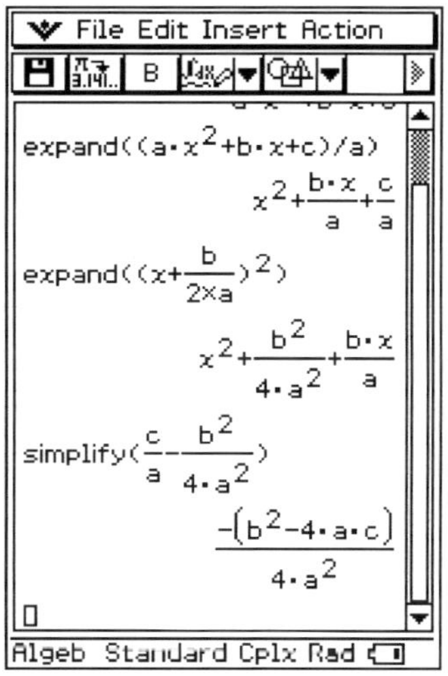

We can construct our whole expression by dragging and dropping the different components. Applying expand(), we can check that the expression is correct:

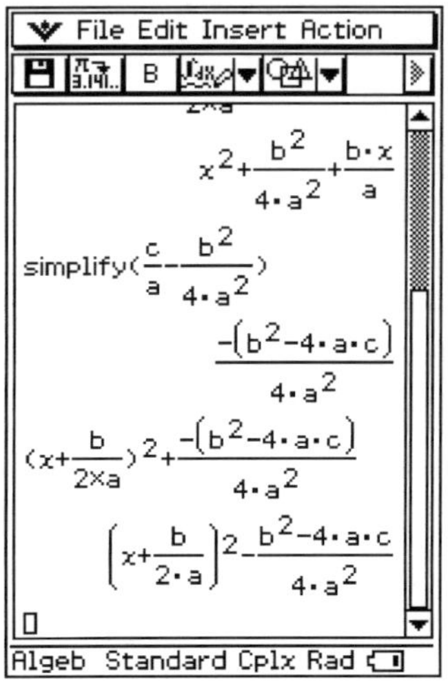

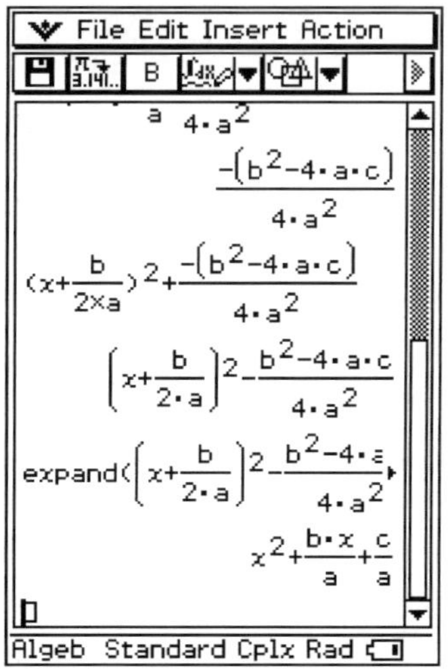

Again, applying the solve() function to the original equation gives the solution. Looking at the expression in completed square format shows where the solution comes from!

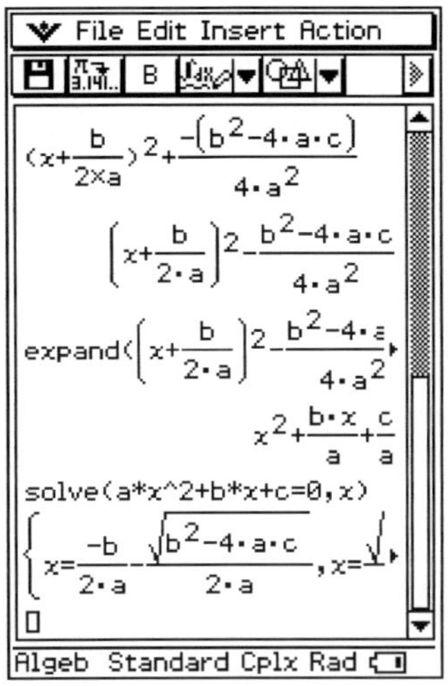

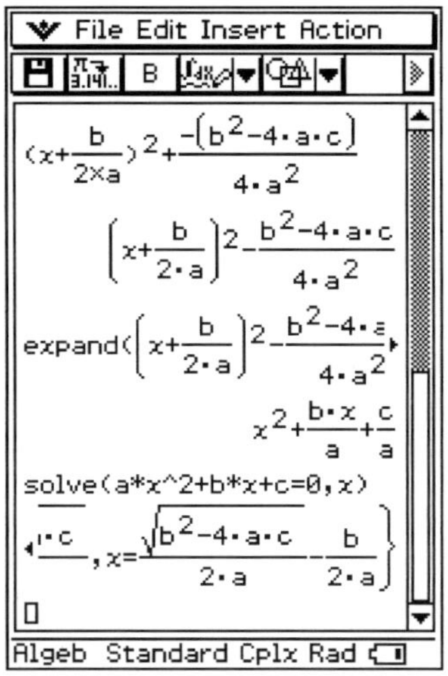

Taylor Series

An important body of mathematics surrounds the local approximation of functions by polynomial functions. Using this technique we can replace a difficult function by one which is easy to deal with and accurate so long as we do not stray too far.

The simplest local approximation of a function is its tangent. Let's study the function:

$$y = \frac{5}{\sqrt{1+x^2}}$$

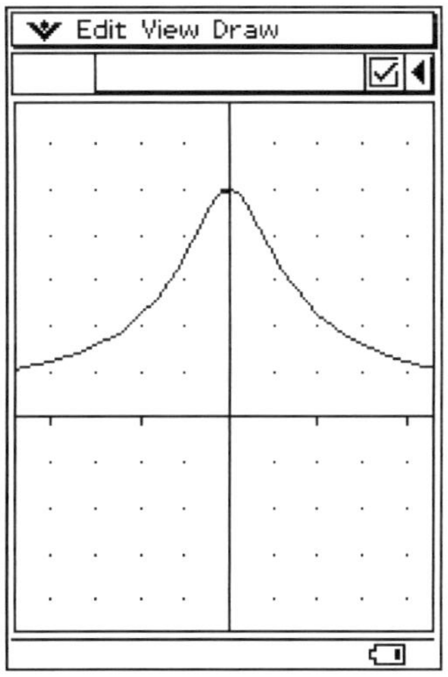

Use the Tangent tool to create a tangent line, and drag the point so that it has x coordinate -1.

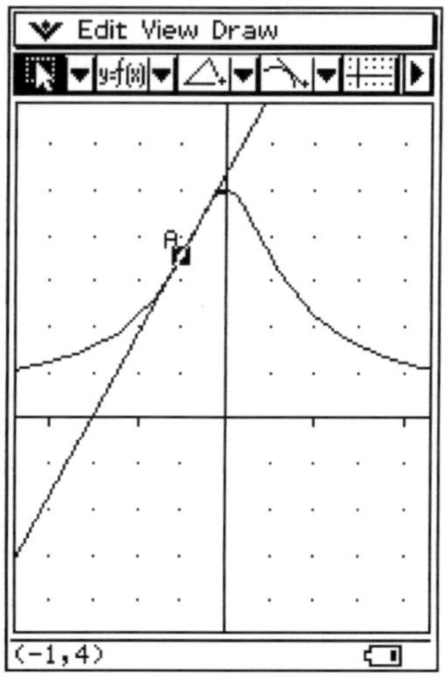

We can see that the tangent line is indeed a good local approximation to our curve.

The function taylor() in eActivity produces a polynomial of given degree which approximates a function at a point. The syntax is as follows:

taylor(**function**, **variable**, **degree**, **x_location**)

This creates a polynomial approximation to the **function** local to **variable=x_location** of the given **degree**.

Let's look at the degree 1 approximation to our function at x = -1:

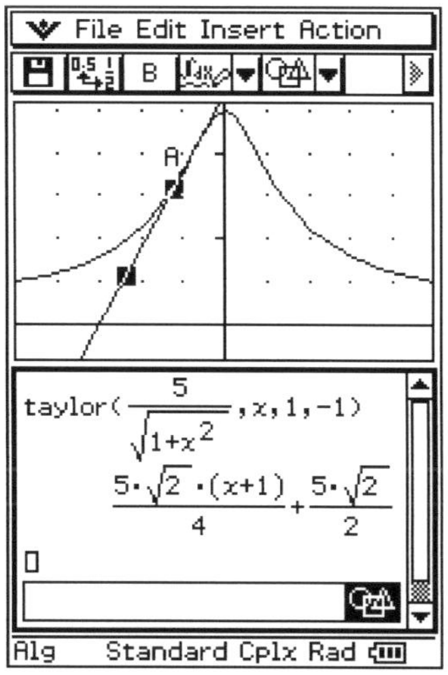

Expanding this, then using the **Fraction/Decimal** button change it to floating point lets us compare with the equation of the tangent in the geometry module and observe that they are the same (given the lower resolution of the geometry system's decimals).

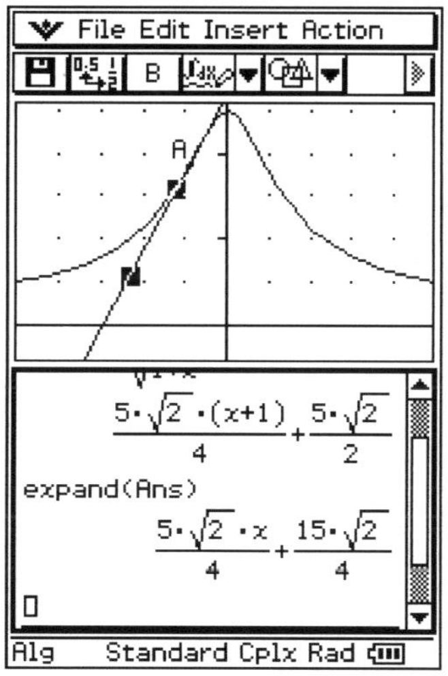

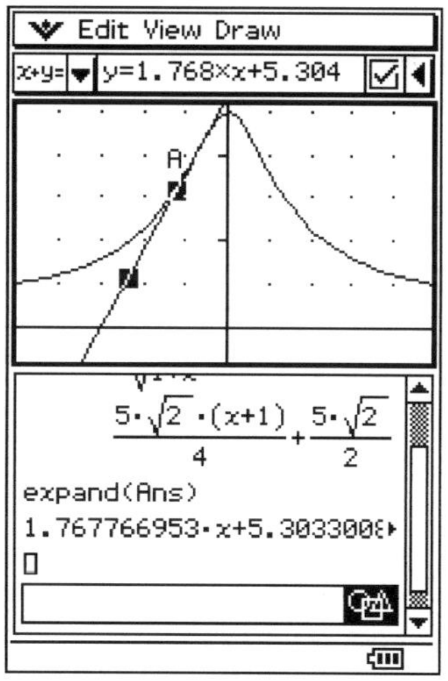

Higher Order Approximation

We can create higher order approximations simply by changing the third parameter of the taylor() function:

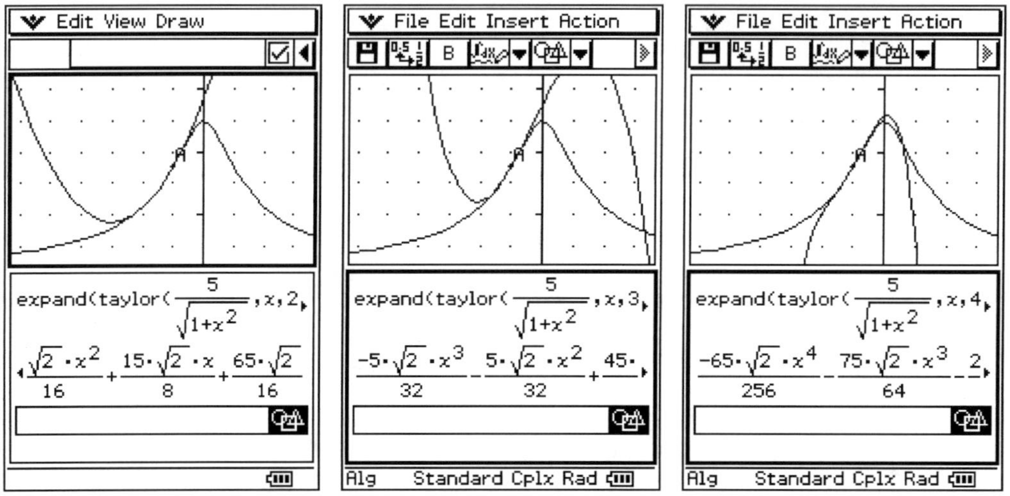

Changing the last parameter alters the location at which the approximation is made:

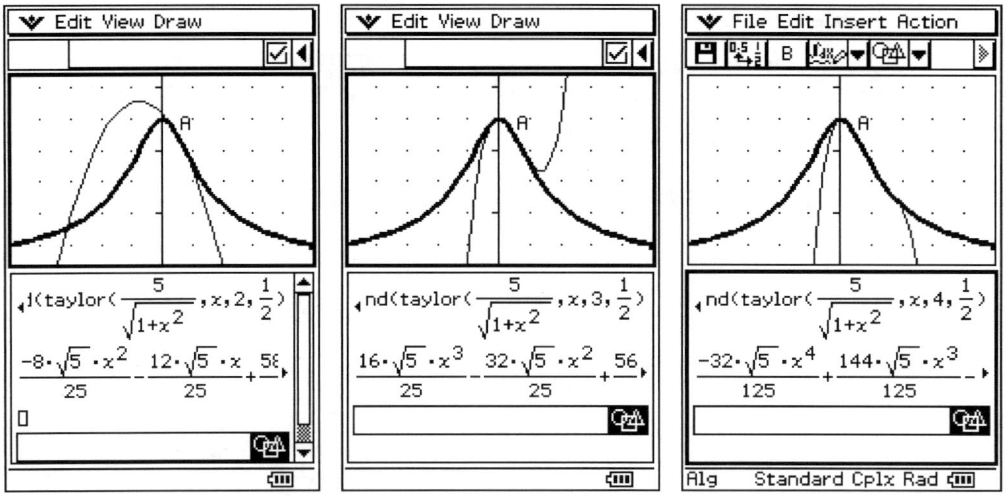

Animation

ClassPad is able to create an animation which runs a point along a line segment, a line, an arc, a circle, a polygonal curve or a function. It is possible to animate several points simultaneously. Hence, creating an animation is a two step process - first you specify which points are to be animated and which curves they are to run along, then you set the animation in motion.

In addition to displaying the model in motion, ClassPad will trace the curve swept out by one or more points in the figure. It can also create an output table for any measurement extracted from the drawing.

A special Animation UI bar is displayed by selecting **View / Animation UI** from the menu. This gives you a slider bar which allows you to single step your model through its range of motion. It also gives you button bar access to the Animation commands. These can also be accessed from the **Edit / Animation** menu.

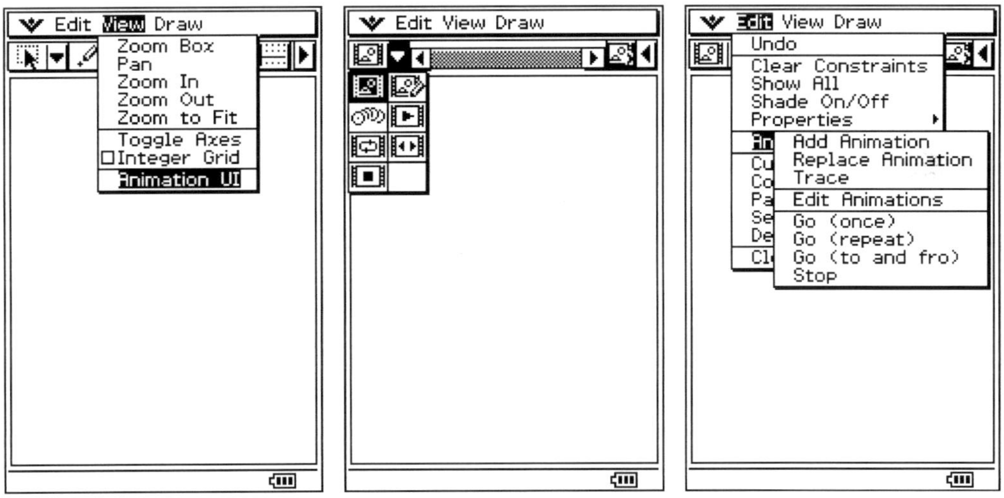

The Animation UI has 3 components:

- Animation drop down – gives you access to all the animation commands
- Slider bar – lets you choose a particular place in the animation

- Close Animation UI button – takes you back to the normal UI (button bar & measure box)

Notice that the button bar is still available while the Animation UI is present, you can access it by pressing the tab arrow.

The following commands are available from the animation drop down button:

Add Animation - adds the selected point / curve pair to the list of animations

Replace Animation - clears all existing animations and replaces them with the currently selected point / curve pair.

Add Trace - sets the currently selected points as trace points (removing any existing trace points). When an animation is run, ClassPad creates a curve for each trace point.

Go (once) - plays the animation through once, then stops.

Go (repeat) - keeps playing the animation until Animation Stop is invoked.

Go (to and fro) - plays the animation forwards, then backwards, then forwards etc. until the Stop button is pressed.

Stop - stops the animation

Edit Animations

The Animation Edit window provides an interface for you to:

- Set the number of steps in the animation.

- Remove animations or traces.

- Change the limits of the animation (limits are specified relative to curve endpoints).

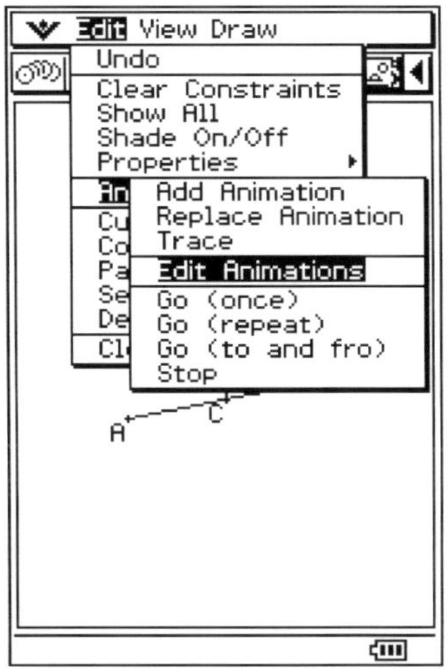

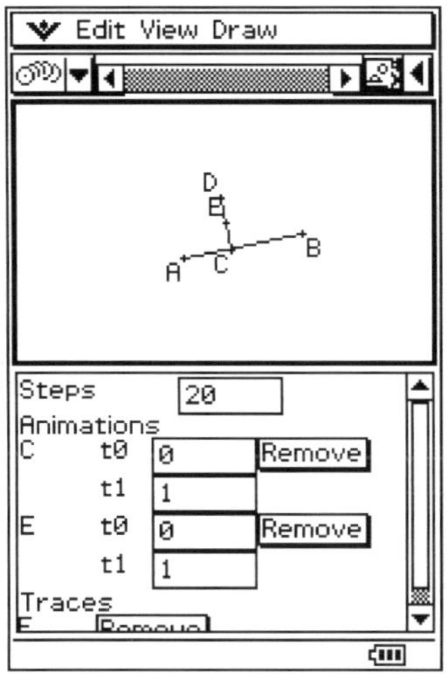

A Simple Animation

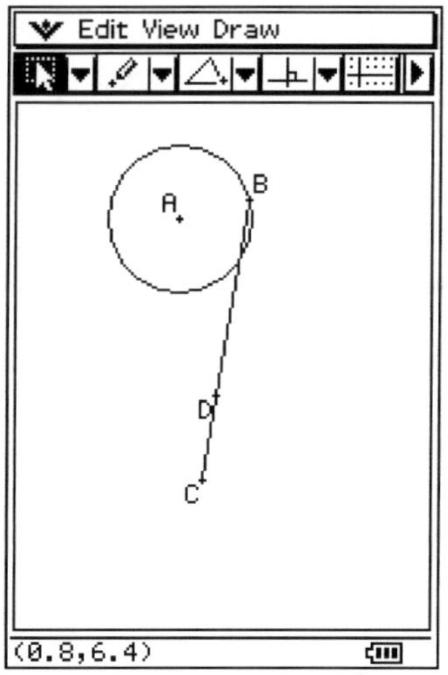

As a simple example, we'll model a mechanism where B rotates around the circle AB, and point D is fixed. BC is a fixed length arm which slides through a collar at D.

First we need to constrain our model to make sure it behaves correctly: we fix the location of A and the radius of the circle:

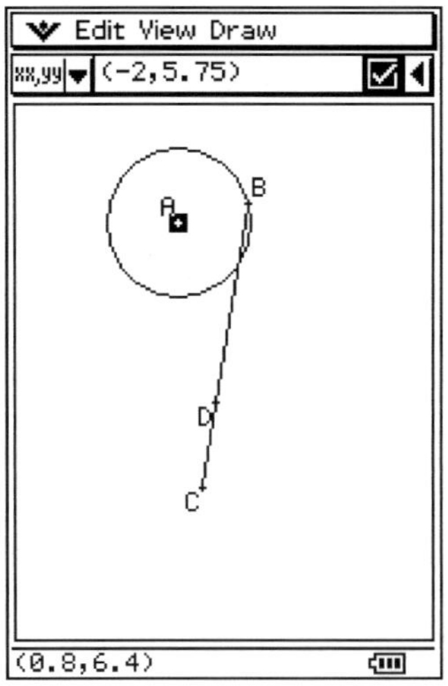

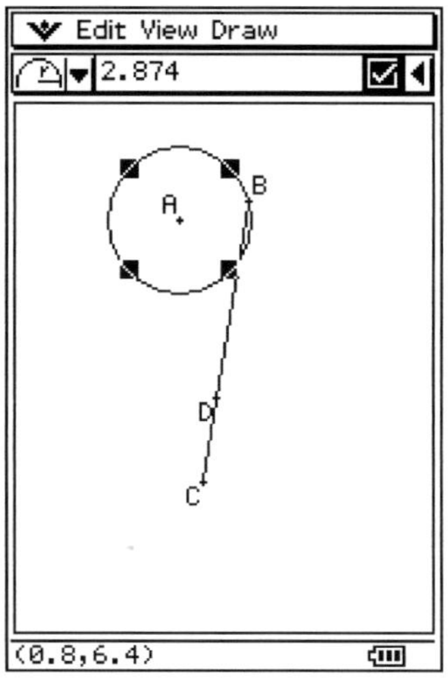

We also fix the location of D and the length of BC:

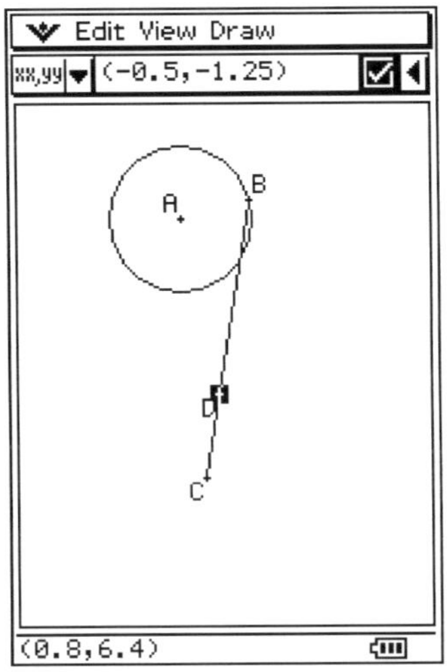

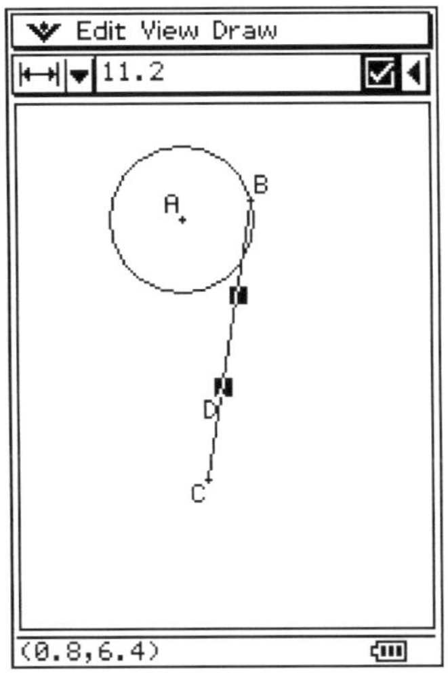

Now we want to add an animation of B around the circle. We select B and the circle and then invoke **Add Animation**:

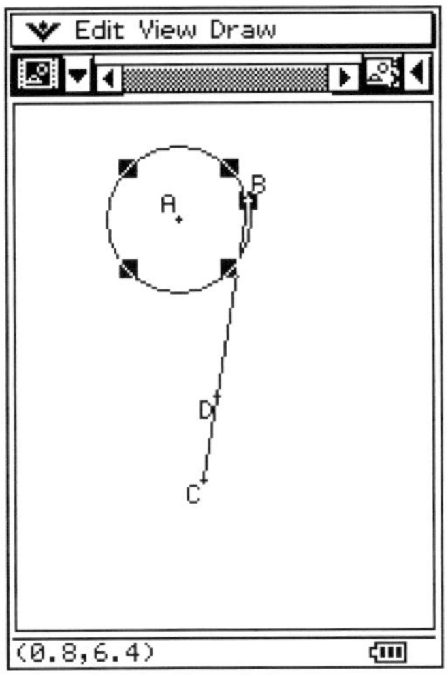

You can now watch the drawing animate by pressing Go, or you can single step through the motion using the Animation Slider bar:

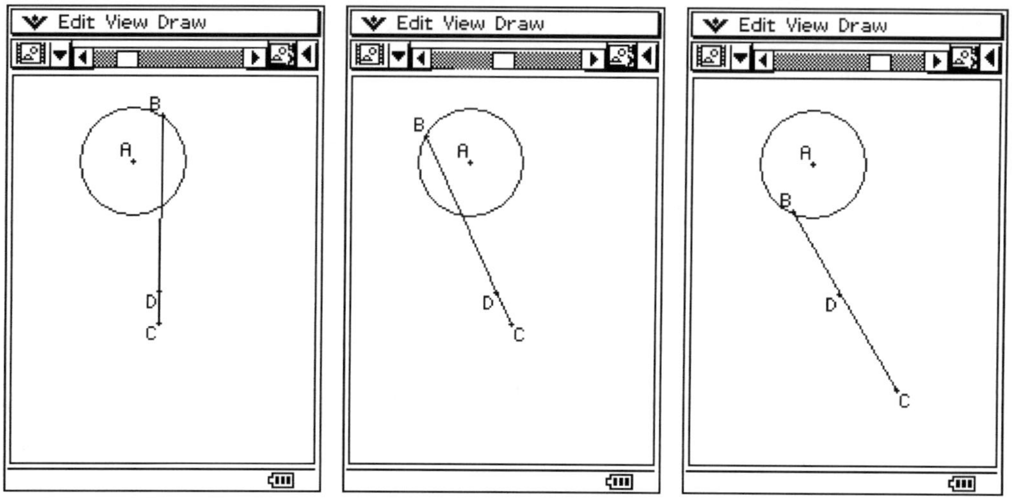

We can observe the curve traversed by point C by selecting it and adding a Trace. (After adding the trace, you need to press **Go** again to see the curve.)

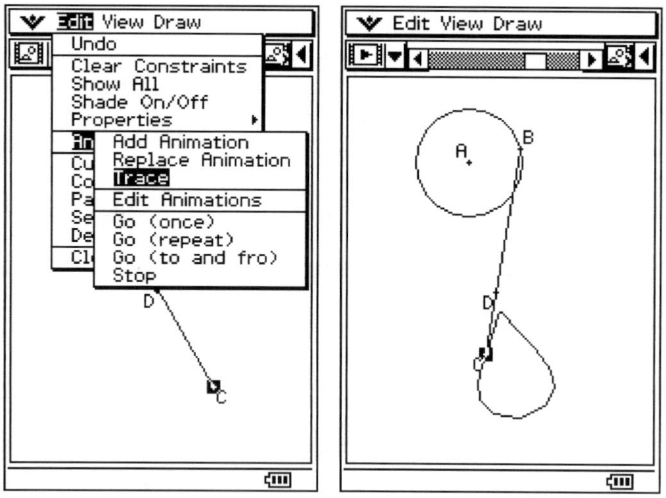

Cycloid Curve

If you mark a spot on the circumference of a wheel, then roll the wheel along the ground, what curve does the wheel traverse?

To model this situation in ClassPad, we need to create two different motions: a motion of the center of the circle, along with a motion of a point around the circumference of the circle.

We draw a segment AB, this is the "track" along which the circle center C will run. The point D will then also be animated around the circumference of the circle.

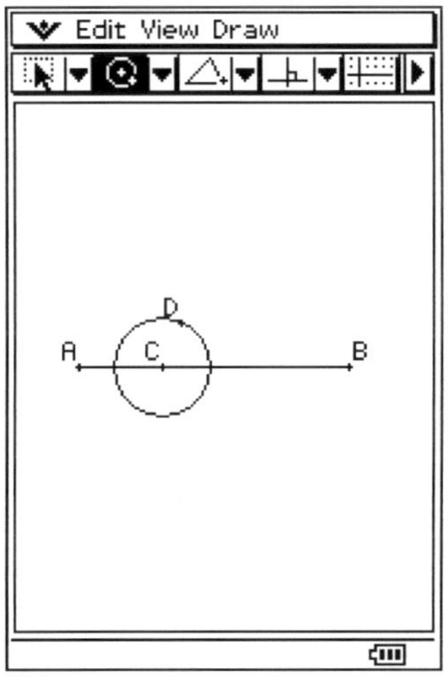

First, however, we need to make sure the model behaves appropriately: We fix the locations of A and B and the radius of the circle:

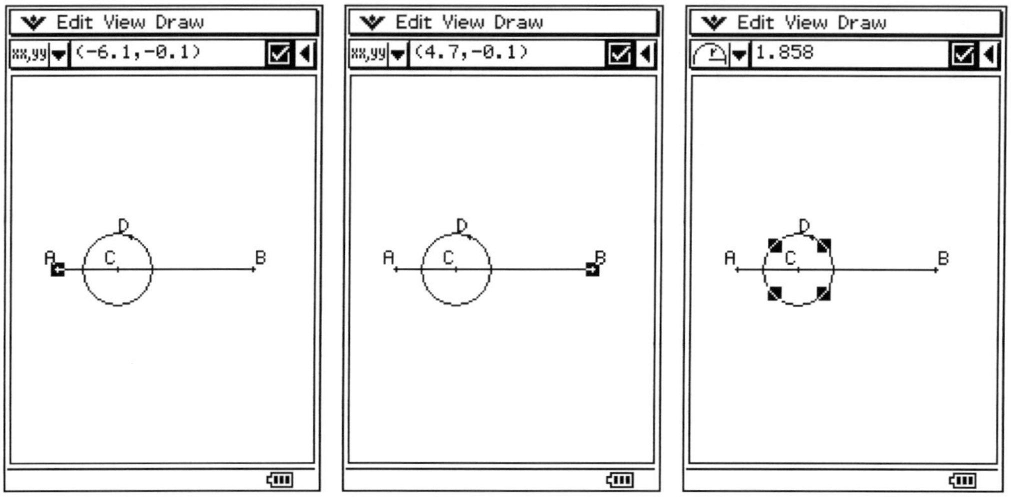

We can now add the following two animations: C on the line AB and D on the circle CD:

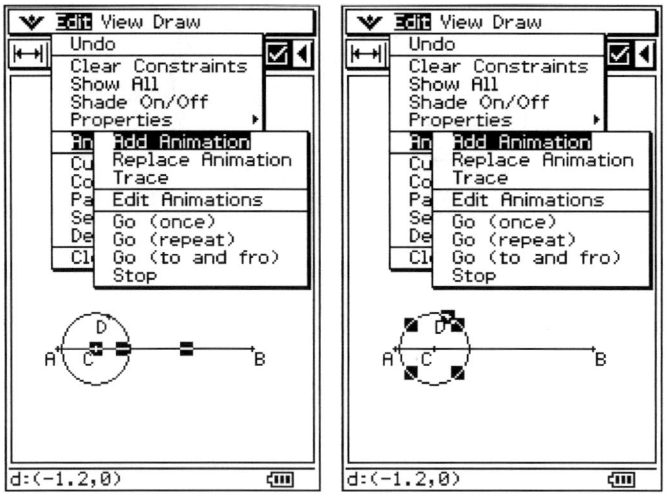

We can specify that D is traced, then make the animation go

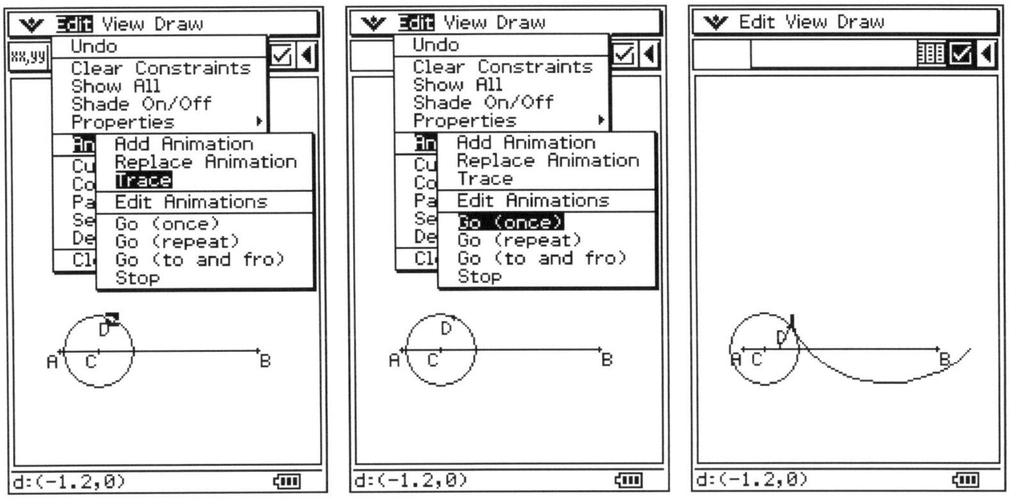

We should see D rotate at the same time as C slides along from A to B.

However, as D is rotating counter-clockwise, this is not a good model of the situation where a wheel is rolling along the ground. To model that situation accurately, we need to make D go clockwise. We can do this by bringing up the Edit Animation window, and resetting the limits of the animation of D. Instead of animating from 0 to 1, we can animate from 1 to 0 - this will make D go backwards (i.e. clockwise) round the circle.

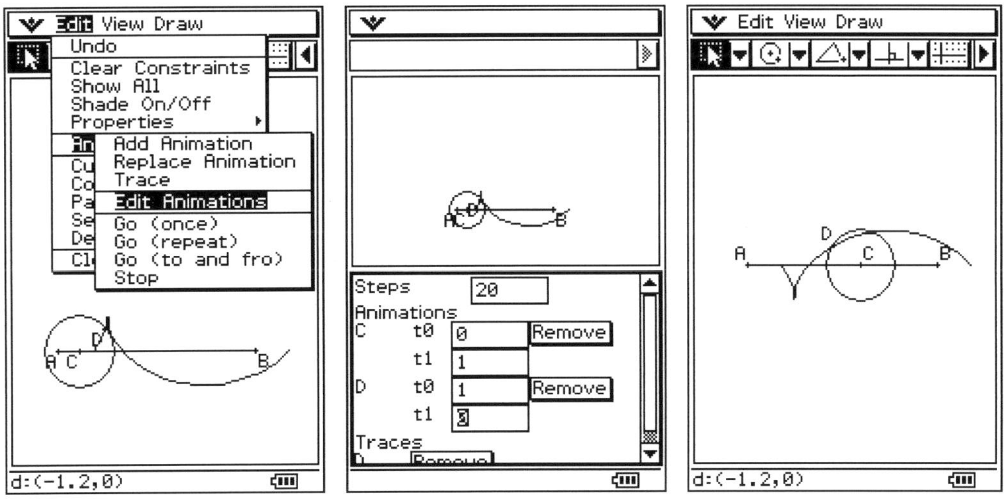

Now we have D going the correct direction around the circle, but we have not established that it is going at the right speed. To do this we should set the distance AB so that it is equal to the circumference of the circle. Then D will perform a complete revolution in the time it takes C to traverse the distance of 1 circumference. However, as we have previously fixed both point A and point B, it is impossible to also specify the distance between them. We have to first relax the constraint on the location of B. Then we can fix the length of AB to be the same as the circle's circumference:

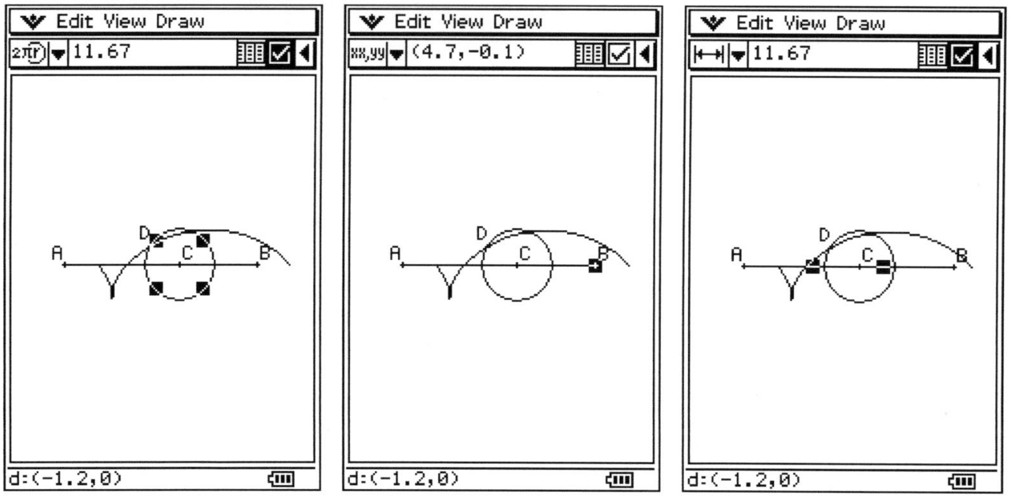

Now an animation will reveal the path of D in a single revolution of the wheel:

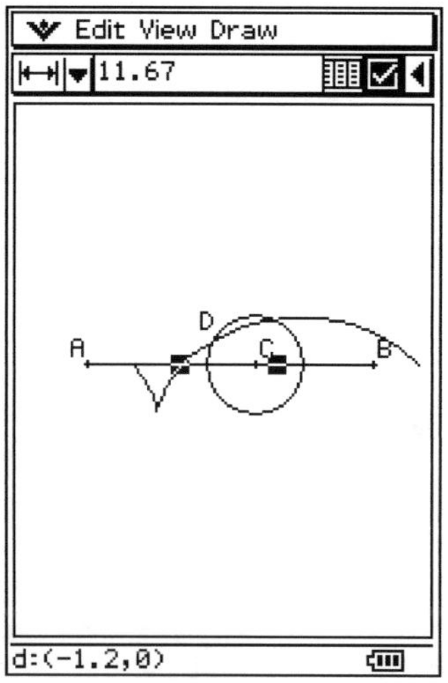

What if we want to see a picture of the motion over the course of 2 complete revolutions of the wheel?

To model this situation, we need to make two changes:

- We need to change the length of the path AB to be twice the length of the circumference of the circle.

- We need to change the limits of D so that it goes from 1 to -1 (or from 2 to 0).

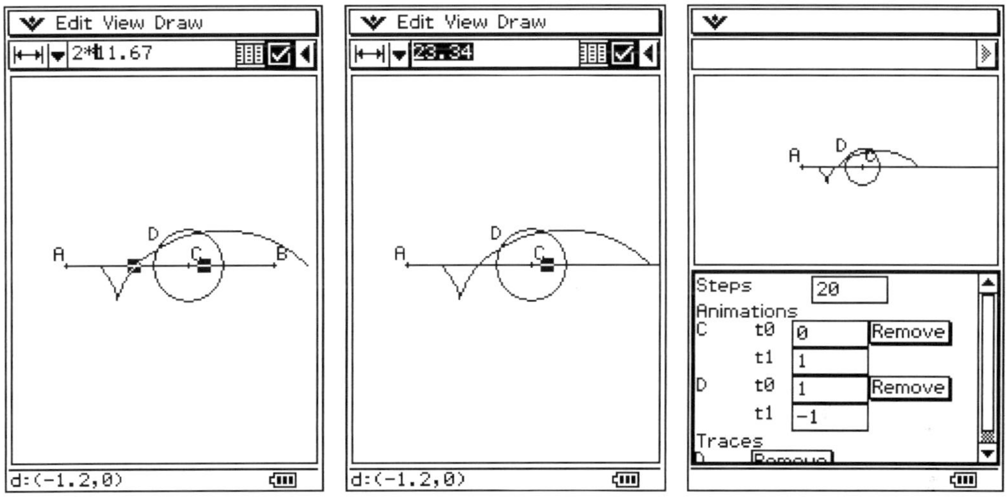

Now (zooming appropriately) we see the new animation:

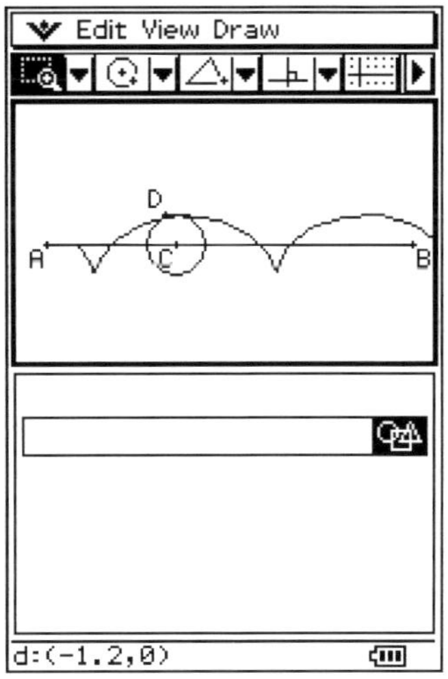

Making Waves

Instead of tracing D, let's imagine we are interested in graphing its height against time.

One way of doing this is to construct a point with the same height as D, but which moves along steadily in the x direction.

First, lets create a line CE and constrain it to be perpendicular to AB. Then a line DF, where F lies on CE, and constrained perpendicular: Now tracing F will do the trick:.

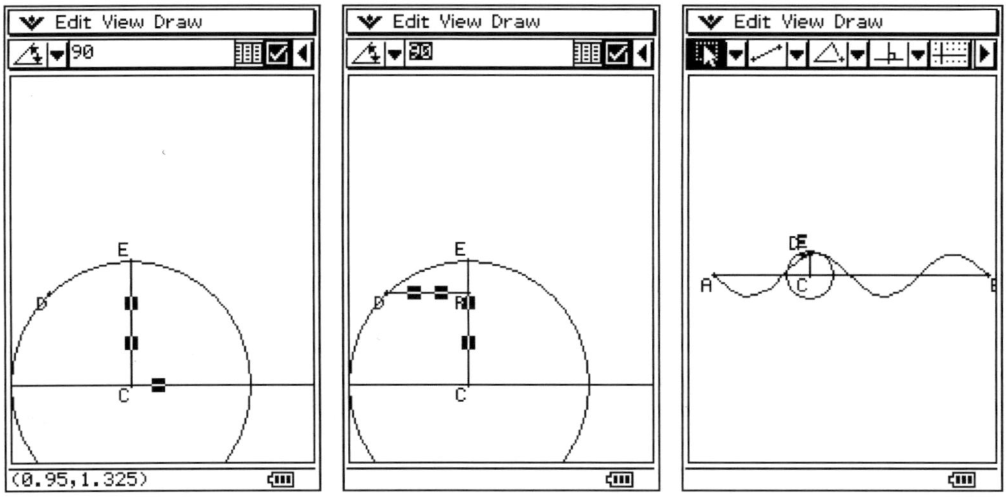

Let's see if we can work out the equation of this wave form:

First, let's set A to be the origin, and recompute the trace:

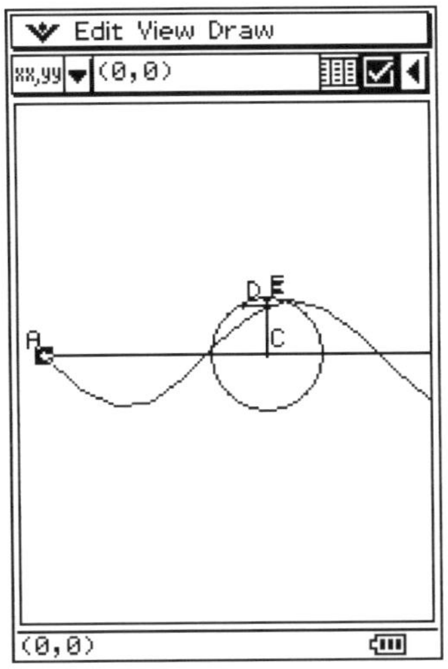

Now, let's postulate that this is a sin wave, so let's add a sin function to the drawing (the superscript ***r*** specifies that we are working in radians):

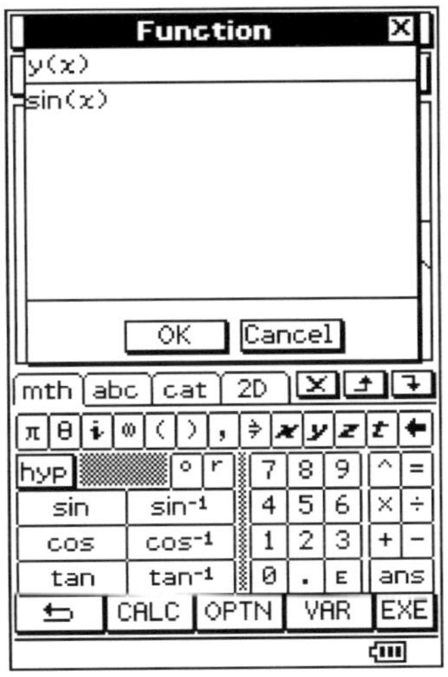

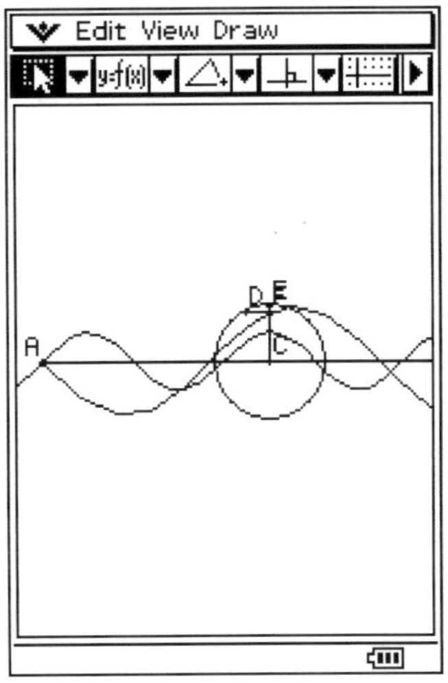

Clearly this does not fit. For a start, the amplitude (height the wave reaches) of the sin function is 1. And the height D reaches is the radius of the circle. Adjusting for this gives:

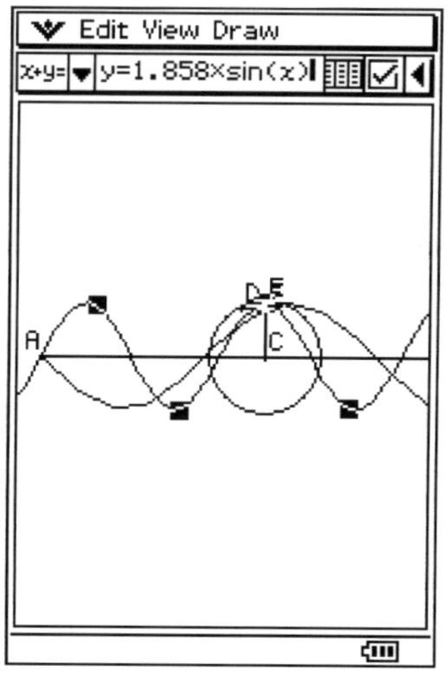

Now the wavelength also needs adjusted. We need to divide x by the radius length:

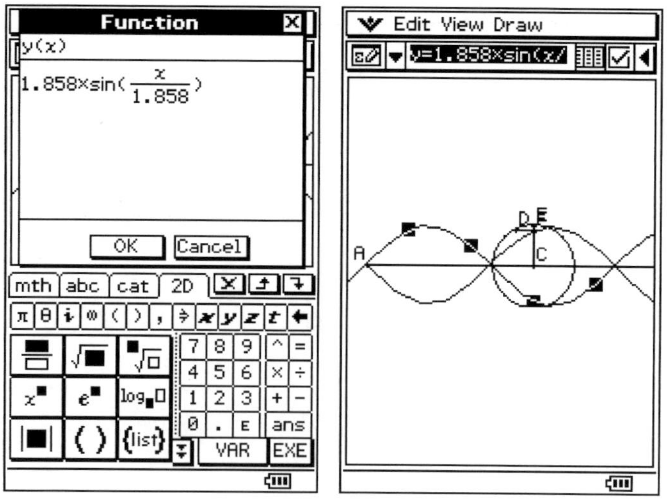

This looks promising, but upside down. Let's invert it: by prepending a minus sign:

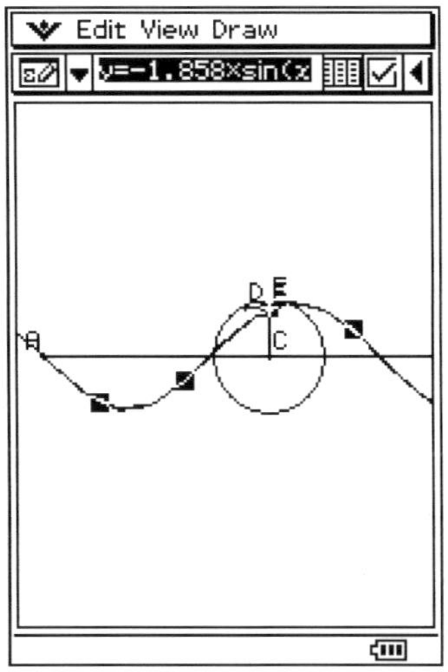

We see that the curve generated is indeed a **sin** curve.

We could have saved ourselves a little bother if we simply set the radius of the circle to 1 in the beginning!

Polar Functions

You can draw a spiral by animating a point around a circle, then animating a second point along the radius line:

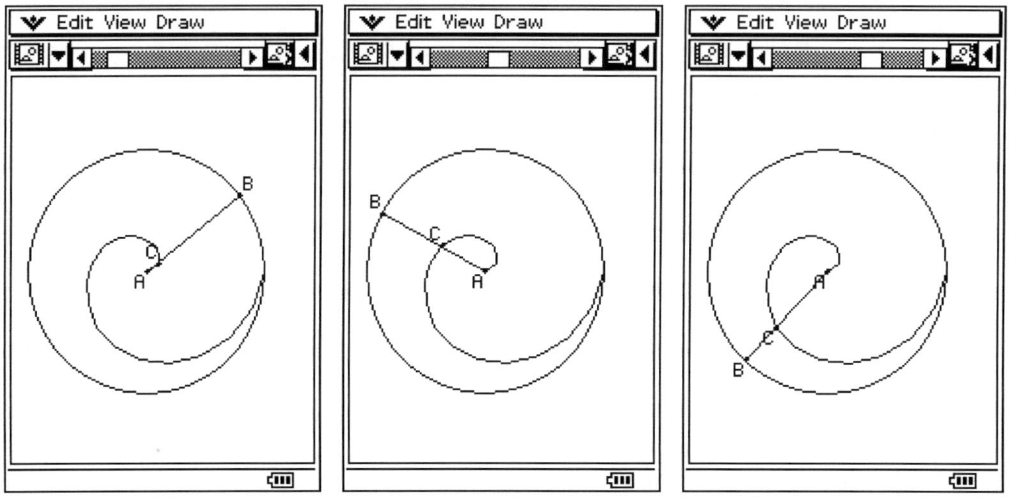

This figure in effect graphs the polar function r = θ.

Can we adapt this technique to graph a more general polar function? Say r = 1.1+sin(θ).

First we create a graph of y=1.1+sin(x). Then we add the line y=0 (otherwise known as the x-axis) as a reference.

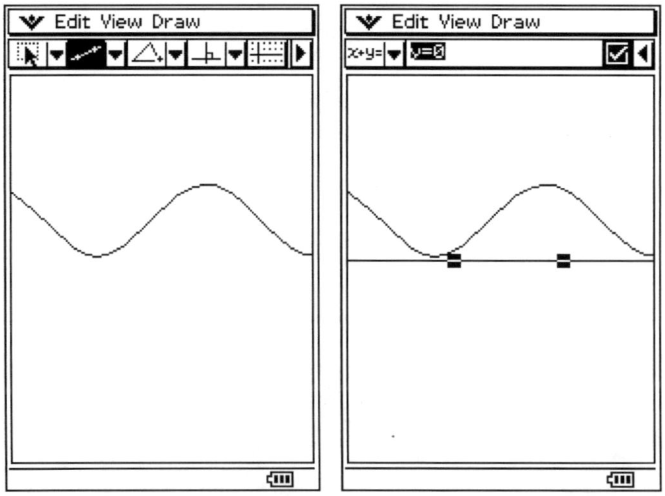

Now we'd like to create a point to run along the function, and a vertical line segment joining it to the x-axis. We will use the length of this line segment in the specification of our polar function.

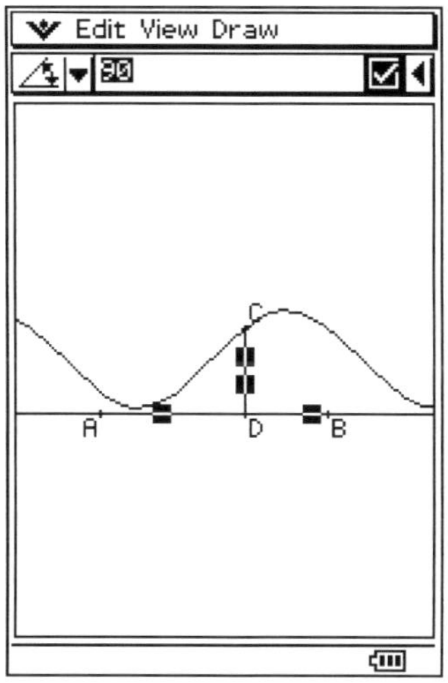

Next we'll create a circle and a line segment one of whose end points is the circle's center. We then constrain the circle's circumference point to lie on the segment:

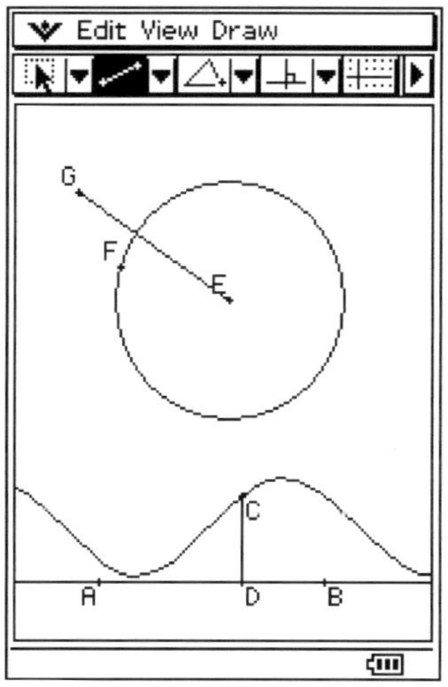

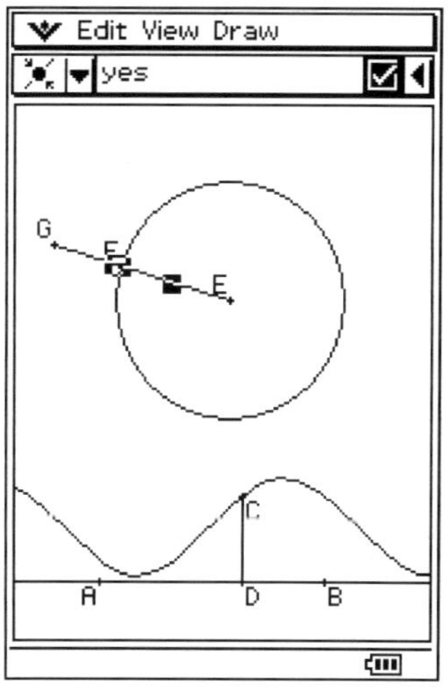

We now set the length of segment EG to be congruent to the length of segment CD.

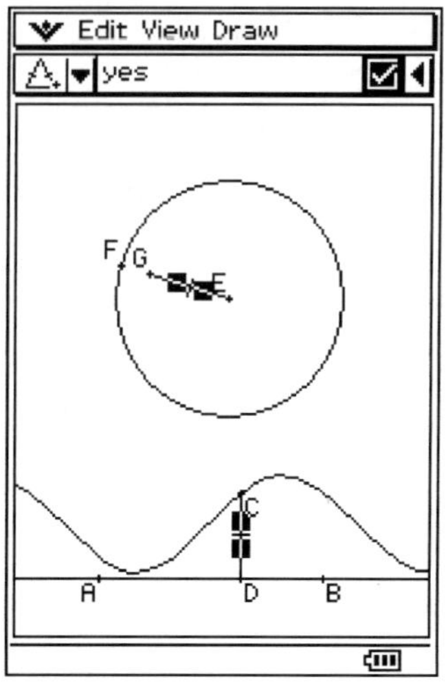

Now we need to add an animation of F on the circle, and of C on the function. Also set G to be a trace point::

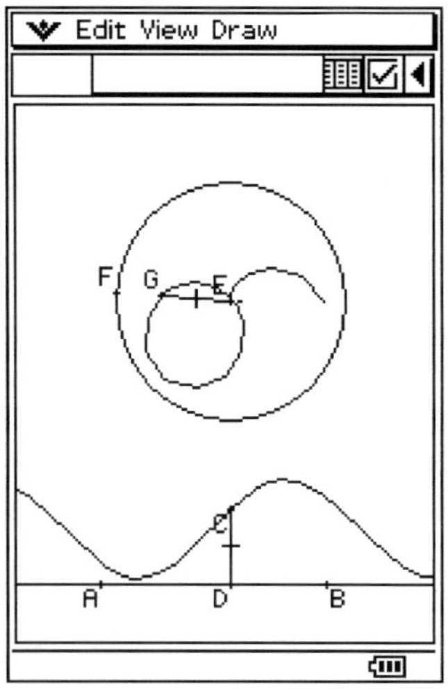

We notice that C is set by default to move along the portion of the function which is visible on the screen. This is not necessarily the range we want. In fact, we'd like our values to range from 0 to 2π. We can specify this in the **Edit Animation** window:

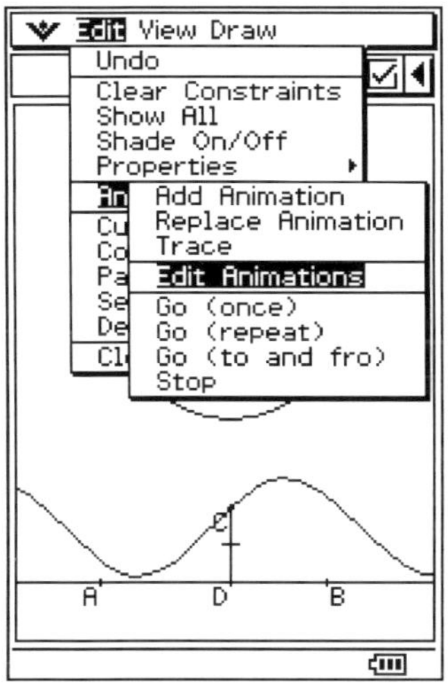

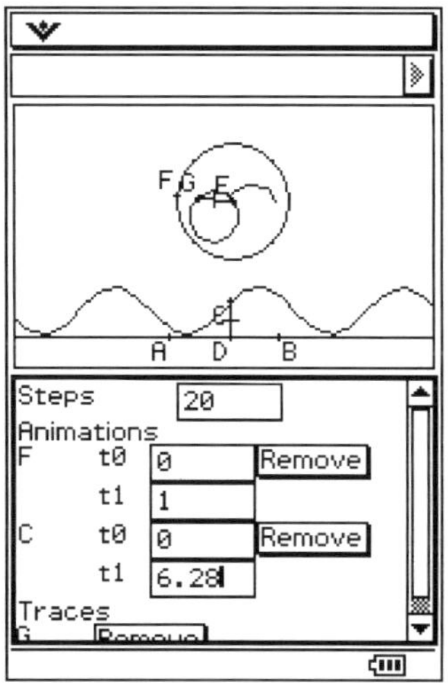

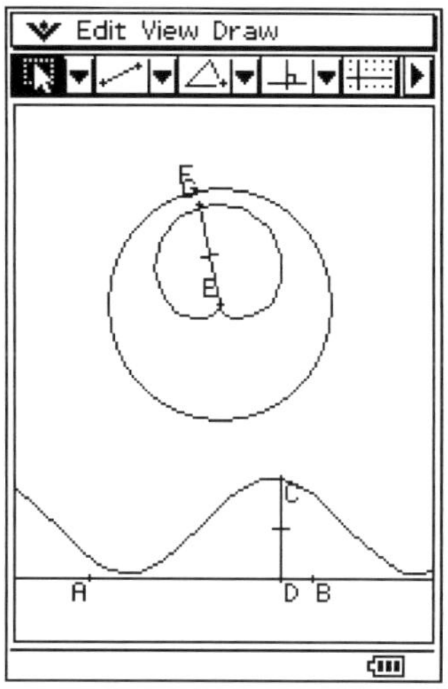

We have used the Animation capabilities of ClassPad's geometry system to draw a polar plot. In the process, we have illustrated some useful geometry techniques. Of course there is an easier way to draw polar graphs, available as you might expect in ClassPad's graphing module. Let's see how to use the graphing module to create this graph.

First, we create a function definition strip in eActivity:

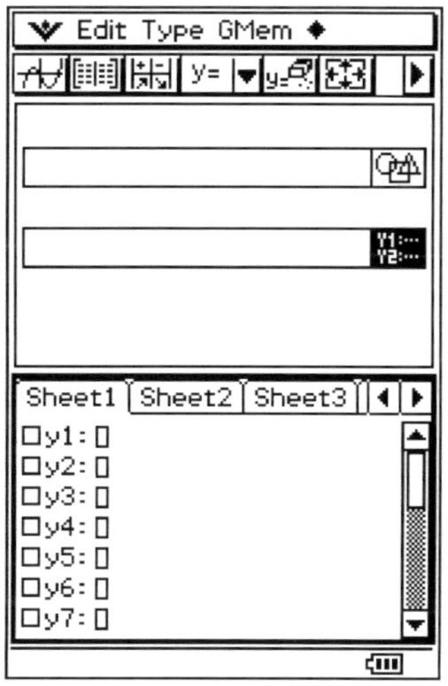

We set the function definition to be r= from the drop down button and enter r as a function of θ

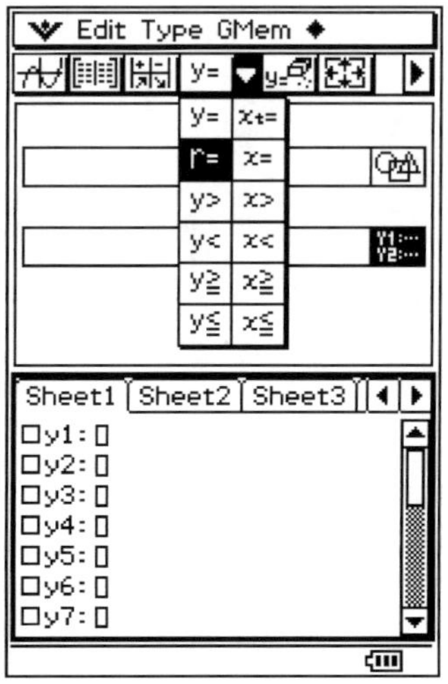

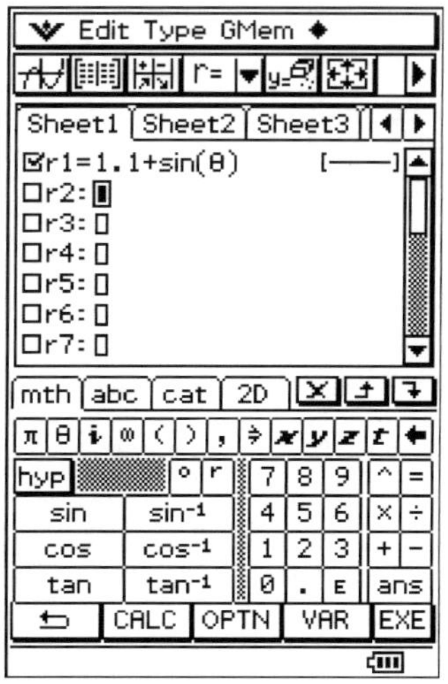

Pressing the Graph button on the extreme left of the button bar causes the graph to be drawn.

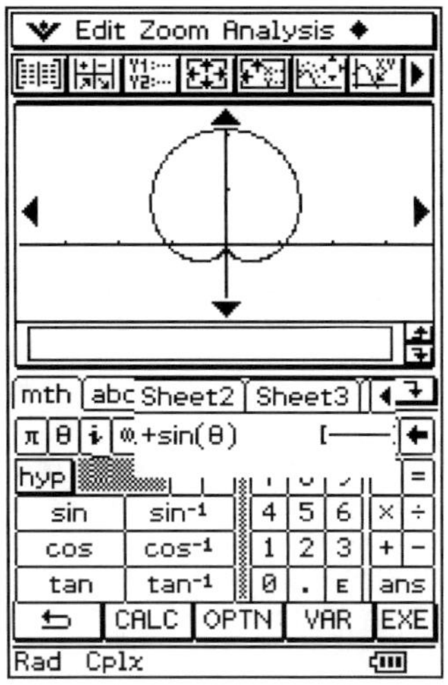

Pantograph

A pantograph is a tool which was used for enlarging or reducing drawings before the advent of the photocopier.

We'll create a pantograph in ClassPad.

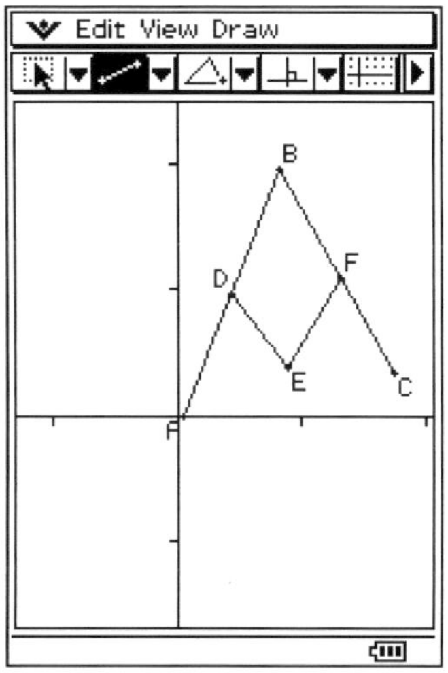

The tool works by fixing point A, then tracing round a design with point C. A pen positioned at point E traces a scaled version of the drawing. In order that our drawing works like a pantograph, we need to establish constraints.

First, let's lock the lengths of AB and BC to 10. Then let's lock the distances DB, BF, DE, EF to 5. Then we need to lock the location of A: let's set it to be the point (0,0):

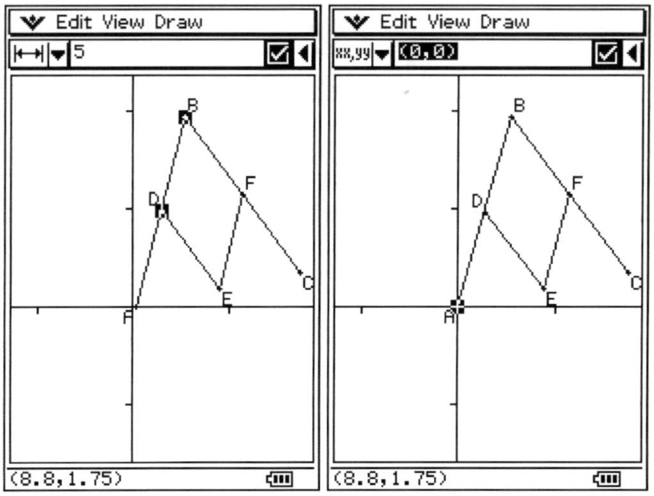

Now, we should be able to drag point C around and let the pantograph follow through:

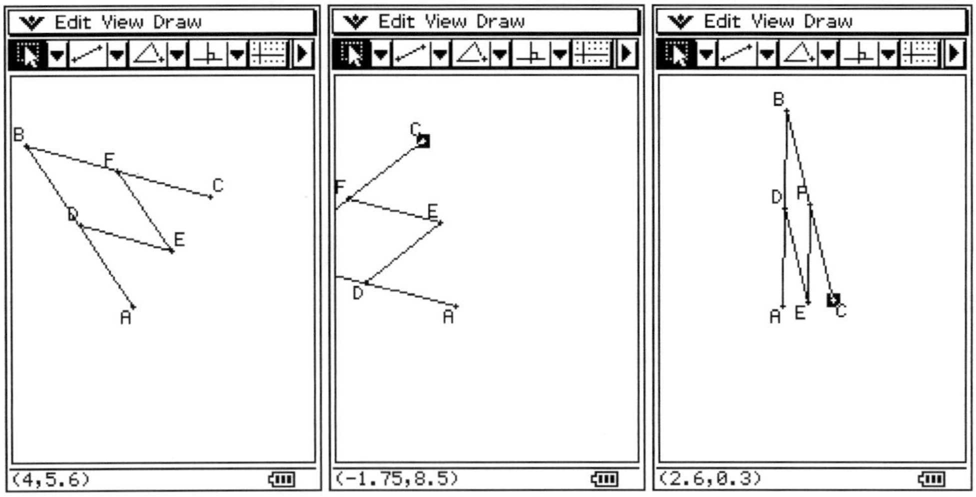

Let's give the pantograph a curve to trace. We can create a piecewise linear curve in eActivity as a 2 column matrix, where each row is the location of a point. (Notice that if we create a 2 row matrix, with each column representing a point, then it is interpreted geometrically as a closed polygon.)

First create a matrix with 5 rows and 2 columns. You do this by first clicking on the 2x2 matrix, then clicking 3 times on the vertical matrix extension button.

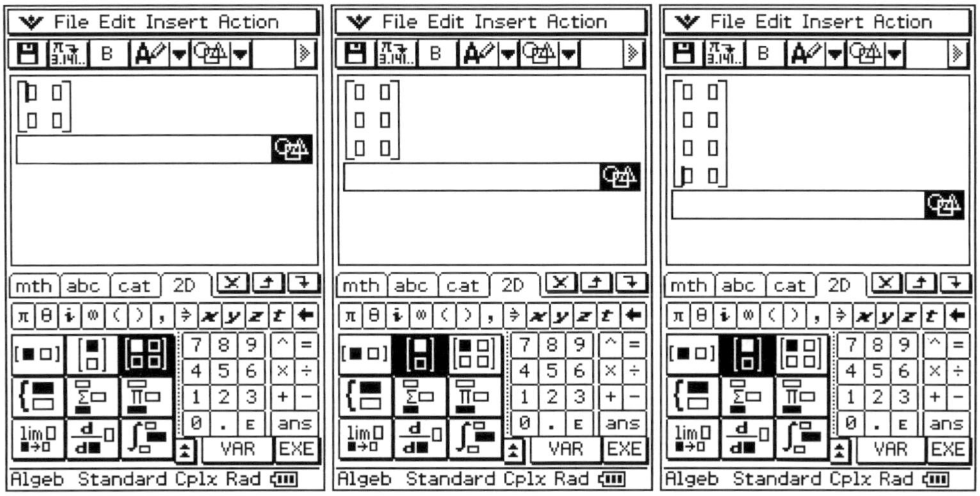

Enter the points shown below, then drag the matrix into geometry:

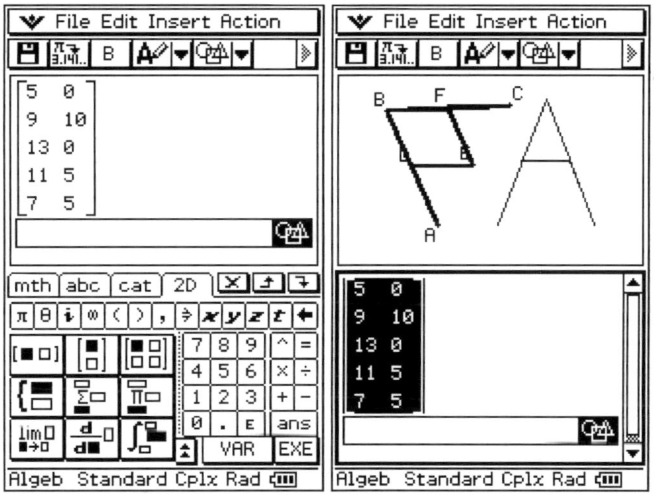

We have drawn the letter A. Now we would like to trace the letter with the point C of the pantograph. Select C and the curve and then pick **Add Animation** from the **Edit/Animations** menu.

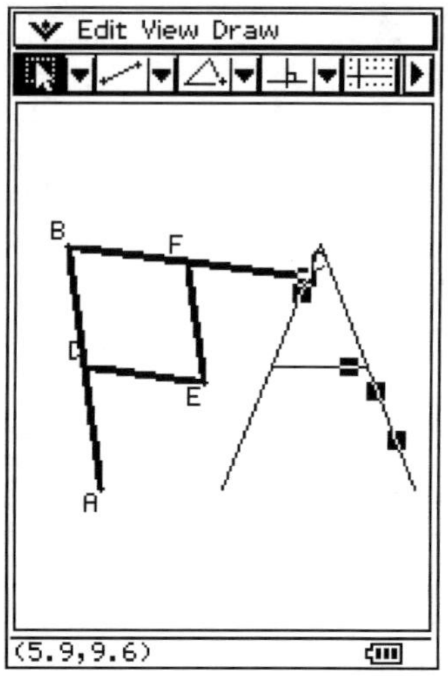

We now need to create a trace at E and make the animation go:

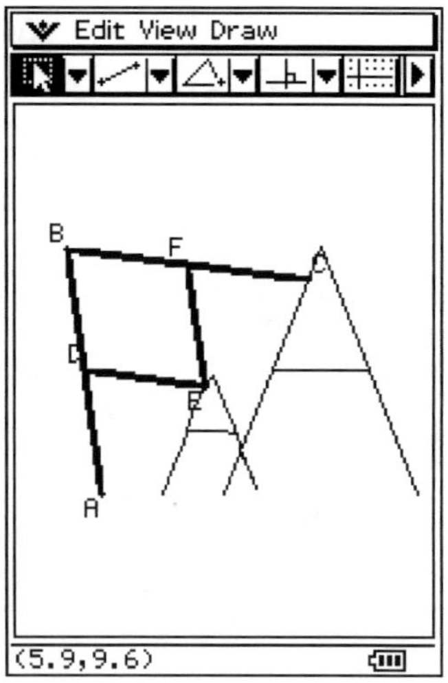

The pantograph has created a scaled down image of our letter. Bringing up the animation UI lets us single-step the pantograph through its paces:

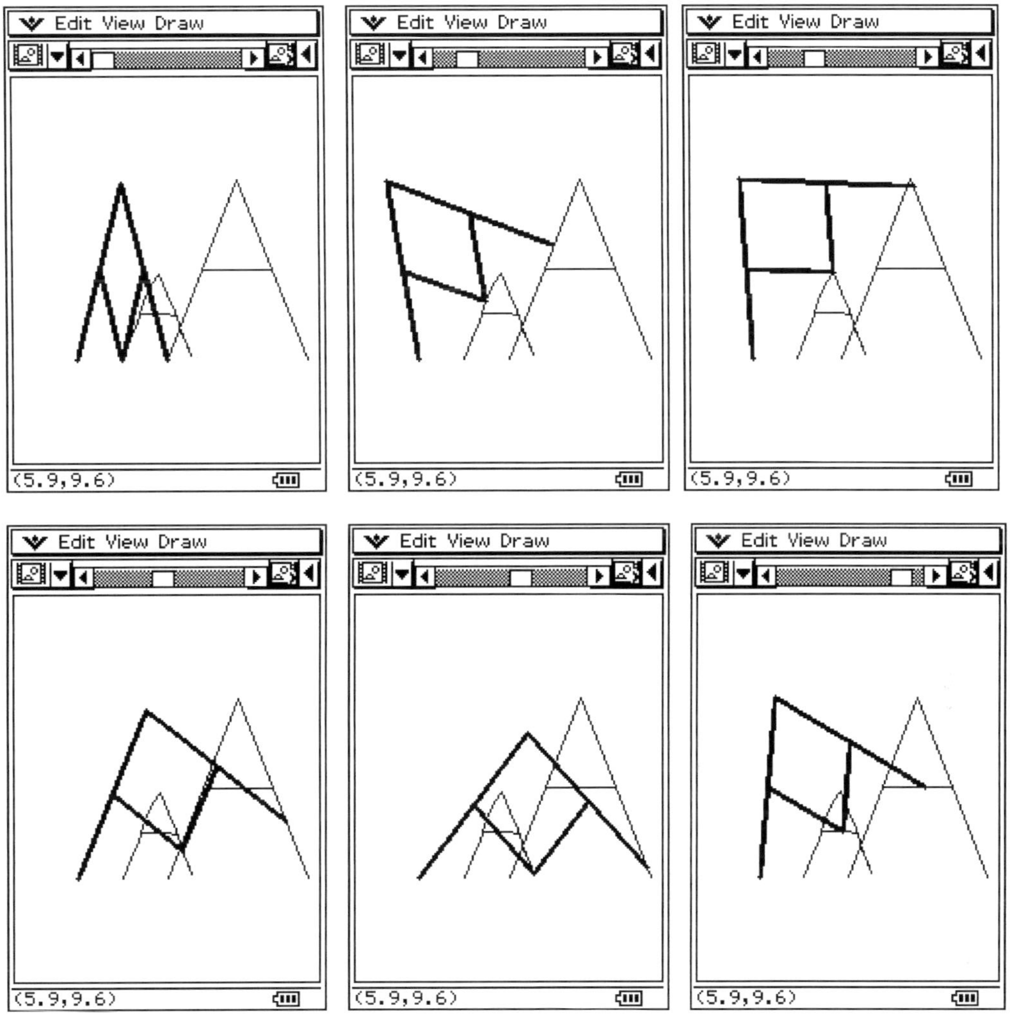

Further Investigation

The pantograph shown here halves the scale of the drawing (or doubles it if you reverse the roles of points C and E.)

Can you design a pantograph which gives you ¼ scale?

Can you come up with a general design which can be used for any scale?

Cubic Spline

The **cubic spline** is a curve which is frequently used in computer aided design. It is useful because the location of the endpoints of the curve and the tangent to the curve at the endpoint may both be specified.

The spline curve is frequently defined in terms of four control points. These four points form a quadrilateral which prescribes the shape of the curve. DeCastlejeau's construction defines the curve in terms of these four points, and can be modeled in ClassPad using the Animation features:

First let's create the four control points along with the lines connecting them. Using the grid is an easy way to make these integers - a fact that will help some later calculations fit more easily on the page!

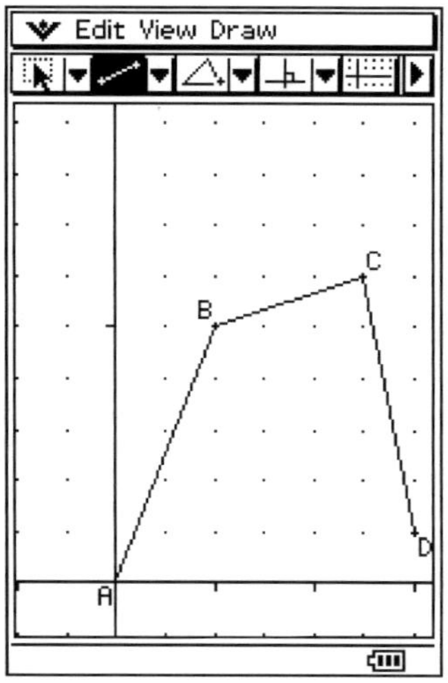

We now create two more line segments EF and FG, with E lying on AB and F lying on BC and G lying on CD:

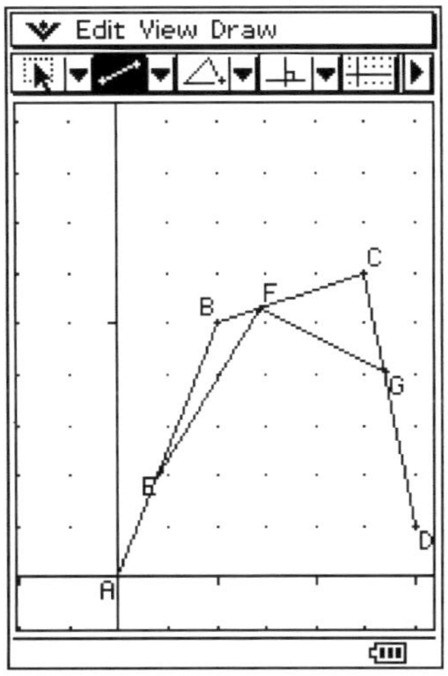

Finally we create another segment HI with H on EF and I on CD. Then we create a point J lying on the new segment:

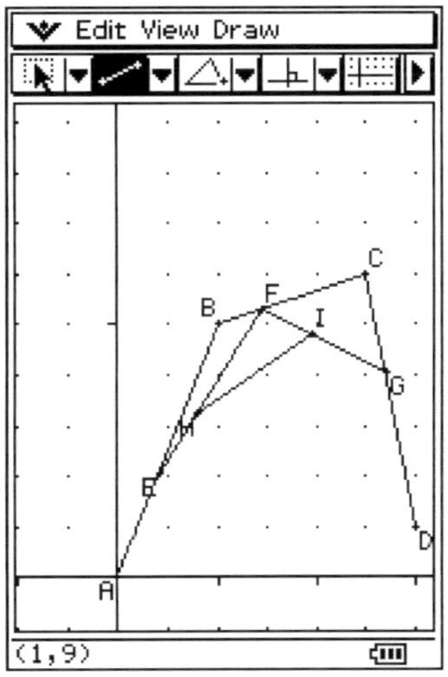

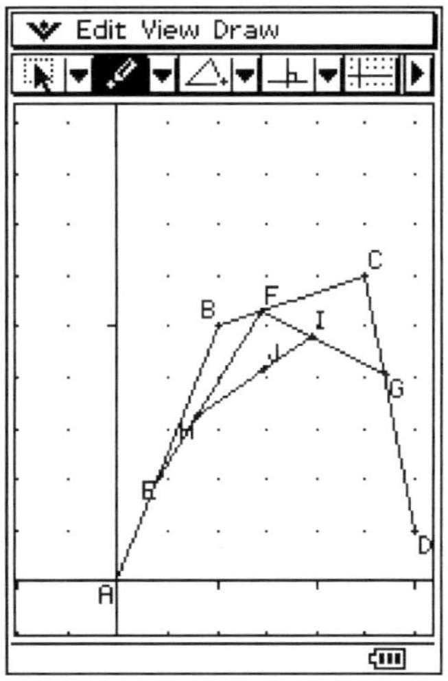

We now need to create a long list of animations:

- Animate E along AB
- Animate F along BC

- Animate G along CD
- Animate H along EF
- Animate I along FG
- Animate J along HI

Finally create a trace at J

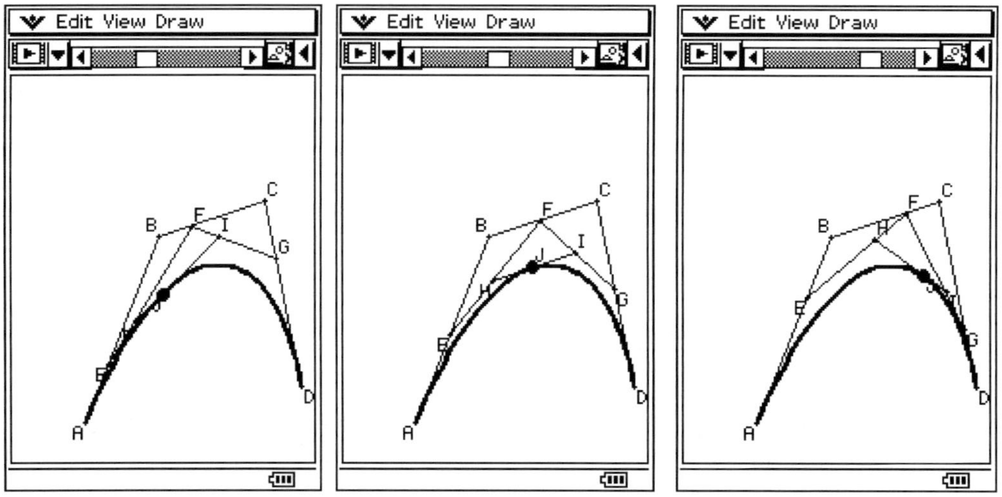

The resulting curve is the cubic spline whose control points are A,B,C,D

Try changing the control points and notice the relationship between the shape of the curve and the shape of the control quadrilateral:

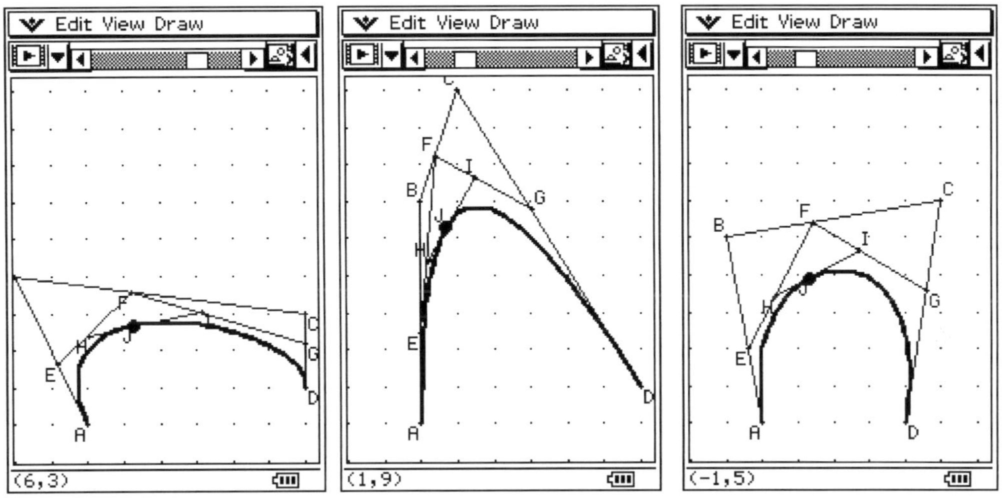

So much for the pictures, what about the equation of the curve?

Each animation is defined by running a point along a line segment, from its beginning to its end. If **a** is the first end point of the segment and **b** is the second end point, then we can express the location of the animation point at time t as **a***(1-t) + **b***t. Let's use this equation to derive the equation for the spline curve of the third example above:

First, we drag and drop point A into a Calculation Row in the eActivity window. We then store this vector in the variable **a**.

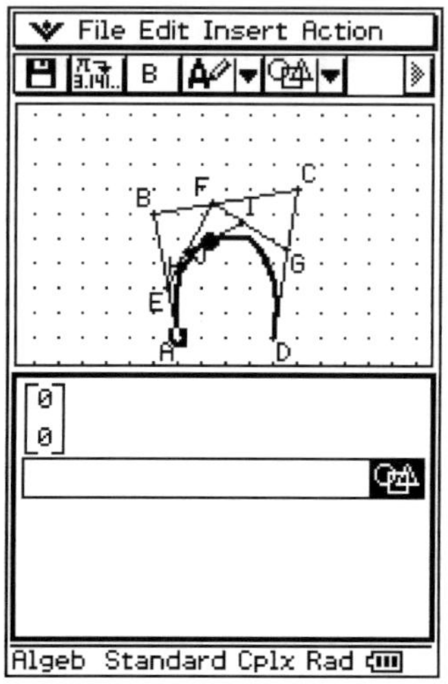

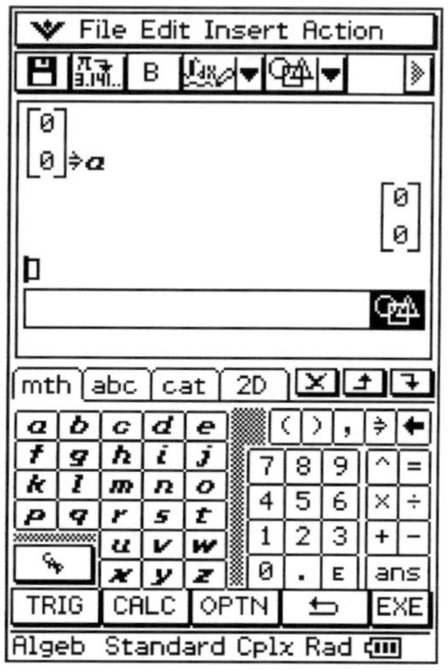

Likewise we can drag and drop point B and C and store them in variables **b** and **c**.

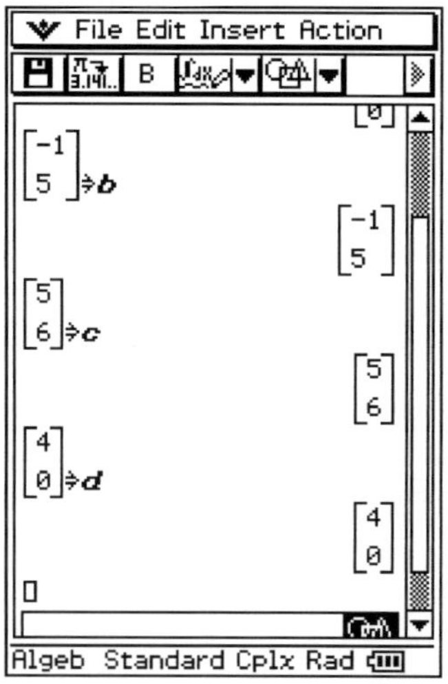

Now, we can define e in terms of **a** and **b**

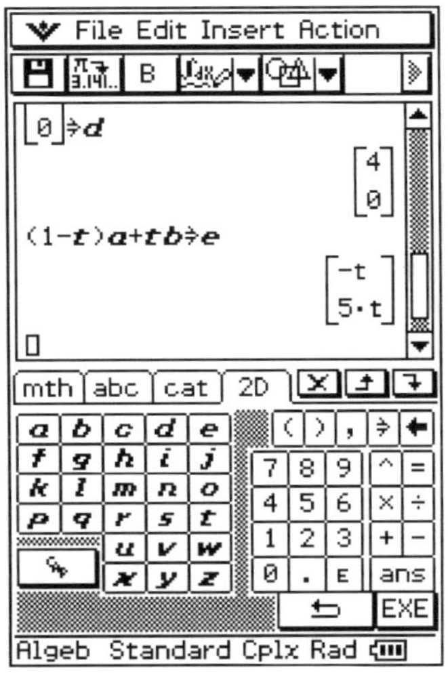

Notice that when using the single letter variables (the bold letters found on the VAR tab of the keyboard), you do not need to insert a product symbol (for example between the **t** and the **b** in the above expression).

A similar expression can be created for **f** . The expand() function can be used to simplify the resulting vector.

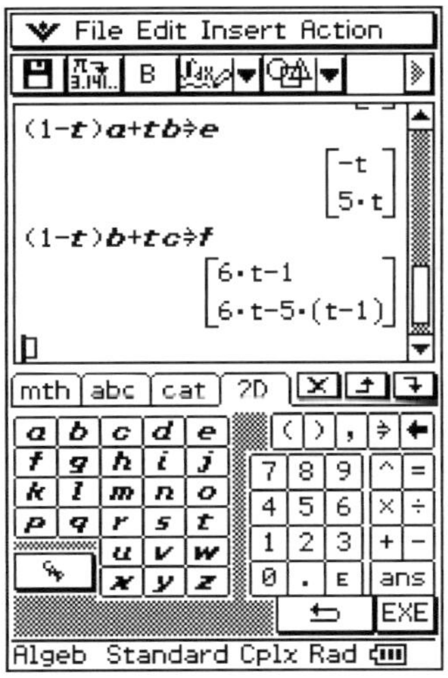

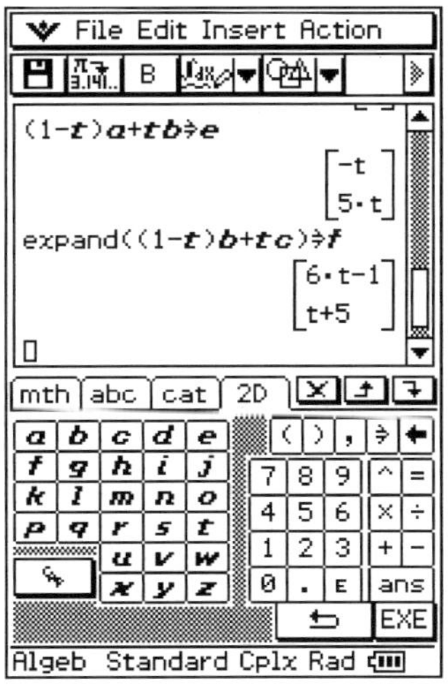

We can continue in this vein to create expressions for the rest of the defined points. As one would expect, they get more complicated as the outputs from previous computations become the inputs to current computations:

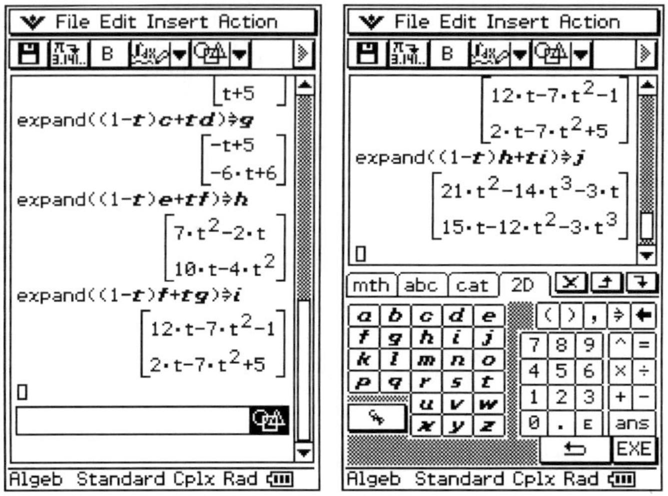

J is the point which we traced, so this is the parametric equation for J.

Let's use the graphing module to plot it:

First we insert a Graph Editor into eActivity, and set it up for a parametric plot:

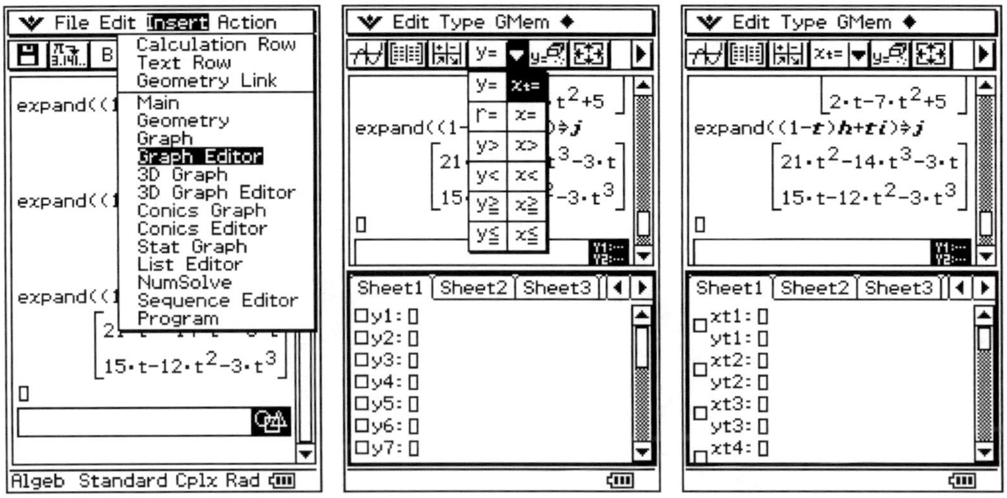

Now we can drag and drop the x and y from our vector into the graph expression editor (remembering to press ENTER to activate the expression).

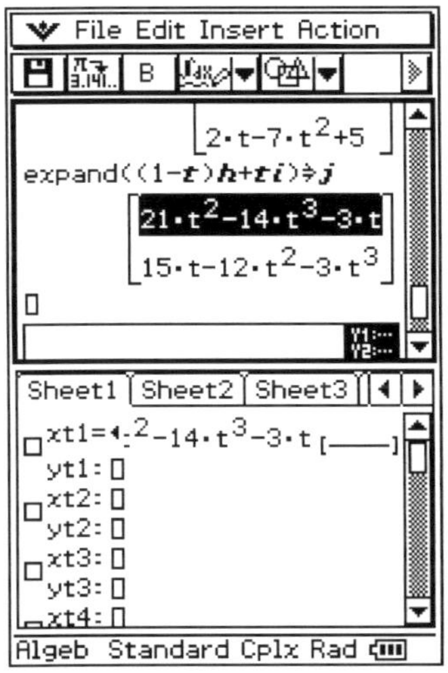

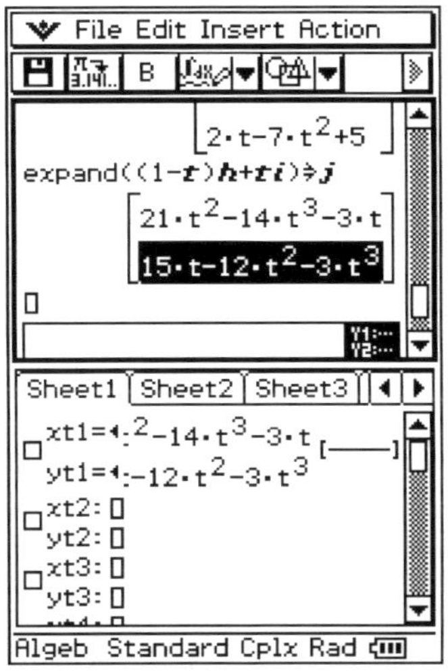

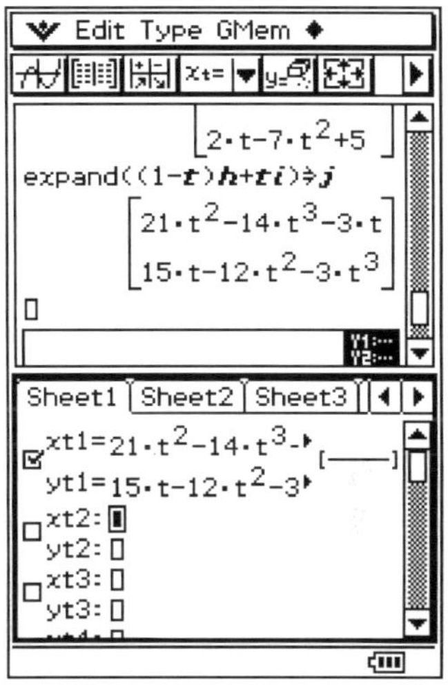

Pressing the graph button displays the plot.

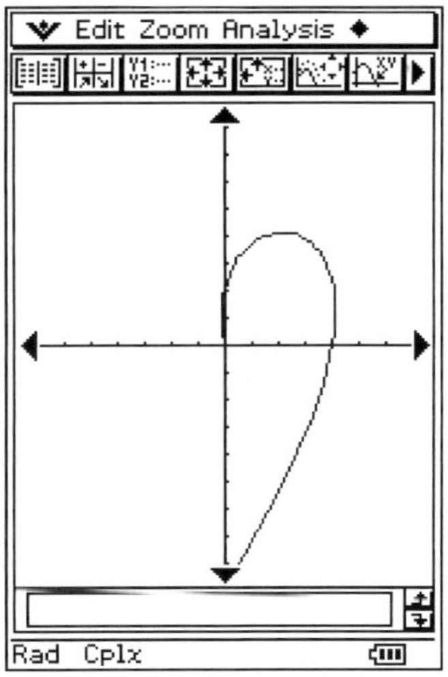

We observe that this does indeed resemble the curve traced by our animation, but it goes on a bit too long. Pressing the View Window button brings up the View Window dialog which enables us to examine the range of t.

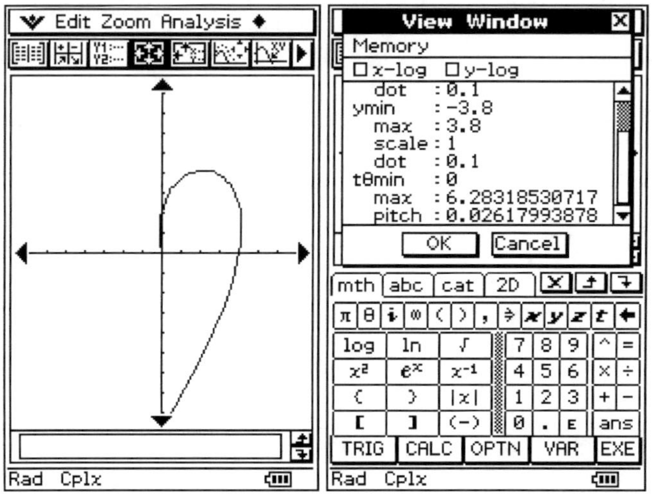

We should set this to run from 0 to 1

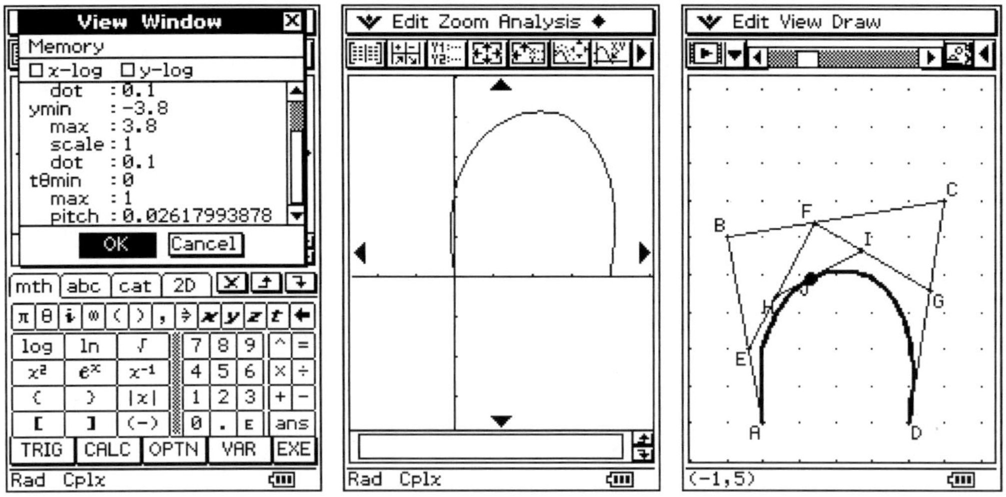

Slopes, Tangents and Derivatives

The derivative of a function at a given point is defined as the slope of the tangent to the function at that point. We can illustrate this concept quite handily with ClassPad. Let's take as our function y = sin(2x).

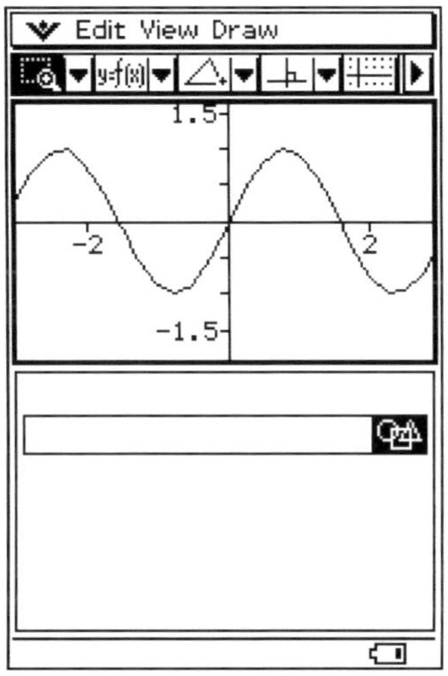

Select the Tangent tool from the construction drop down button and click on the function. This creates a point on the curve and the tangent at that point.

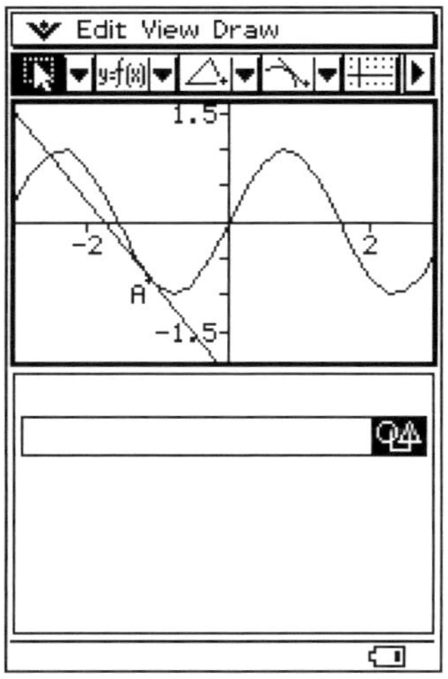

Select point A and the function and create an animation. We can now watch the tangent traverse the curve:

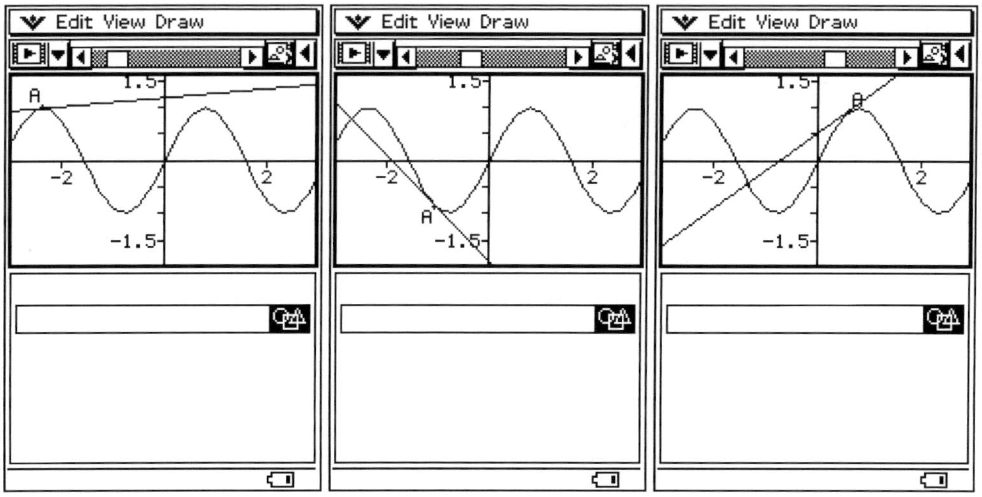

The slope of the tangent can be displayed in the measurement box:

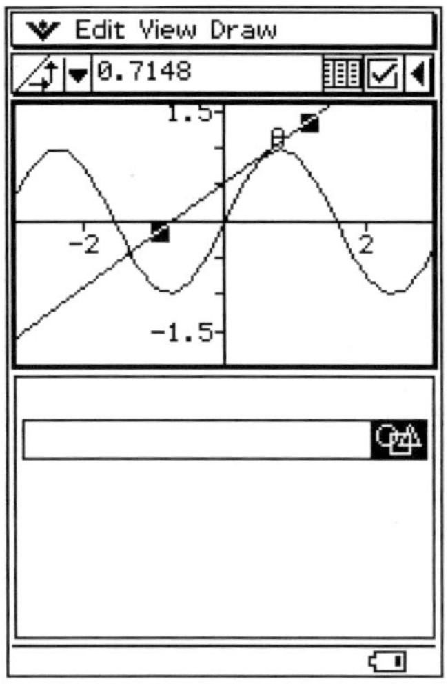

We can tabulate the slope through the course of the motion by clicking the **Table** button to the right of the Measurement box. Notice that the Table button is present if an animation has been defined. Pressing the button tabulates the current measurement over the course of the animation:

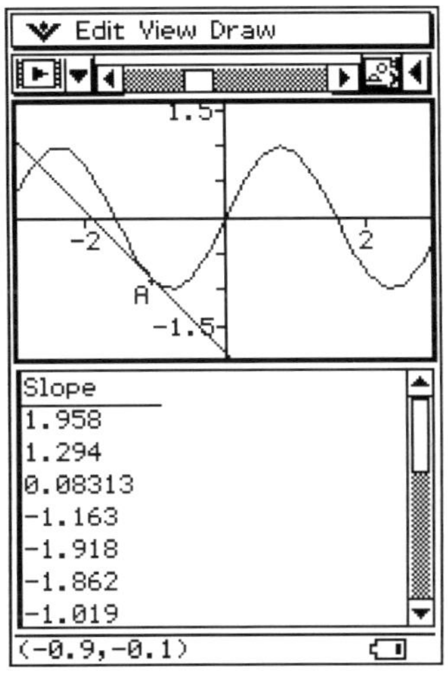

In order to create a graph of the slope over the course of the animation, we'd like to have the x coordinate corresponding to each slope value. Select the point A and display its coordinates in the measurement box, then press the Table button.

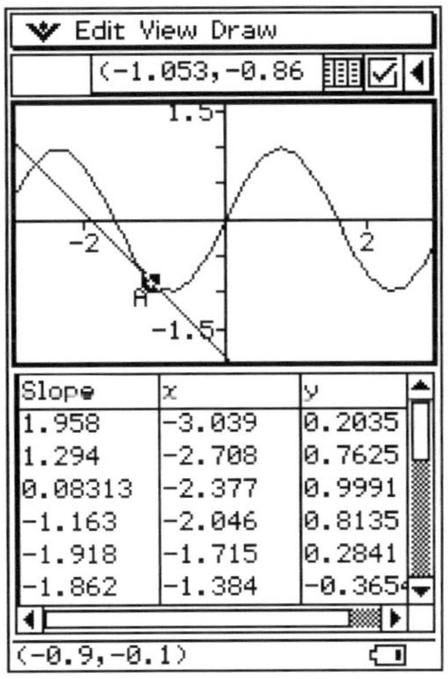

To create a graph of the slope, we need the tables reordered so that x comes before the slope. Do this by selecting the x column then **Edit / Move to Front** from the menu.

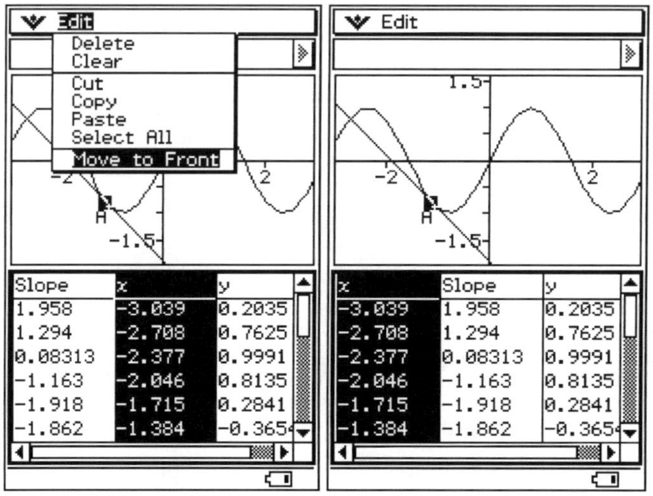

We can now select the two columns x and Slope and drag the pair into the geometry window. A graph of Slope is created:

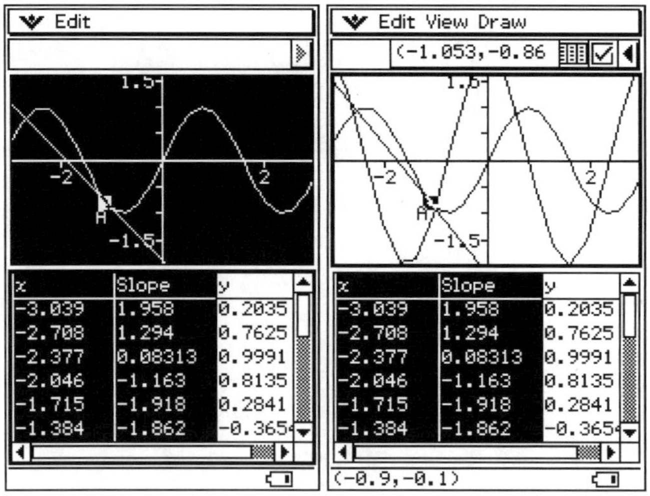

We can illustrate the correspondence between the new function and the tangent by animating a point along the new curve. As the animations run along at the same speed, the point B will illustrate the location on the derived curve corresponding to the slope at A.

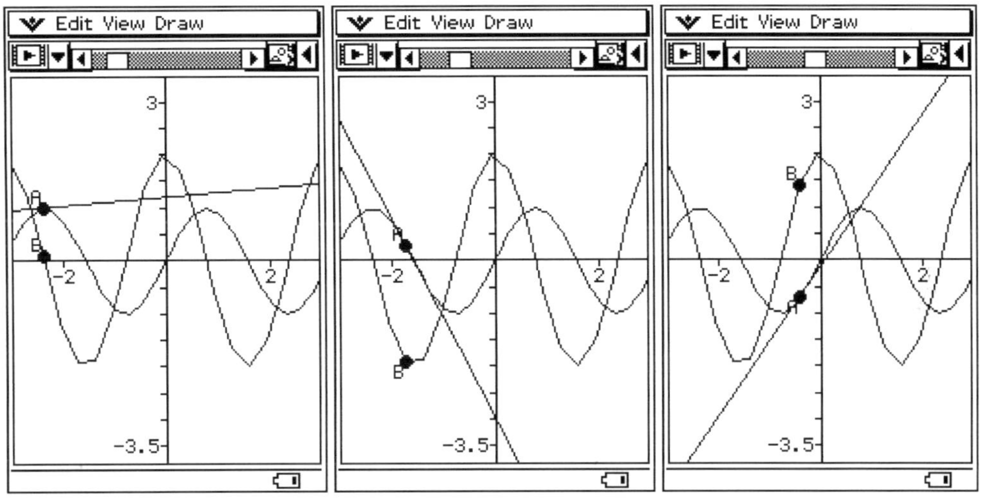

Name that Function

Can you name the new function? It looks like the original sin(2x), but shifted over so that its maximum is at 0. Let's look at cos(2x):

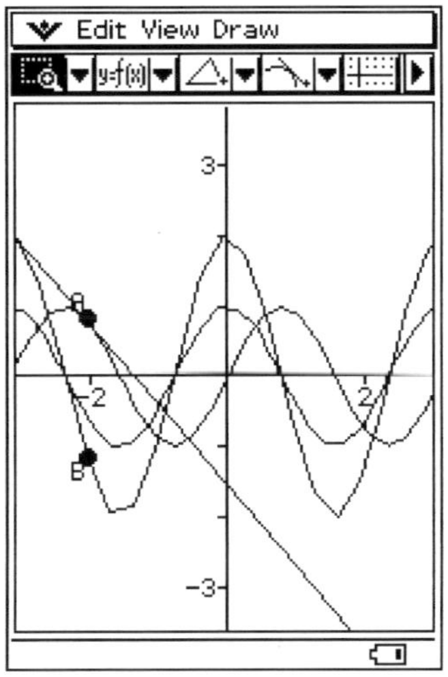

We need to increase the amplitude: let's multiply by 2:

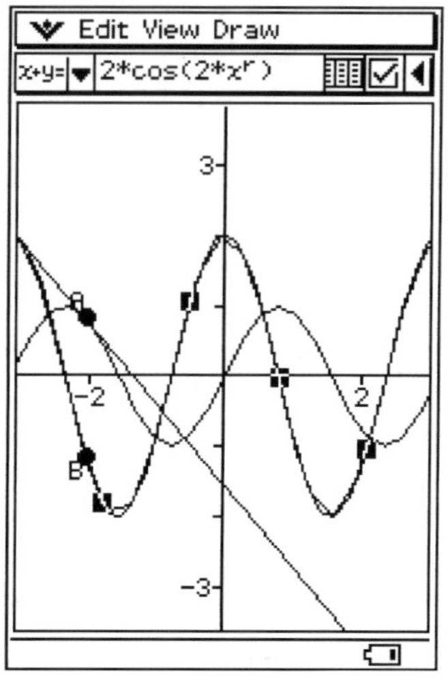

It looks like 2cos(2x) is our derivative. We can verify this using the all-knowing algebra system in eActivity:

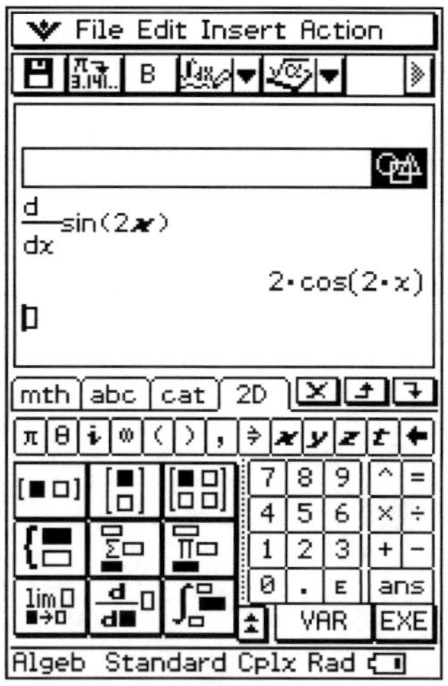

You're right, it would have saved time to have done this from the very beginning, but it wouldn't have been as much fun, would it?

Conics

In previous sections, we have split out different aspects of the ClassPad's geometry functionality, with examples pulled from a variety of different mathematical areas. Here we will look at a single mathematical area, but apply a variety of different ClassPad techniques.

Parabola

One way of defining a parabola is in terms of a focus (point) and directrix (line). The parabola is defined to be the locus of all points whose perpendicular distance from the directrix is the same as their distance from the focus.

Here is how to create this model in ClassPad:

First we draw a line AB for the directrix and a point C for the focus.

Then we create a point D and join it to C and to the directrix:

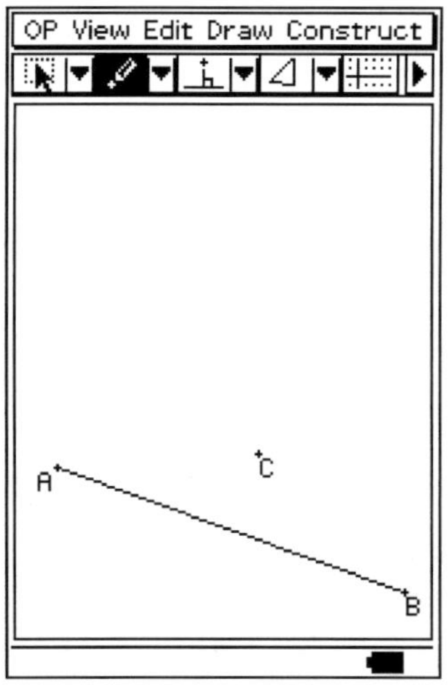

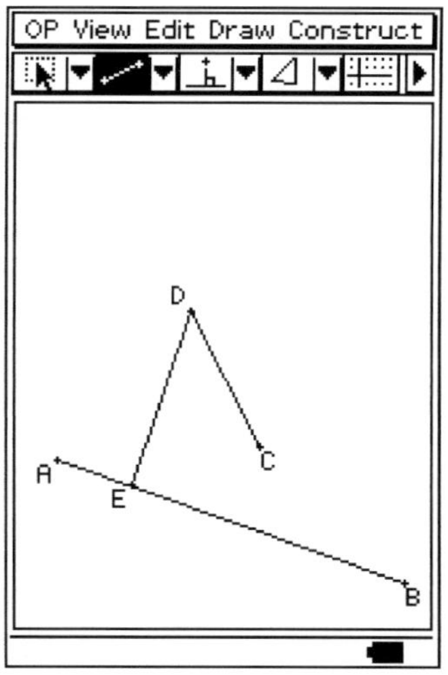

We now need to specify the following constraints:

- DE is 90 degrees to AB
- DE and DC are congruent.

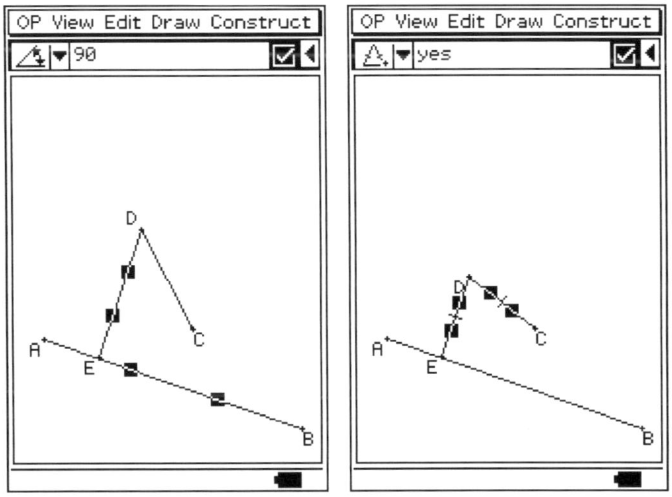

We should now specify that E should slide along the line AB, while tracing D. We do this by selecting point E and line AB and clicking **Edit/Animation/AddAnimation**. Then we select D and click **Edit/Animation/Trace**.

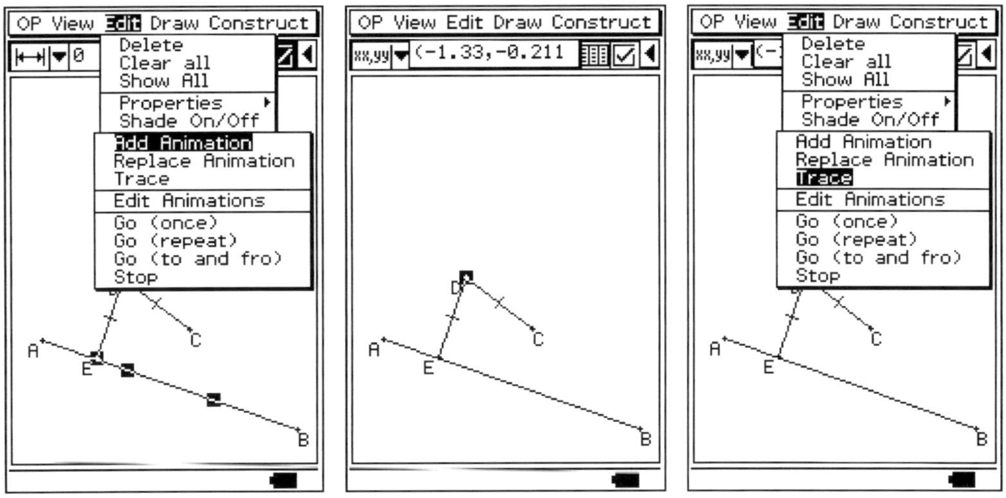

Start the animation by clicking Edit/Animation/Go (once):

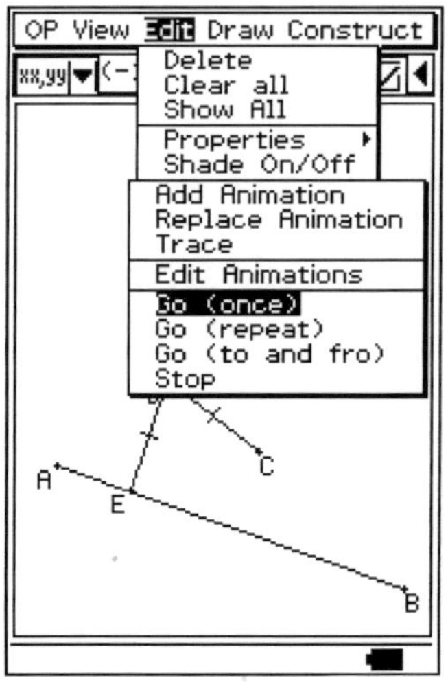

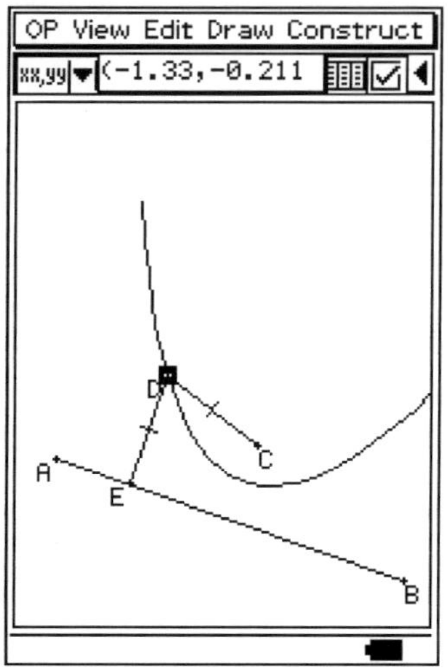

Thicken the curve using **Edit /Properties/Thicker**

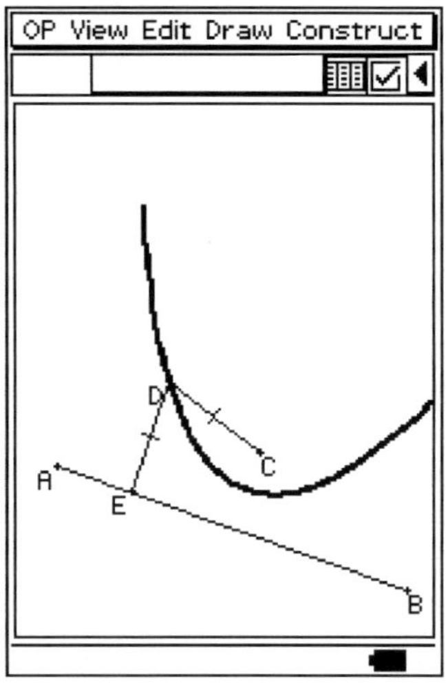

Ellipse

One definition of the ellipse is the locus of points , the sum of whose distances from two focal points is constant. This definition admits the simple pencil and paper construction where a slack piece of string is pinned to two foci and a pencil draws round the ellipse.

We can mimic this construction in ClassPad:

We first create a circle whose center is one of the foci of the ellipse, and whose radius is the length of the string:

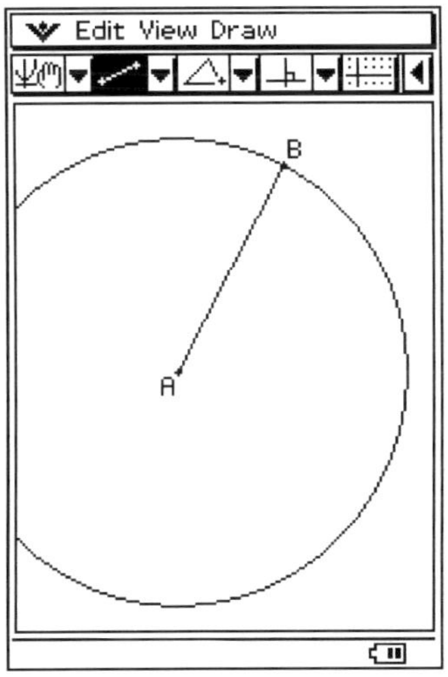

We then draw the second focus C and join it to a point D on AB. We draw the line segment DB, which will lie entirely on top of AB, but which we need in order to take the next step – which is to select DC and DB and make them congruent. DB lies on top of AB, so there is a danger of selecting the longer line. Drag the pen till it is on the center of DB and you will see the selection handles on the shorter line. Then lift the pen to complete the selection.

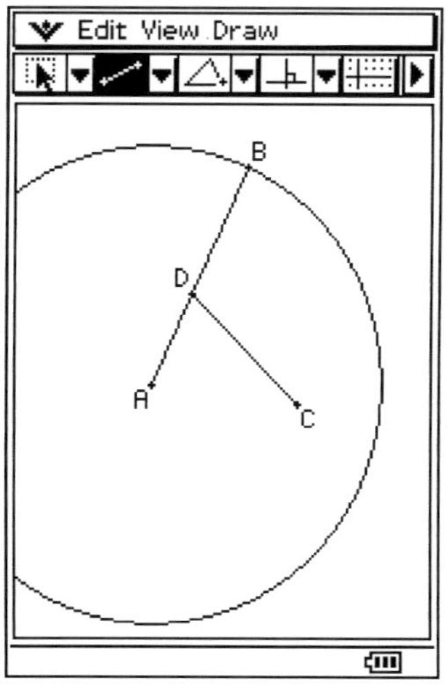

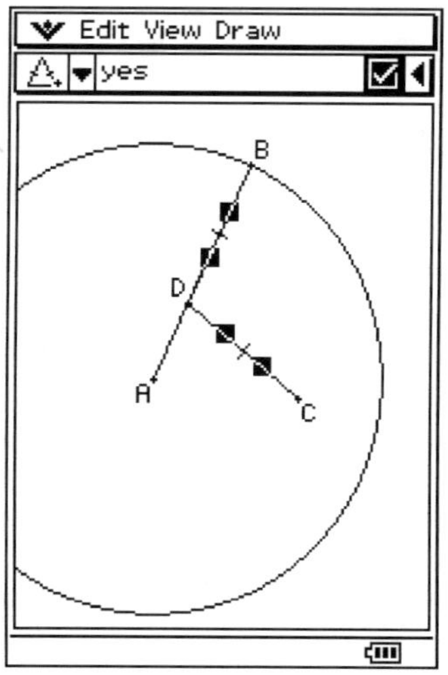

As BD and DC are now the same length, the point D should trace out a path such that $|AD|+|DC| = |AB|$, which is constant.

To see the path, we need to select point B and the circle and add an animation:

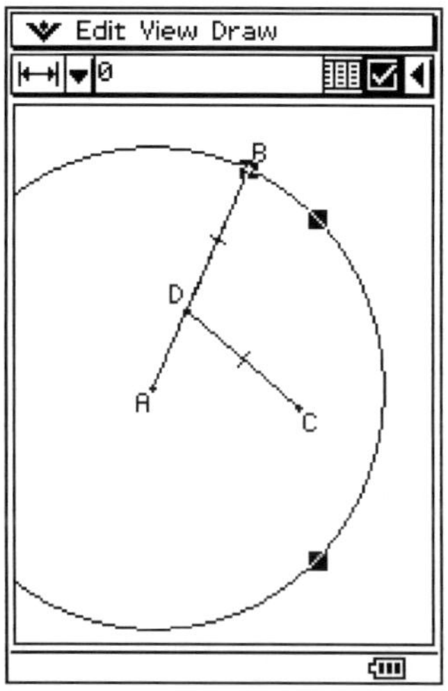

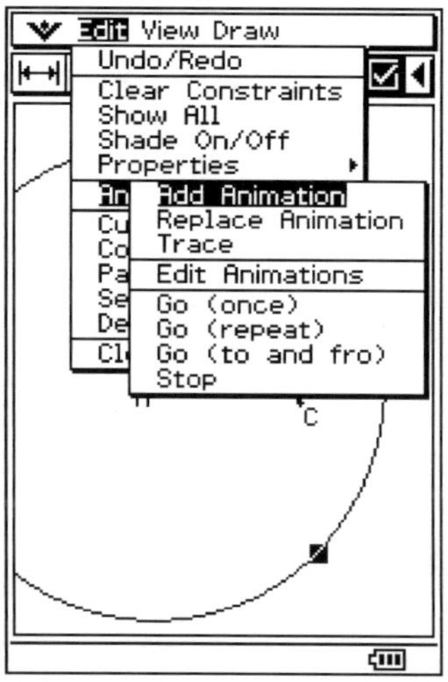

We then need to select point D and create a trace of this point:

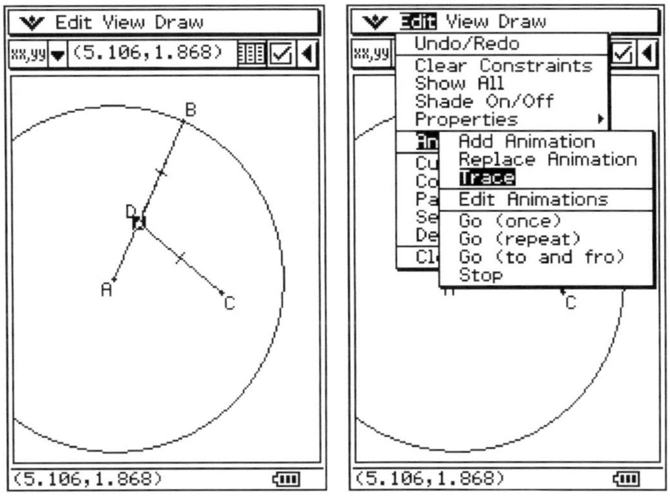

Running the animation will cause the trace of point D to be drawn:

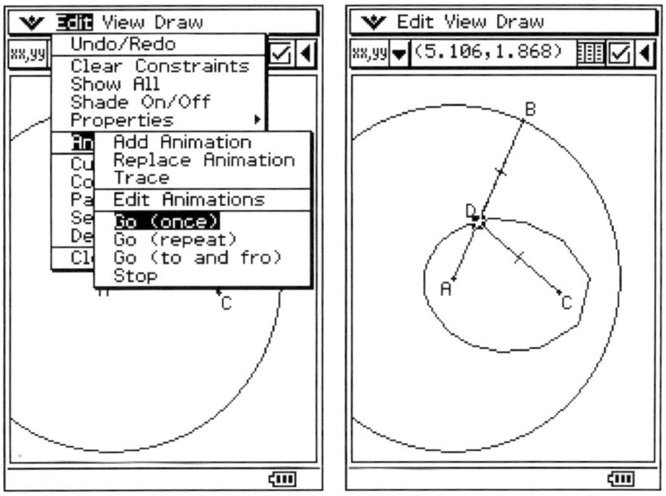

The right hand side of the ellipse is a little angular: this is due to the step size of our animation. We can increase the number of steps by bringing up the Edit Animation window:

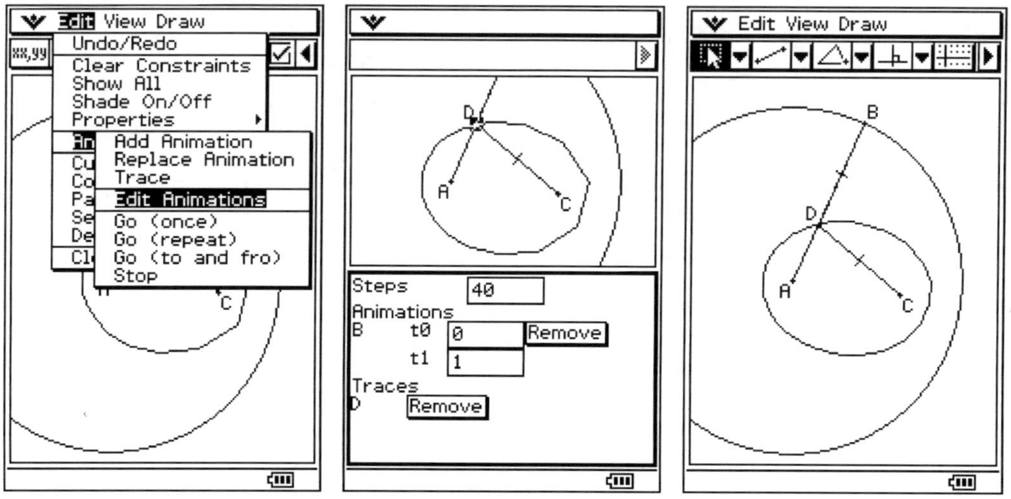

Hyperbola

The hyperbola can be defined as the locus of points such that the difference of the distance from the point to each of two foci is equal to some specific constant.

Again we start with a circle, although in this case the radius of the circle |AB| represents the fixed difference in distances. We now create a third point C on AB extended and a second focus D. One way to do this is to create segment BC, then lock the angle of AB and BC to be 0.

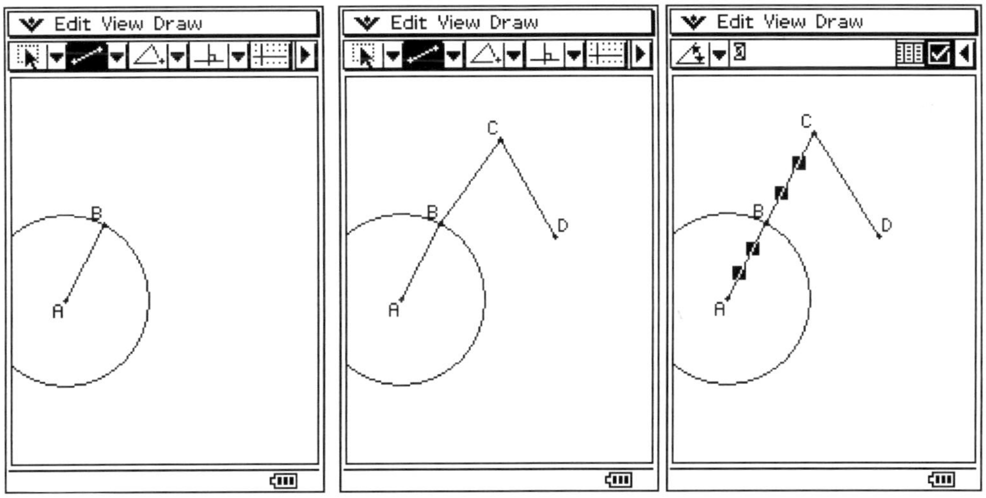

We fix the location of the focus D by selecting it and locking its coordinates with the check button. We then specify that segment CD and BC are congruent. This guarantees that |AC|-|CD| = |AC|-|BC| = |AB| which is the (constant) radius of the circle AB.

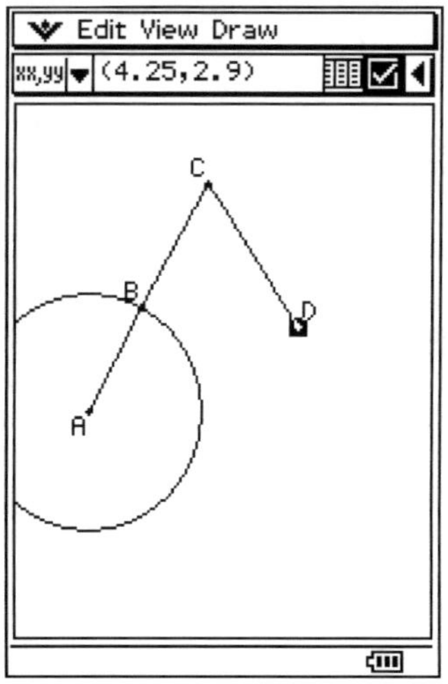

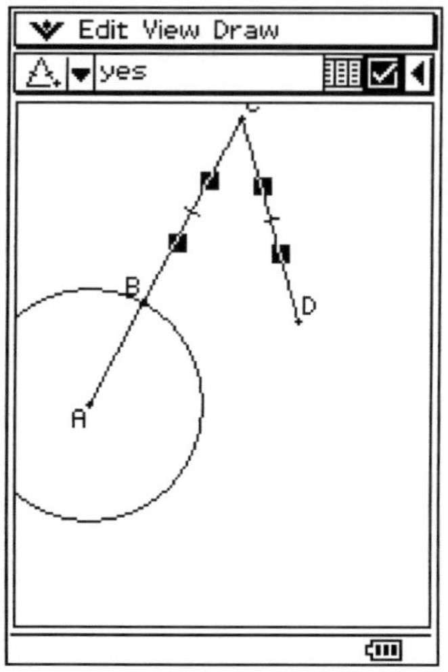

We now just need to add an animation of B round the circle AB, and a Trace of point C. Then run the animation to draw both the hyperbola and its asymptotes:

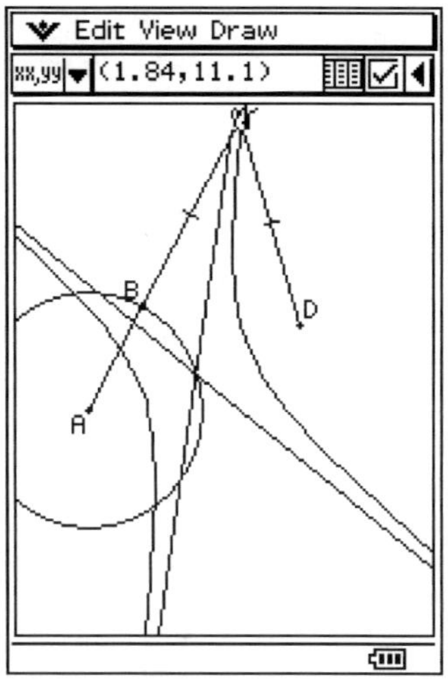

Changing the radius of the circle AB changes the characteristics of the hyperbola:

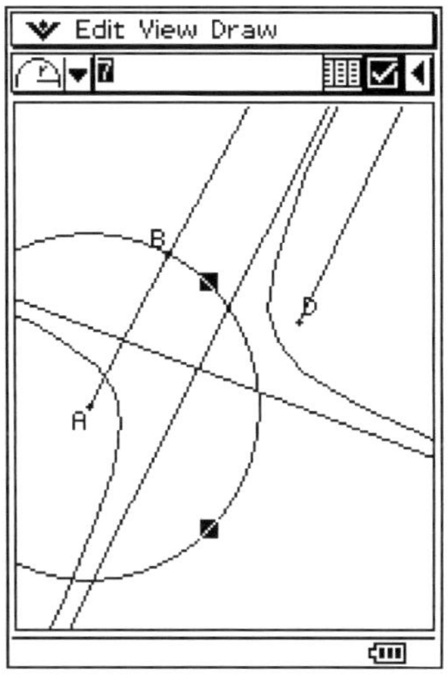

Focus On Ellipse

Here's our assignment:

> Draw the ellipse centered at the origin, with semi-major axis length 5 in the direction of the x-axis, and semi-minor axis length 4. Show that if a ray of light emanating from the point (-3,0) reflects from the surface of the ellipse, it passes through the point (3,0).

As usual, with coordinates defined so specifically, we should turn on the axes and the integer grid. The ellipse tool lets us position first the center of the ellipse, then the end of one of the principal axes, then the end of the other. (The principal axes are the axes of symmetry of the ellipse.) Locate the center at the origin, the end of the first principal axis at the point (5,0), and the end of the second principal axis at the point (0,4). Also create the points (-3,0) and (3,0).

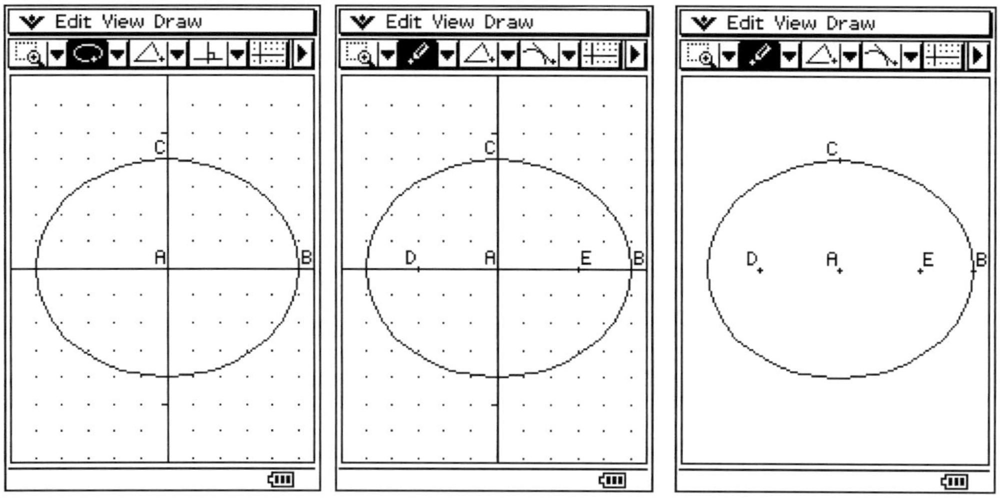

Once the figure is drawn, you can turn off the axes and grid points to relieve screen clutter.

Our next step will be to create a tangent line to reflect in. To do this, select the tangent tool from the construction drop down button, or from the menu, and click on the ellipse at the position where you would like to see the tangent:

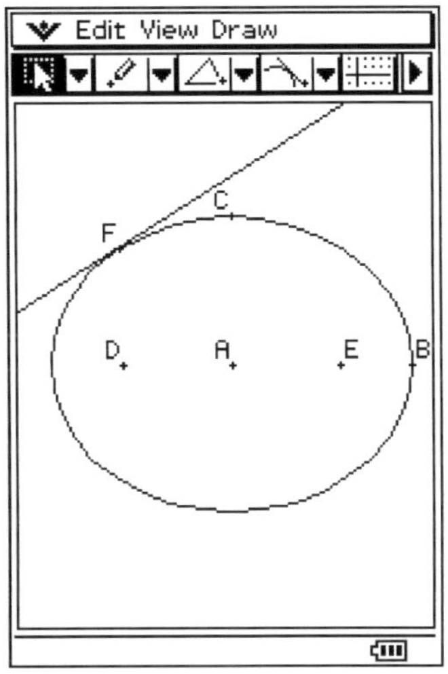

We now create a line between point D and F, then reflect this line in the tangent:.

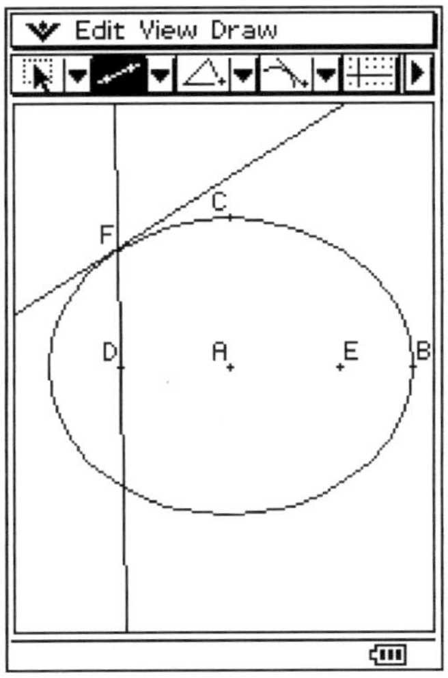

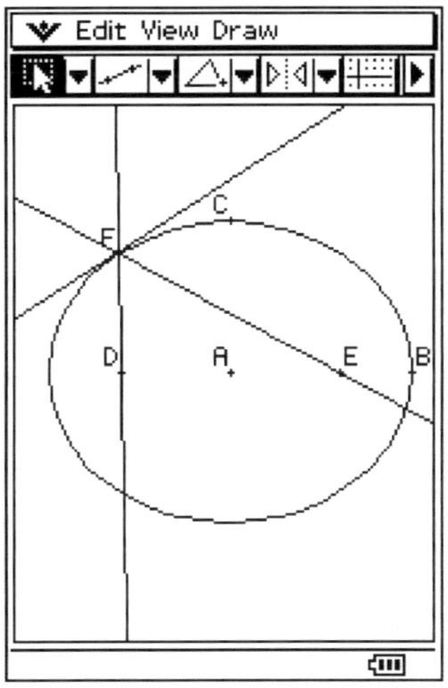

We can observe that the reflected line appears to go through the point E (3,0). To ascertain if this is more than just an optical illusion, we can select the point E and the line EF and check that the point on line relationship is satisfied.

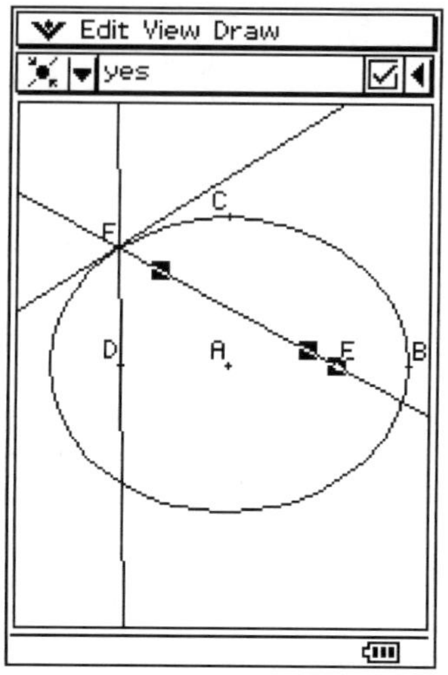

Dragging point F around the ellipse shows that this is not due to luck in locating F:

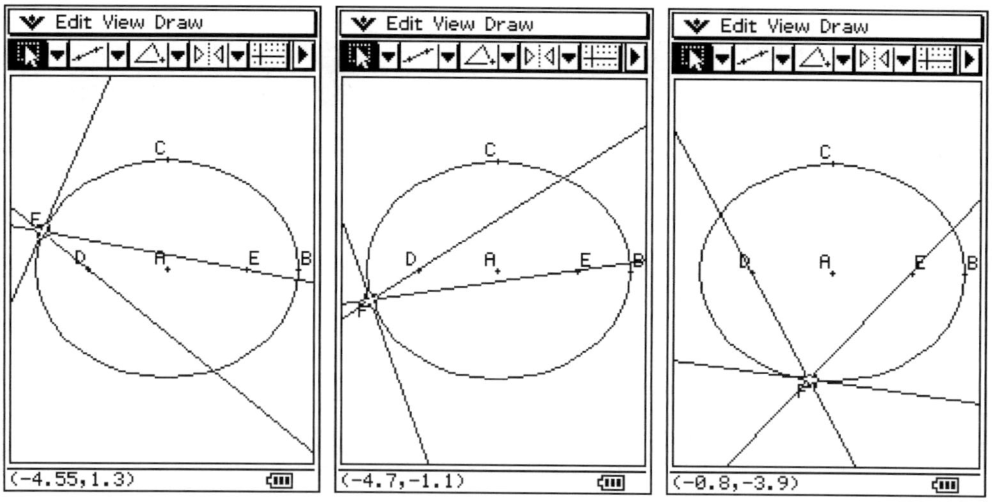

Strings & Pins & Foci

The points D and E in the drawing above are called the foci of the ellipse, because light produced at one such point will be focused by the ellipse onto the other point. It turns out that these foci are also the points you can use for the familiar pin and string construction of the ellipse. Let's investigate this a little further.

For the moment, we'll hide the tangent and the light rays (just select them and pick from the menu **Edit / Properties / Hide)** .

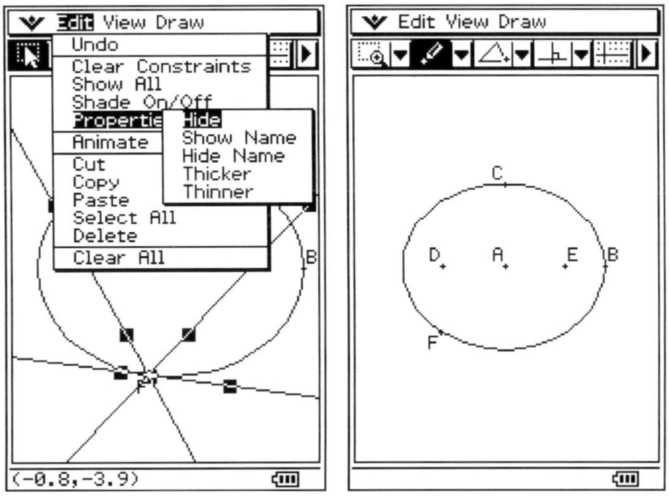

We then select point F and the ellipse and add an animation. Now we can select point F and D and display the distance between them. Pressing the Table button (to the left of the check button) creates a table of values for this distance. (If the table button does not show up, then you did not successfully add an animation.)

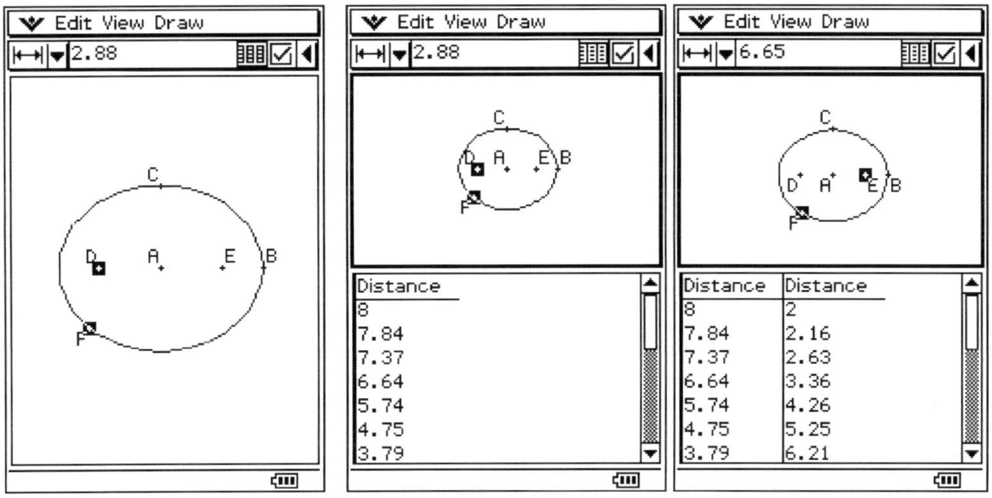

Inspecting these distances, we notice that they add up to 10. The points D and E are therefore the locations we would use for the string and pin construction of the ellipse.

Algebraic Focus

The algebra required to show that the relationship above holds is not too complicated, but does get a little messy. ClassPad can help with the messy stuff:

First we should copy the equation of the ellipse into eActivity (do this by dragging and dropping the ellipse into a Calculation Row in eActivity:)

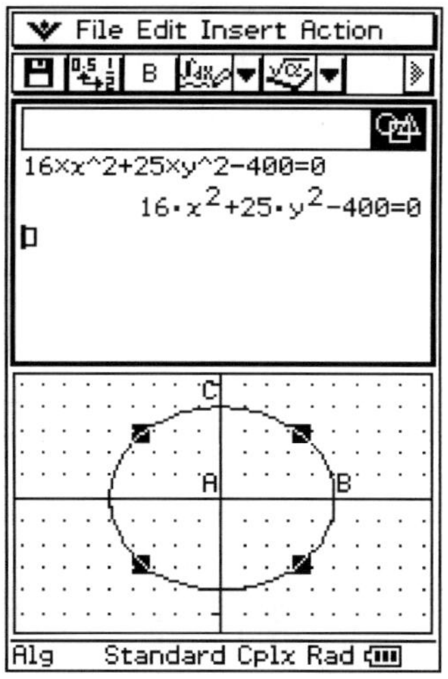

For later reference, we'd like to solve for y^2. However, if we try solving for y^2 we get a syntax error - we get an error (we have to solve for a variable). We can work our way round this problem by replacing y^2 by a dummy variable u, and solving for u:

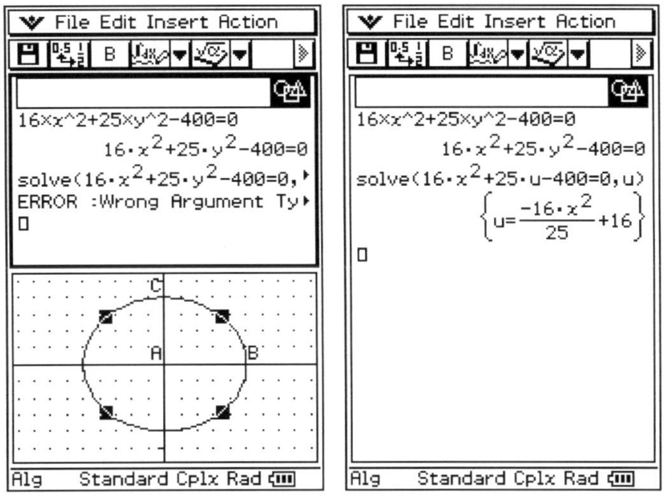

We now want to construct the expression for the sum of the distances |DC| +|EC|:

This is the expression

$$\sqrt{(x-3)^2+y^2}+\sqrt{(x+3)^2+y^2}$$

However, we can substitute the variable u for y^2:

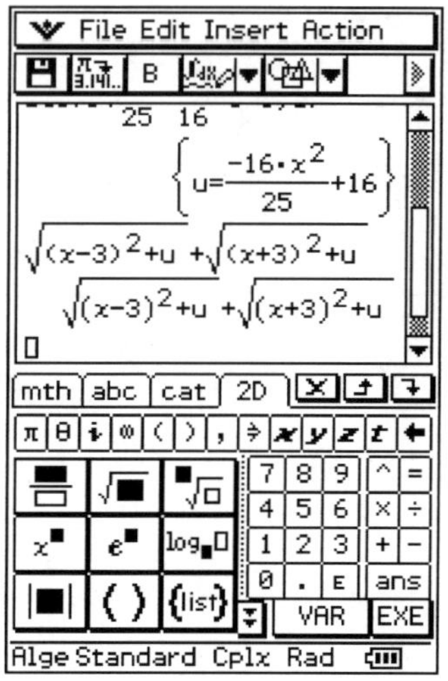

This expression does not look too promising as it is the sum of two square roots, and not as such amenable to much simplification. When roots are present, it is often a good idea to square the expression. Let's do that, and then expand the result:

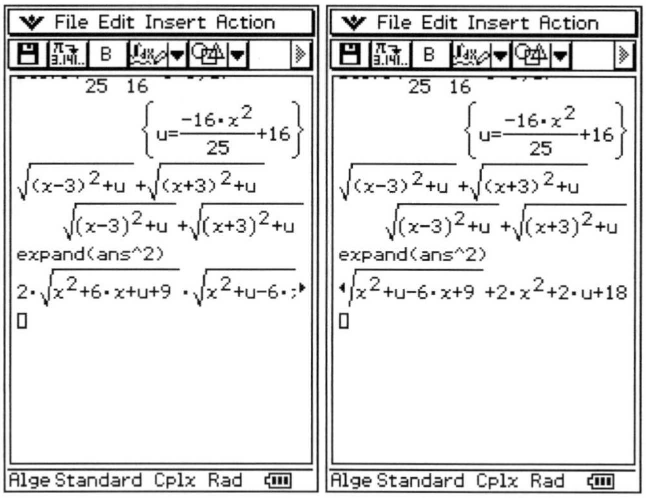

The result, of course still has square roots in it, but at least they are multiplied by each other. Let's examine the inside of the roots. The two expressions look similar and promising for simplification. Let's substitute for u from the above equation in each of the terms inside the roots. The vertical bar line (read "with") causes the equation after the line to be substituted into the expression before the line.

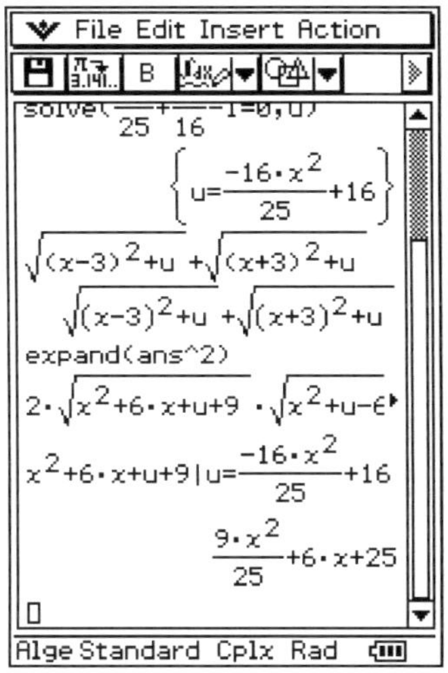

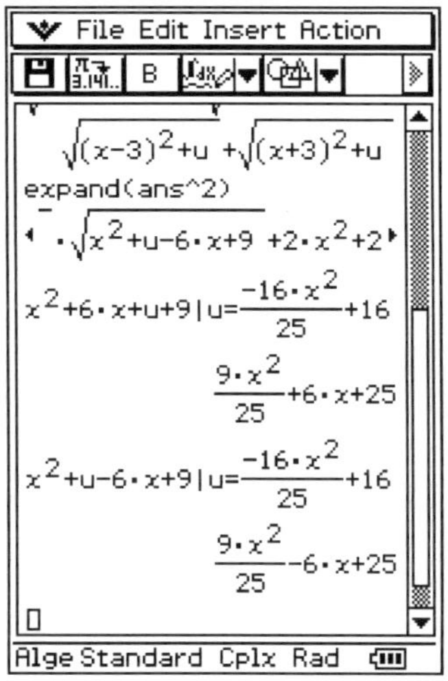

Now let's multiply these terms and simplify the result:

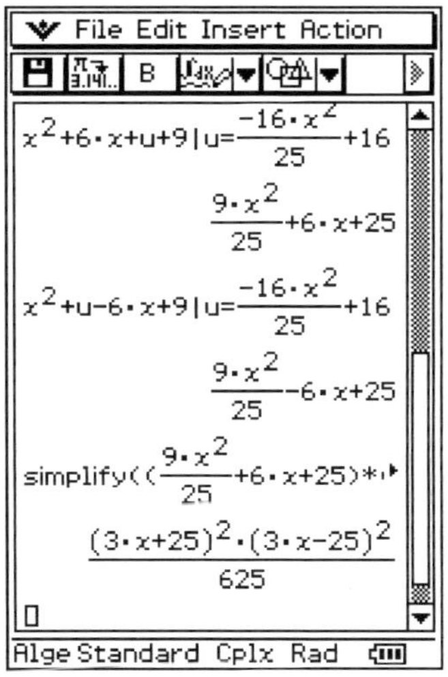

This looks very promising indeed. Remember, this is the term inside the square root, so we should be able to get rid of those squares. However, we need to be careful: $\sqrt{a^2}$

simplifies to a if a>0 and to -a if a<0. Now we know that our x values lie between -5 and 5. So the term 3x+25 is guaranteed to be positive. However, the term 3x-25 is likewise guaranteed to be negative. So when we strip off the square root, we are going to need to change it to -(3x-25). Here then is the term under the root:

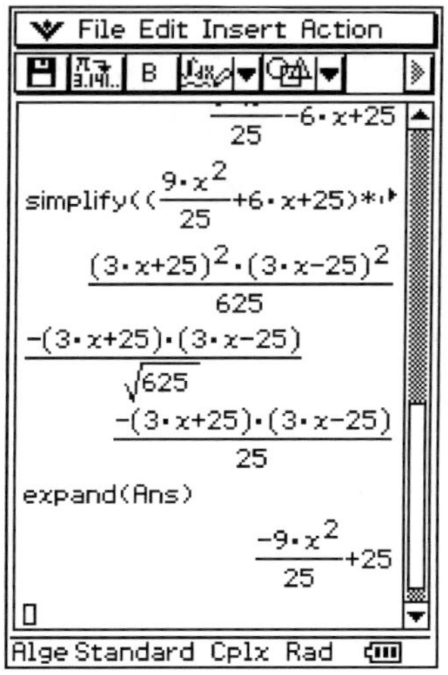

Now, we can copy and paste the original expression for the square of the sum of the distances into a new input area, and replace the product of the roots with the above expression:

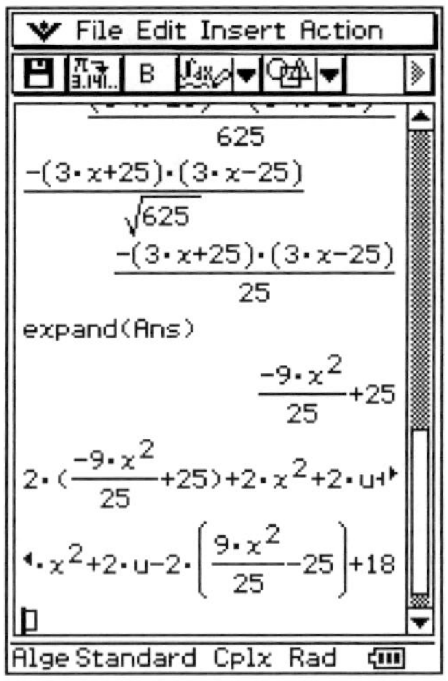

We now need to substitute for u and simplify:

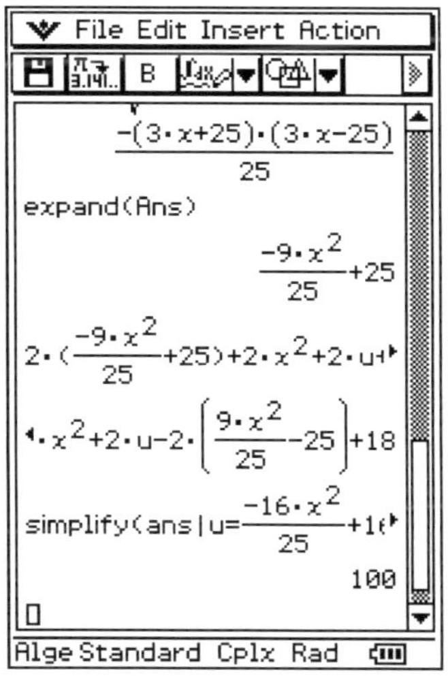

Our big messy algebraic expression simplifies down to the constant 100. Remember we squared the sum of the distances from the circumference point to each of the foci. So the sum itself (the length of the string) is shown to be 10, which agrees with the numerical results obtained in the previous section.

Eigenvectors, Eigenvalues and Ellipses

Eigenvectors and Eigenvalues are inherent properties of matrices, which can seem a bit mysterious in the abstract. However, their significance becomes clearer when connected up with the geometry of ellipses. We'll try to shed light on this subject with ClassPad.

First draw your ellipse.

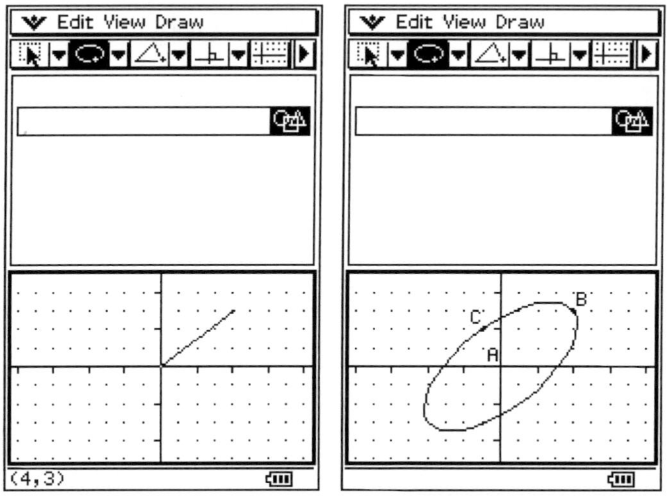

We can drag the ellipse into eActivity to look at its equation. (Your ellipse's equation may have different coefficients to ours. Don't let that worry you – follow through with the numbers you have.)

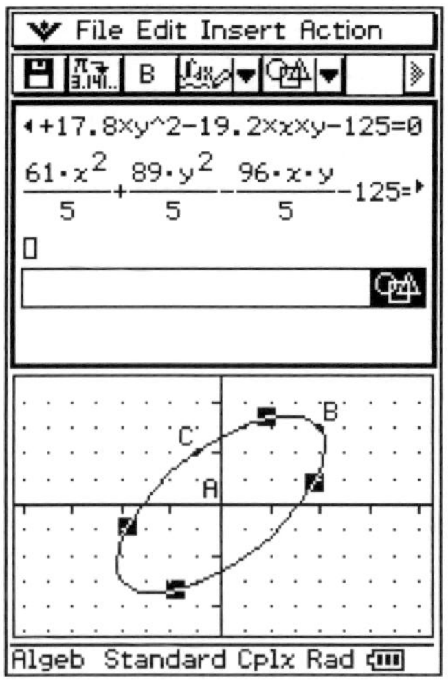

Now an ellipse whose center is at the origin will always have an equation of the form:

$$ax^2 + by^2 + cxy = d$$

We can express this as a matrix equation:

$$\begin{bmatrix} x & y \end{bmatrix} \cdot \begin{bmatrix} a & \frac{c}{2} \\ \frac{c}{2} & b \end{bmatrix} \cdot \begin{bmatrix} x \\ y \end{bmatrix} = d$$

We can verify that these expressions are the same using ClassPad:

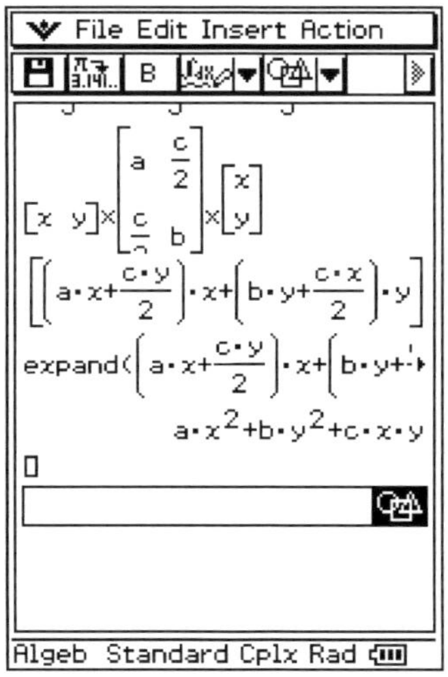

We can create the matrix for our ellipse simply by copying the coefficients into a 2x2 matrix template, remembering to divide the coefficient of xy by 2.

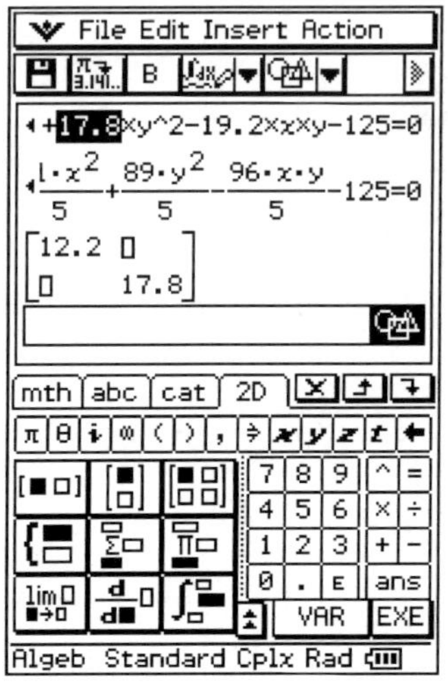

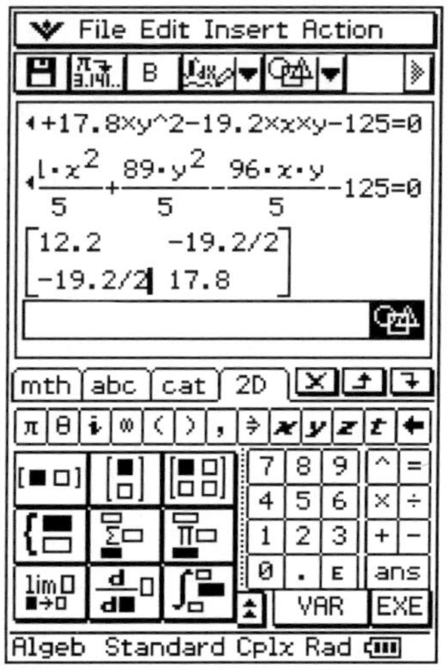

Now eActivity has a function EigVc() which computes the eigenvectors of a matrix. This is just the sort of function abbreviation that it is difficult to remember: is it EigVec() or EigenV() or Evector()? Fortunately the ClassPad Keypad has a function look up list built in so you can skim this list to find the exact correct spelling of the function, then press **INPUT.**

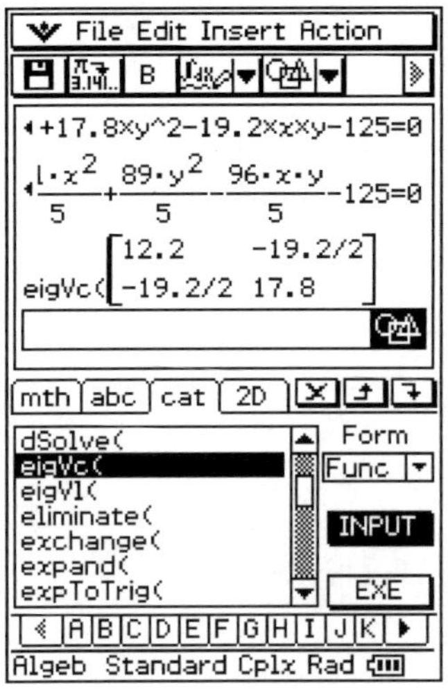

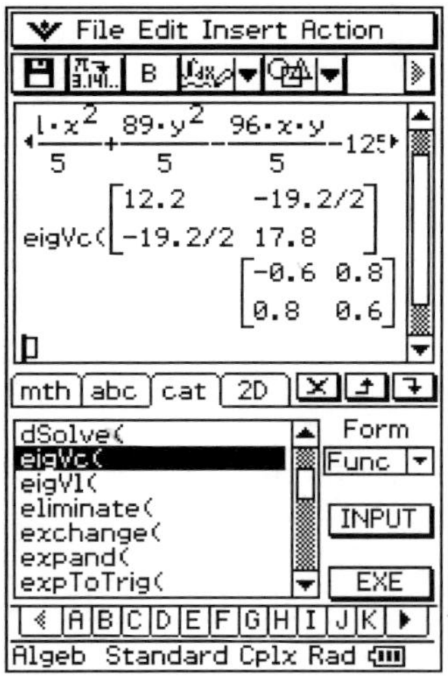

What we get back is in fact not a vector at all, but a matrix! This is a little confusing, but we can interpret each column of the matrix as a vector:

We can create these vectors in ClassPad by dragging coefficients from the matrix into empty vector templates. Now try dragging the vectors onto the drawing (first turn off the integer grid and zoom in a little so you will be able to spot the new points.)

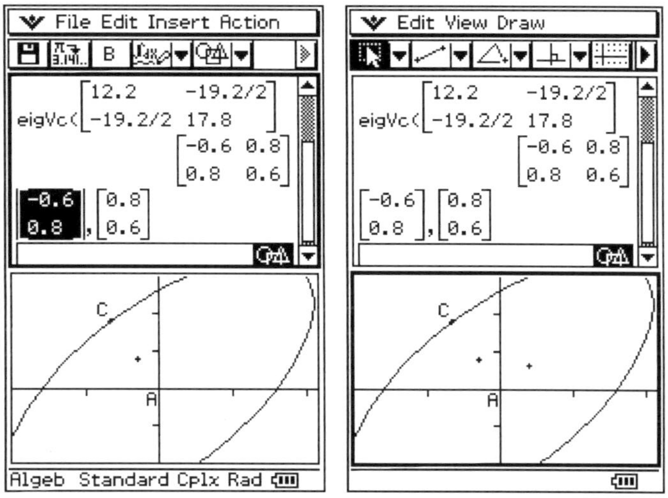

Now draw lines through the origin (A) and each of these new points.

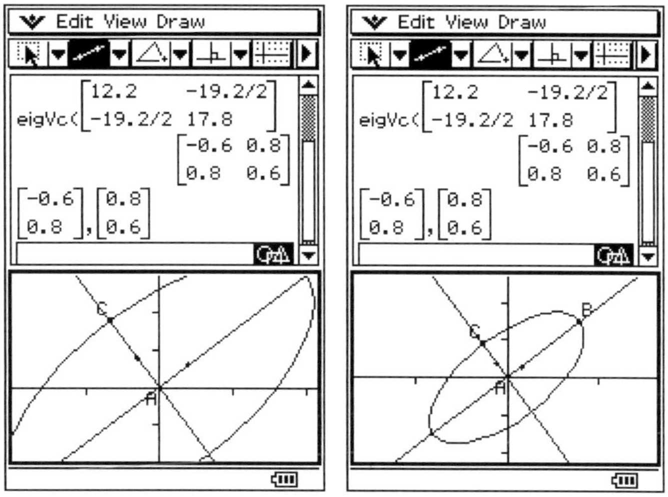

We see that the eigenvectors give the direction of the axes of symmetry (otherwise known as *principal axes*) of the ellipse.

So much for eigenvectors, how about eigenvalues?

ClassPad's eigenvalue function is EigVl()

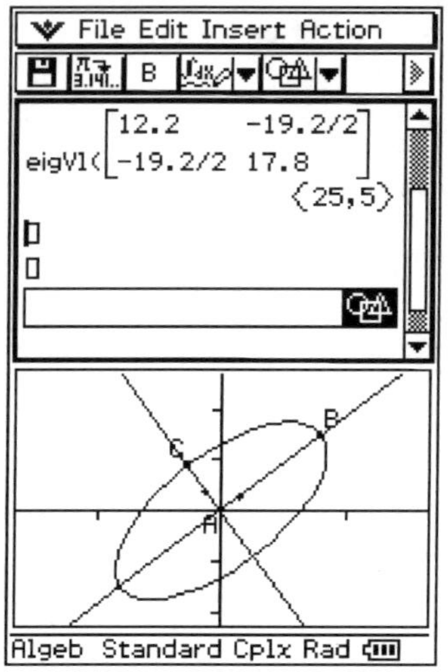

The result of this function is a pair of numbers - the eigenvalues of the ellipse. How do we interpret these?

If we type the original ellipse equation, but leave out the x*y term and replace the coefficients of x^2 and y^2 with the two eigenvalues, we have a new ellipse equation. Drag this into the geometry:

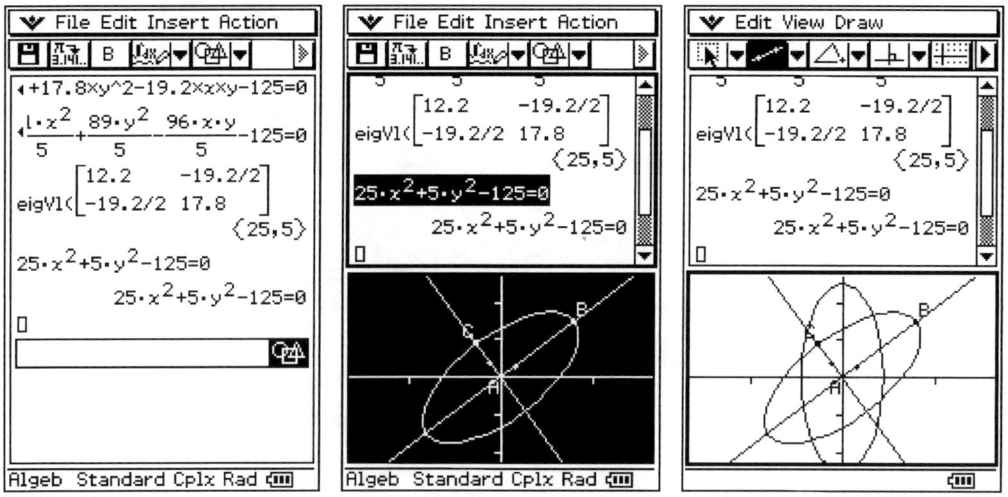

We see that this looks like the original ellipse, but rotated to align with the axis.

We can verify this by performing the rotation. First we check the direction of the line AB so that we know what angle to rotate:

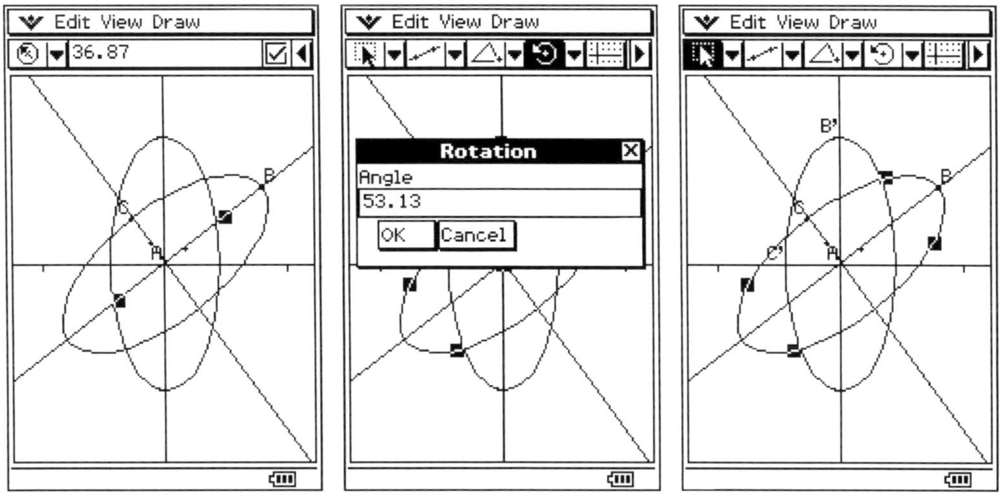

The ellipse rotates into place on top of the existing one.

Legacy Applications

Up to this point, we've only explored the cActivity and Geometry applications. Browsing through the launcher, we find a number of other applications advertised –graphing applications, the Solver, Statistics, and more. Those that have used other Casio graphing calculators will discover that these programs are eerily familiar. Even if you haven't used calculators with similar features, you'll master these applications in no time, with a few helpful hints and examples contained in this section.

Graph and Statistics

As an example we'll investigate the effect on frequency of a couple different functions – e^x and 1/x..

The graph application can be a powerful tool for inspecting functions. Here the eActivity will serve as a good spot to record our measurements. Launch the eActivity – if you're not there already just hit the menu icon on the bottom of the screen, then hit the eActivity icon.

Now insert a graph strip into the eActivity – tap the insert menu and then graph editor.

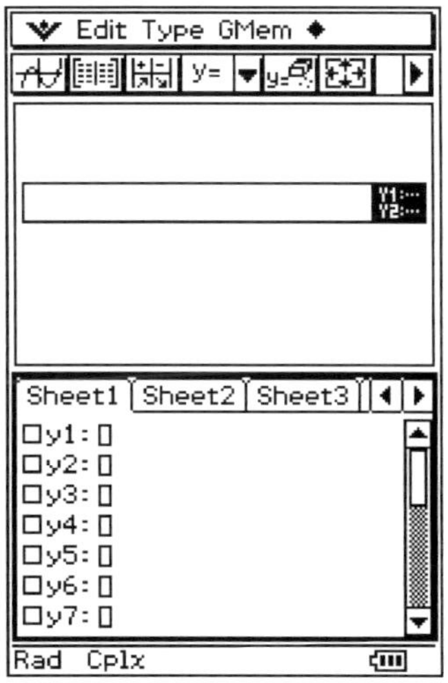

Click on the box after y1 to start the function entry. Then press the keyboard button – it has all we need to type in the expression. Type in sin(x)/x.. You can find 'sin' under the trig layout button – the rest is at the top of the keyboard or you can use the main calculator buttons.

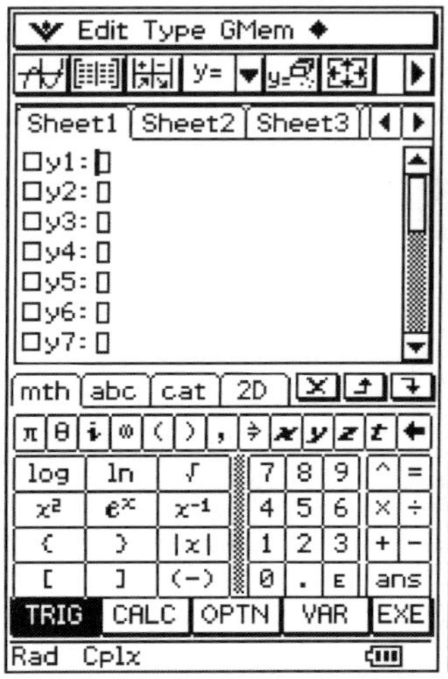

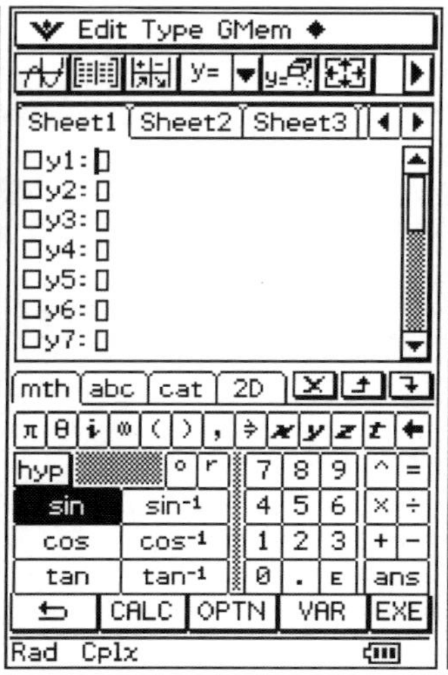

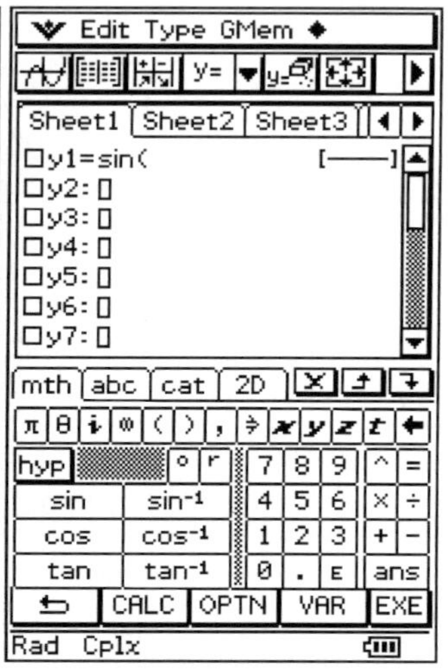

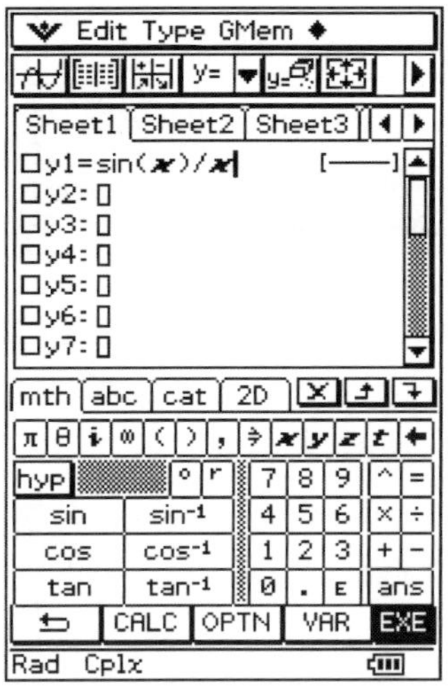

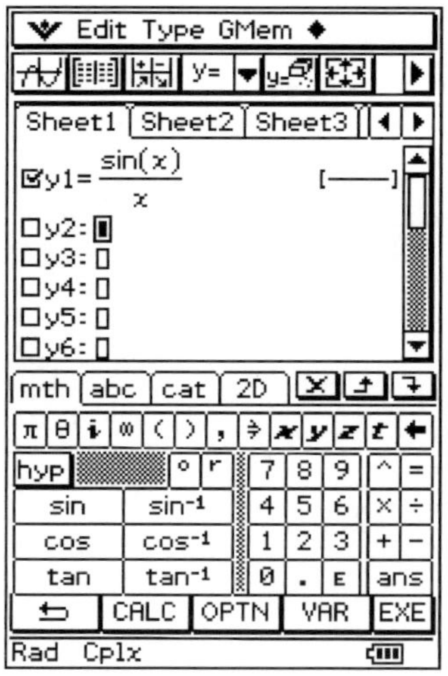

Now press the graph button in the upper left hand corner of the screen. After a few moments, we can see the function.

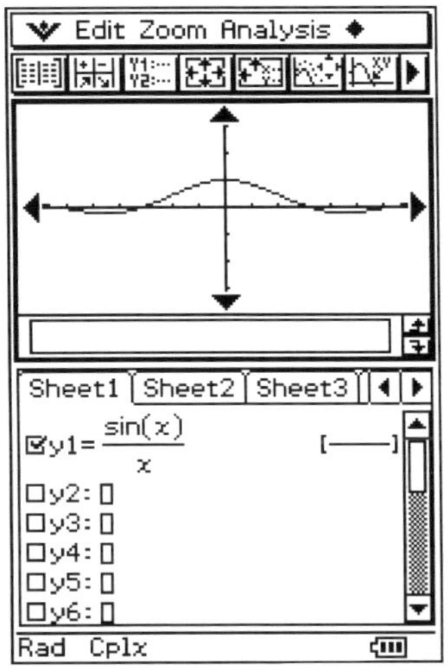

If we want to see a few more cycles, we can adjust our zoom. Tap the zoom menu, and select zoom out.

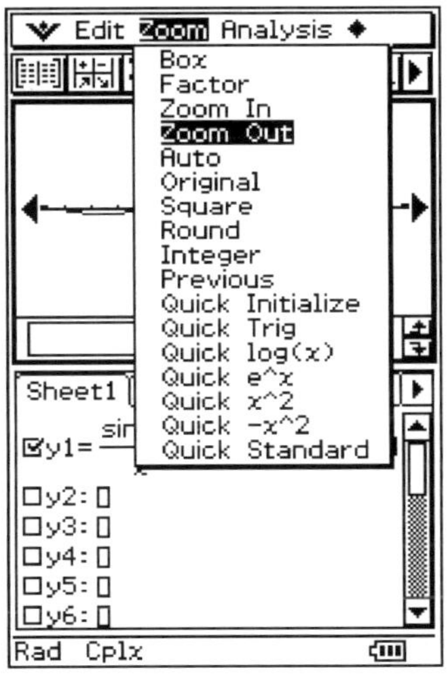

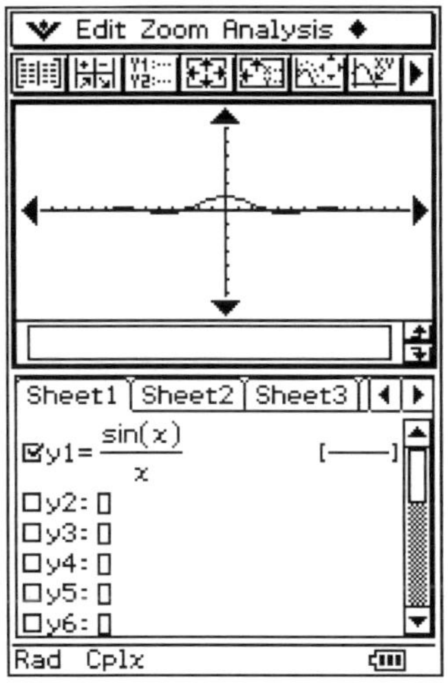

That's not quite what we wanted – now we can hardly see the function since the amplitude is too low. We need to zoom in just in the y direction. Hit the button with the four arrows in the center of the button strip.

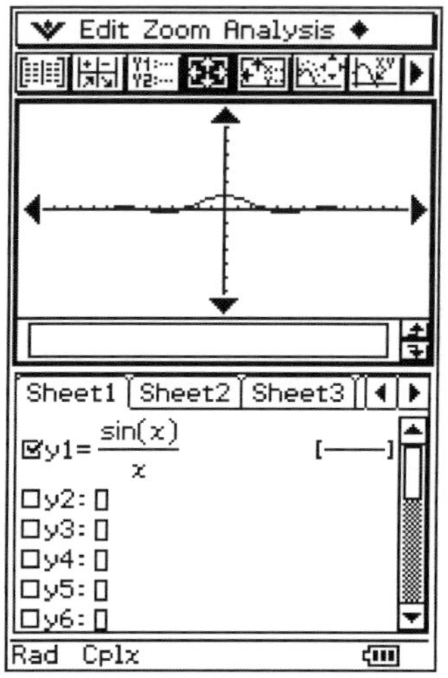

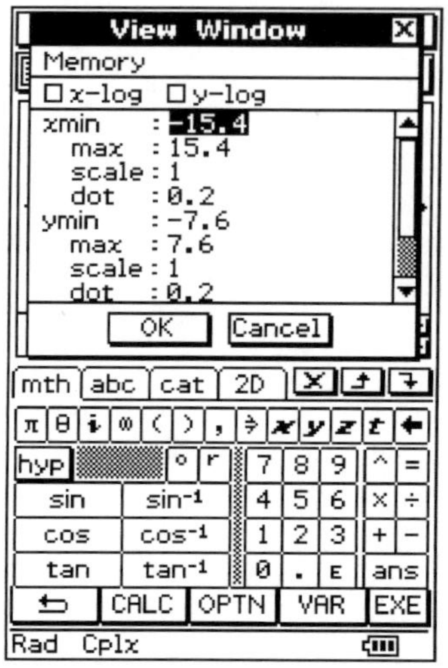

Change the ymin and ymax to –.5 and .5. Then press OK

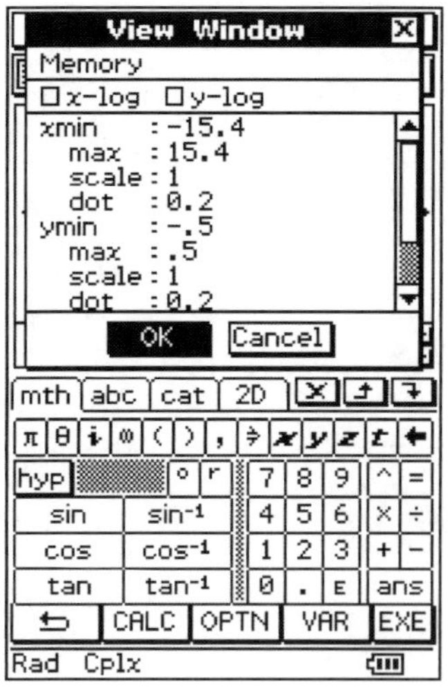

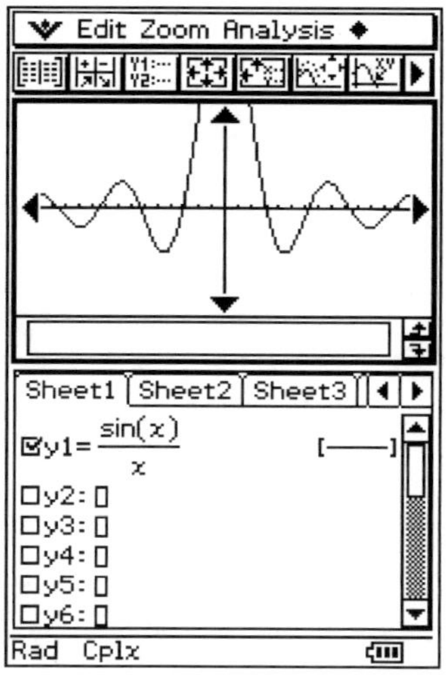

It doesn't take much effort to look at different parts of the function – just tap on the pan arrows on the sides of the graph. Tap on the right button a few times so just the positive x part of the graph is visible.

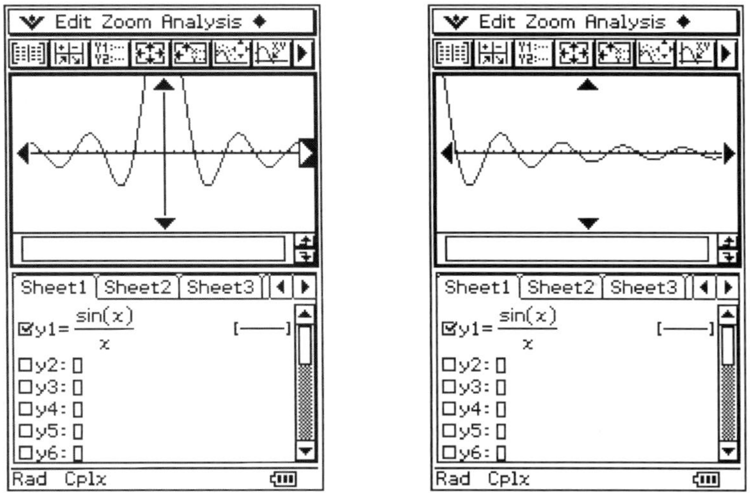

One way to determine the frequency of our modified sin wave is to measure the distance between the peaks. The max graph tool will come in handy. Hitting the arrow in the button bar at the upper right corner will reveal more buttons. One of these is the max button. Tap this, and some information on the first peak will be displayed. It starts searching for the max from the first visible x position. In this case, the first max at (7.72,0.128).

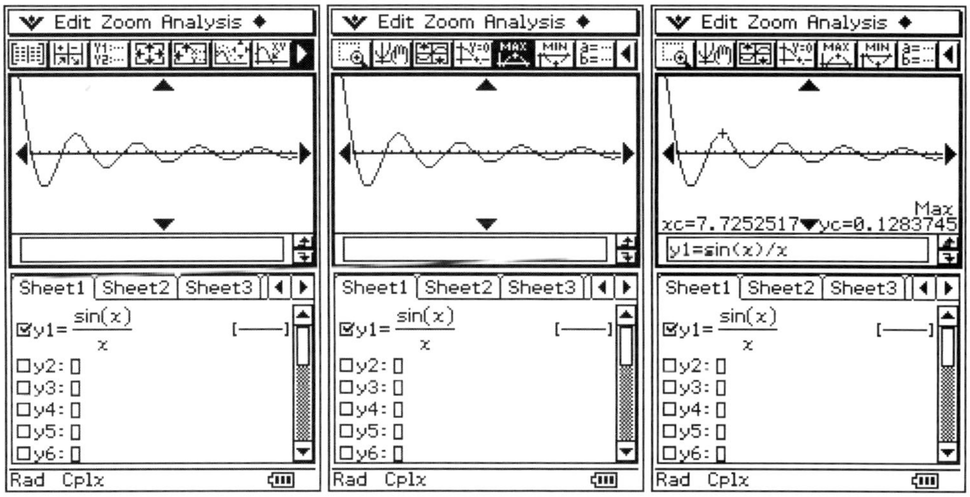

We now have the starting position of one of the cycles. We'll save this by dragging and dropping the x value into the eActivity. The eActivity still waits beneath the graph windows. For now, we don't need the graph editor, so click on the graph editor, and close it.

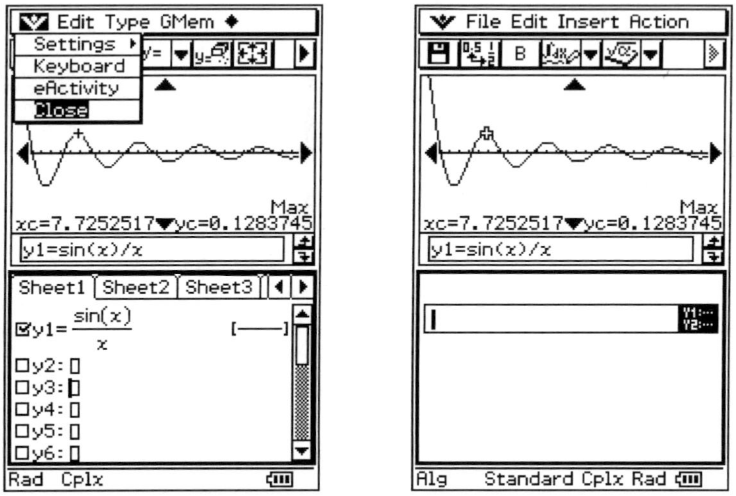

Make sure empty lines are available in the eActivity – just click in the eActivity window below the graph strip, and press EXE a few times.

Click on the 'xc' in the graph window. It will appear in the box at the bottom of the graph. Highlight the value of the x point. Make sure to lift up the pen – then press down again on the selection and without lifting, drag into the empty line. Once the cursor appears in the empty line, let the pen up, and the value will jump into place.

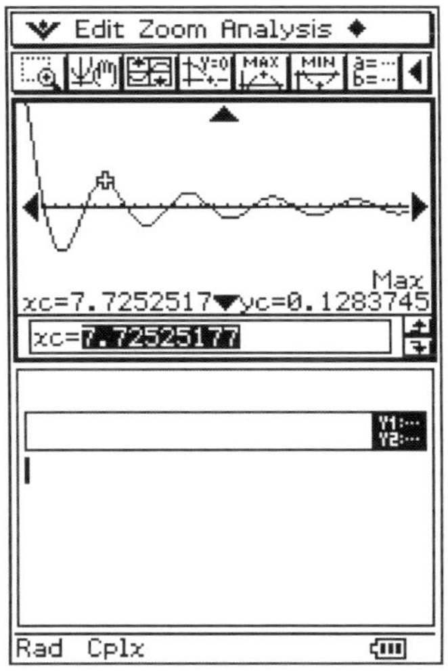

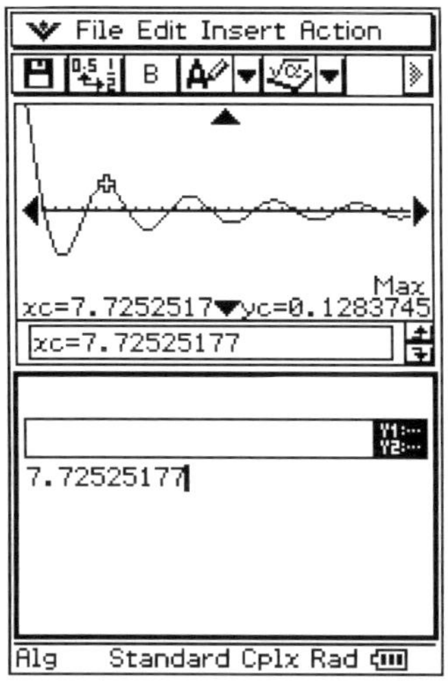

Now on to the next max. Click on the graph to make it active. We're still in max mode, so the left and right pan arrow keys move us from max to max. Tap once on the right arrow, and we'll have the next max point.

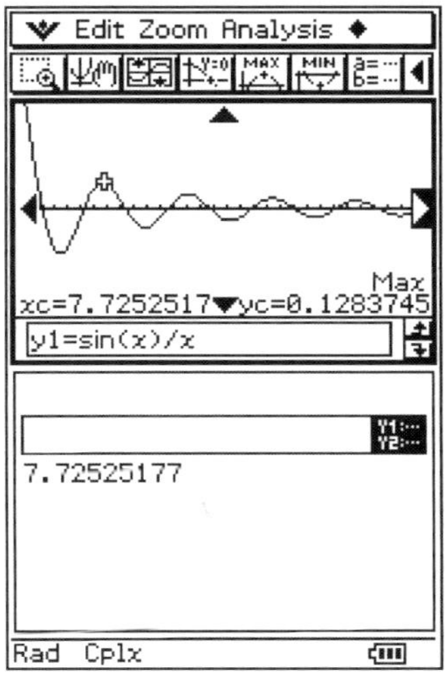

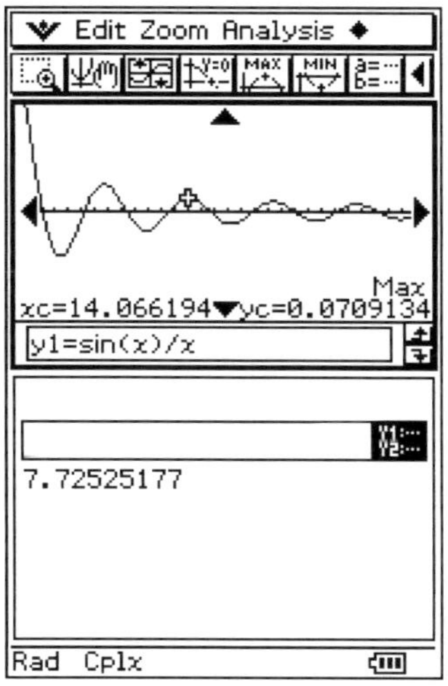

We can grab the x value of the max like we did before. Repeat this once more for three points (or four if you're ambitious).

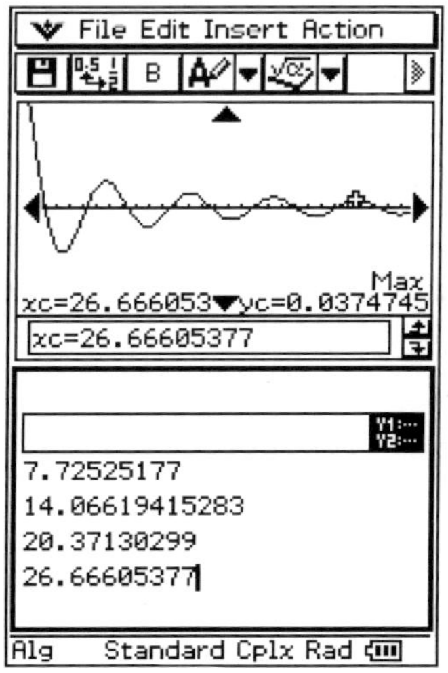

Now that we have the points, we can determine whether the frequency is constant. If the distance between adjacent x points is the same, our frequency isn't changing. Another way to look at this is whether the x points are increasing at a constant rate. If we consider the x points as a data series, we can use the statistics tools to tell us whether these values are increasing linearly.

We can insert a statistics strip below the data points. Just click on the eActivity window at or below the last data point. Then tap the insert menu, and select the list editor.

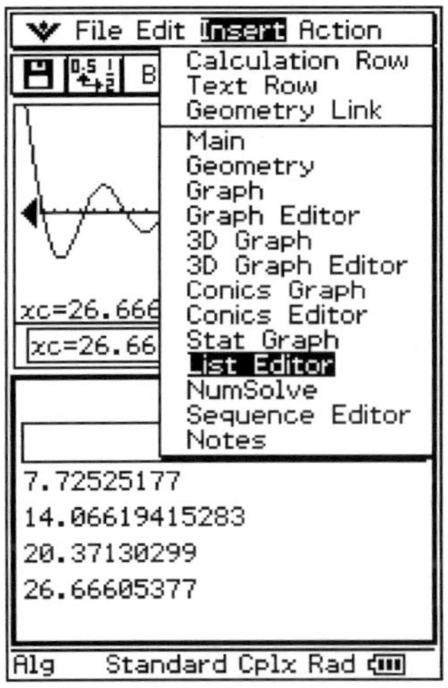

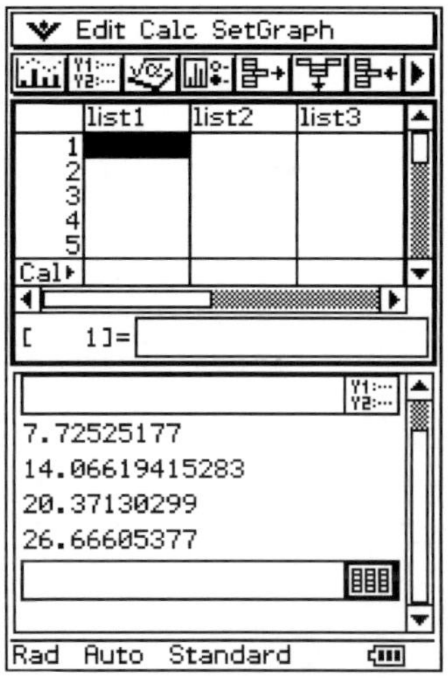

Start out by putting each data point at regular intervals – so fill in the first column with sequential integers – 1 through 3 (or four). The first row of the first column is already highlighted, and EXE commits each cell and moves to the next row. So just hit 1, EXE, 2, EXE etc.

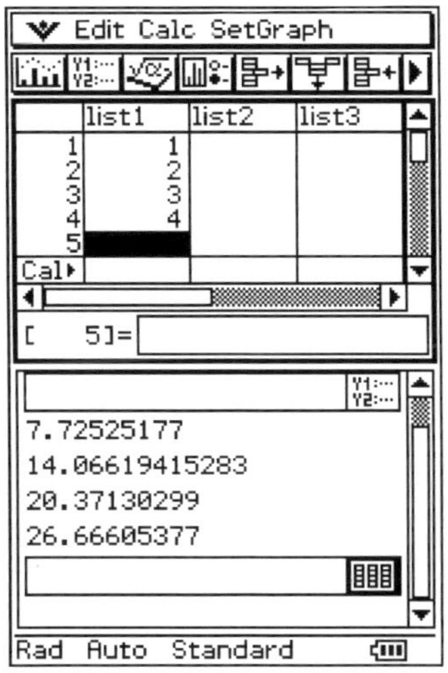

Now we can move the data points into the second column of the list editor. Unfortunately, the list editor does not accept drag and drop. But copy and paste will almost be as convenient. Tap on the eActivity – and drag over the first input to select it. Copy is under the edit menu.

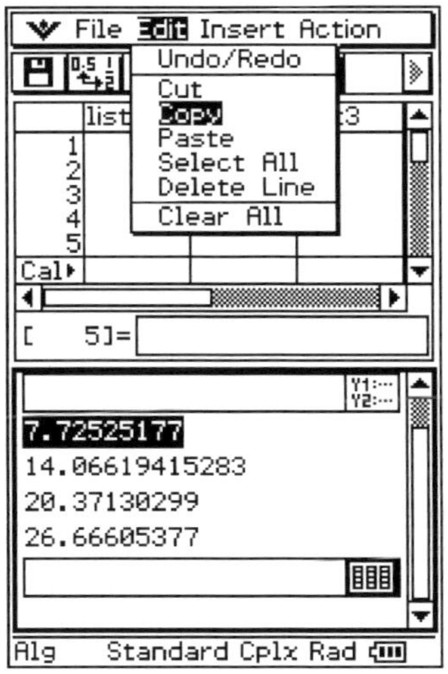

Hit the copy menu item, and then click in the first row of the second column of the list editor.

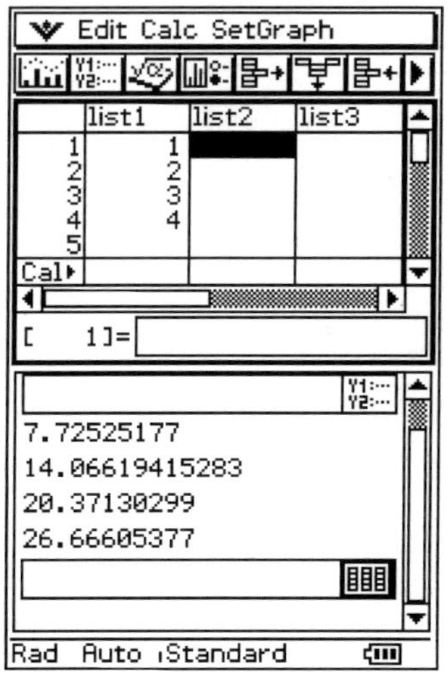

Then hit paste from under the same edit.

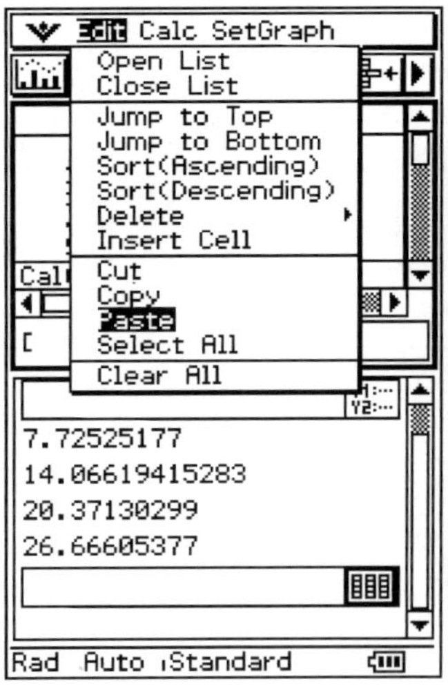

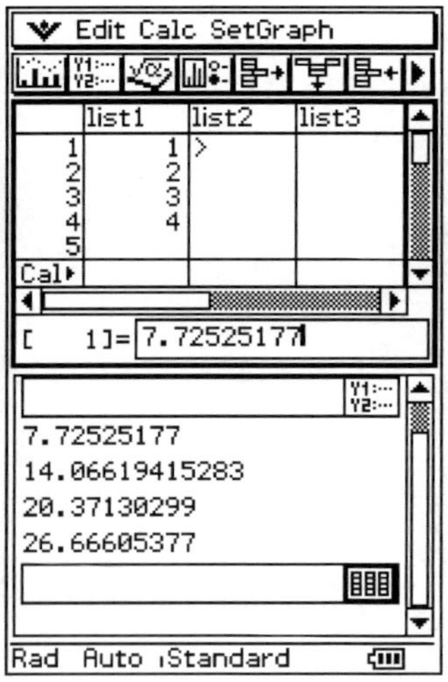

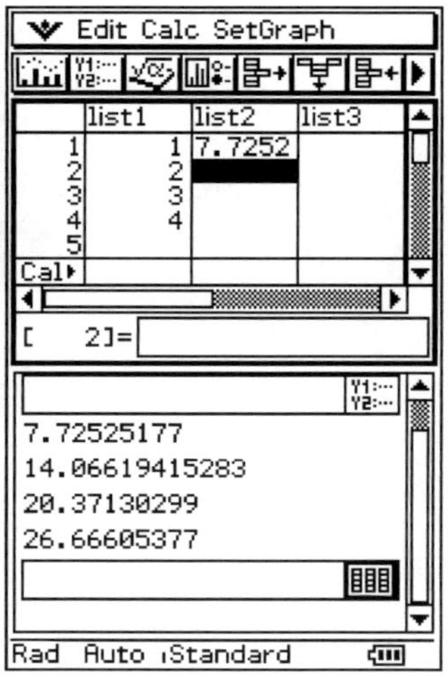

Copy and paste the rest of the data in eActivity into the list editor.

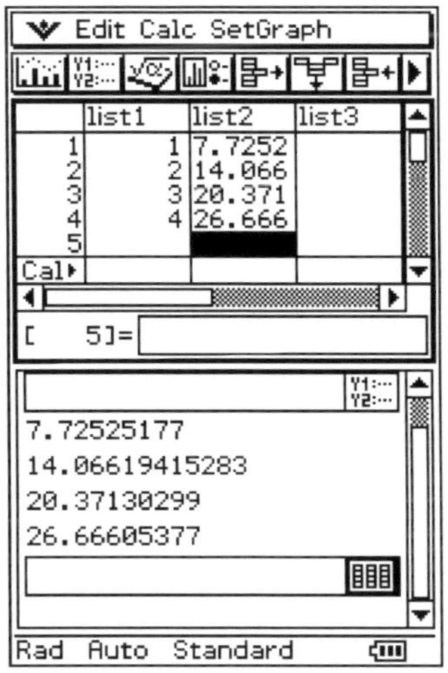

Now hit the graph statistics button – it is the first button on in the button bar.

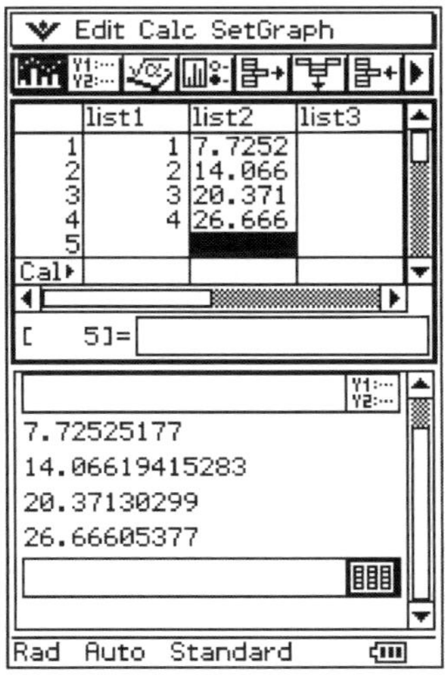

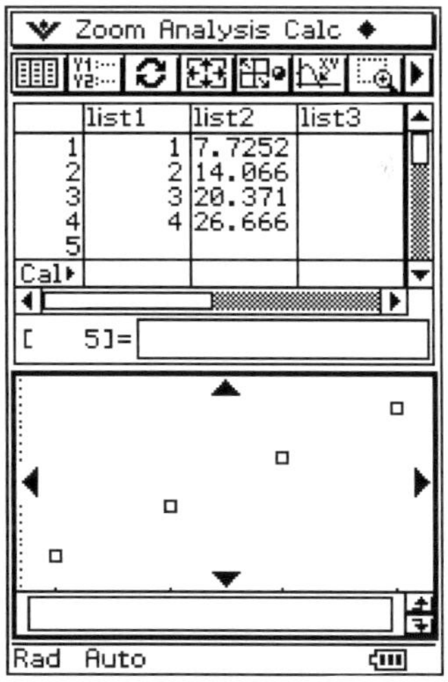

The dots look like they form a perfect line – or the frequency doesn't seem to be changing. But there's an effective way to make sure. We can ask the statistics suite to find the best curve to fit the data points. If it only has a linear component, then we know the data points are perfectly in line.

From the Calc menu, select the Quadratic Regression (Quadratic Reg). Press OK, and it will give us the various components of the best fit quadratic.

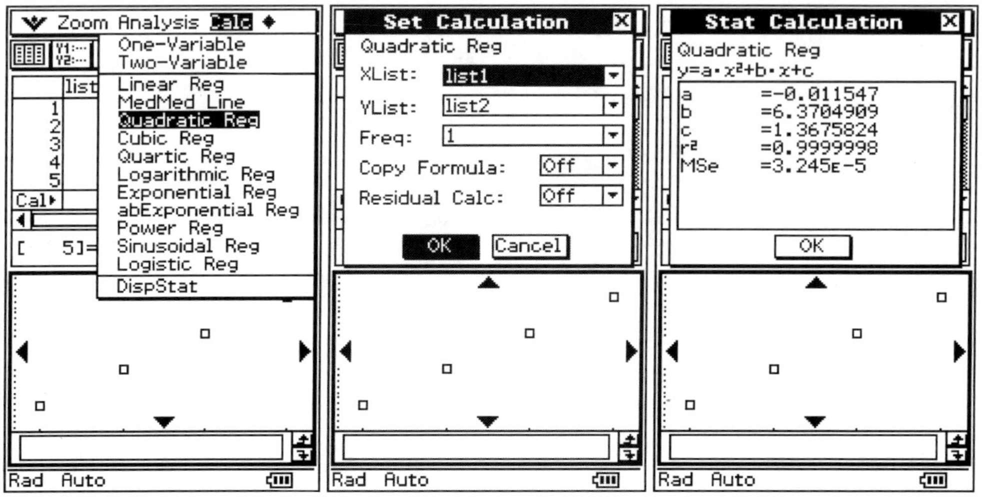

Notice that while the non-liner component (a) is not large, it is not zero. This indicates the frequency is being affected by 1/x.

We can repeat the same process for e^{-x} – we can insert a new graph strip, captured data points, and statistics strip below the ones for 1/x.

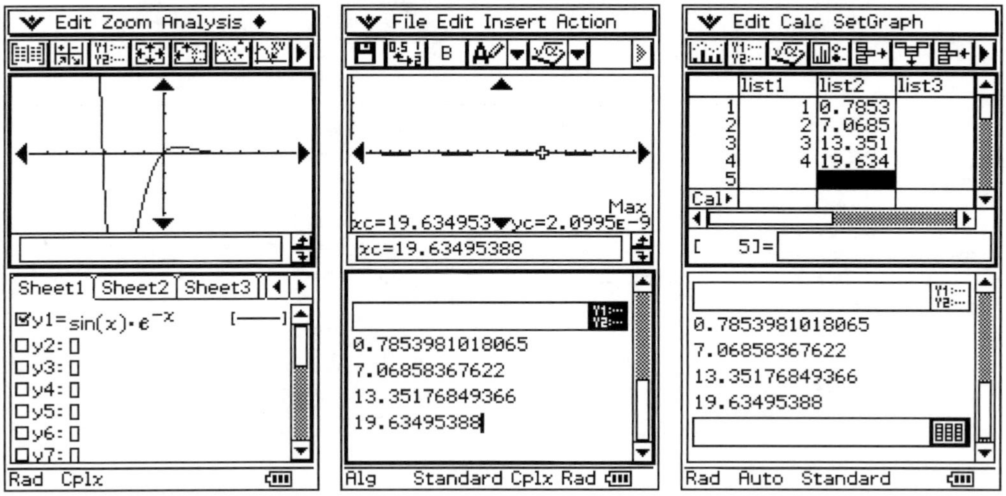

If we pick out Quadratic Regression from the Calc menu, we can see whether the frequency is changing.

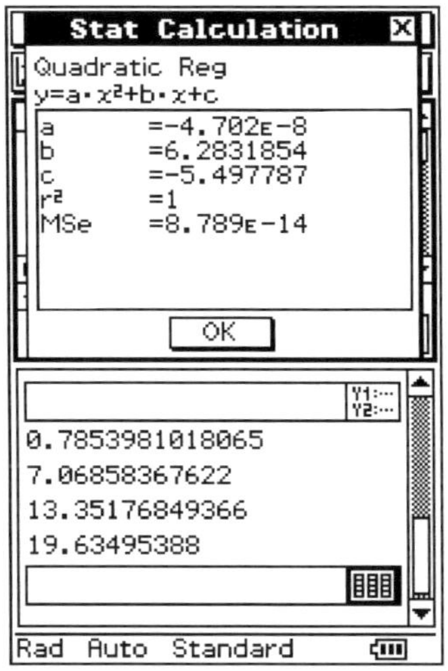

Now this time, the non-linear portion (a) is miniscule. In fact, it is near the precision of the numbers we captured, so it is reasonable to conclude that multiplying by e^{-x} does not cause the frequency to shift over X. In fact, (b) looks like exactly 2π – so the frequency seems completely unaffected.

3D Plot

Of all the applications in the ClassPad, few can match the visual impact of the 3D plot. Equations in x and y are transformed into landscapes - the hills, cliffs and deep ravines reveal the nuances of the expression.

To launch the 3D graph application, go to the launcher by tapping the menu icon at the bottom of the screen, then select 3D plot.

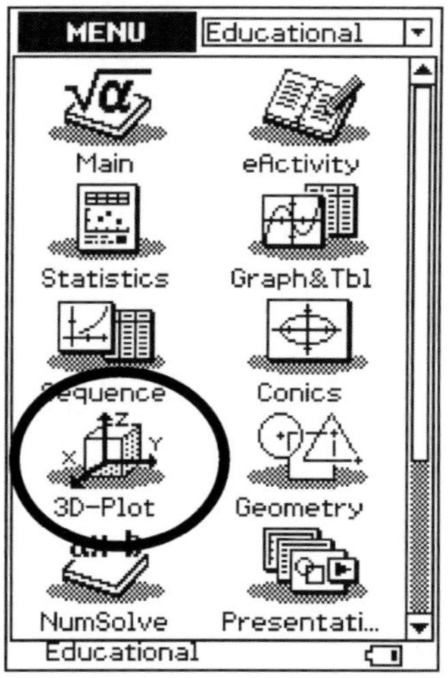

Immediately we see a couple windows – the 3D plot Z editor and the 3D graph itself.

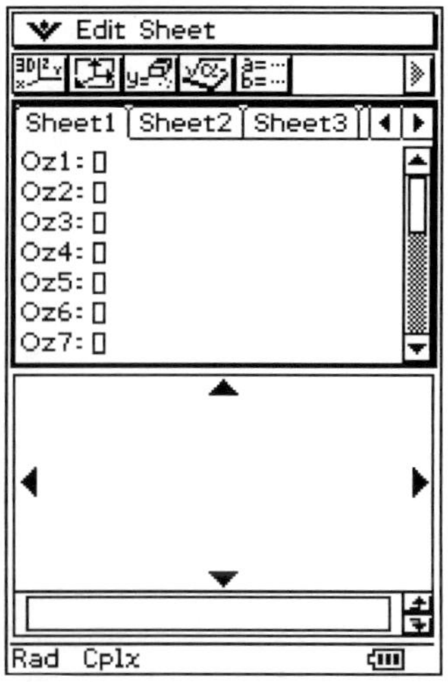

A good exercise that puts the 3D plot through its paces is showing the interference pattern produced by two same-frequency point sources.

The equation for a single point source is straightforward – sin(r) where r is the distance from the point. Let's place our first source at (15,0):

sin(sqrt((x-15)^2+y^2))

Press the keyboard button to bring up the keyboard – this has a convenient square root key. Tap on the box after z1: - then type in the equation. Use that square root key instead of "sqrt": When you're done, hit that EXE key to commit the equation. You don't have to add the final closing parentheses – the calculator will add them for you.

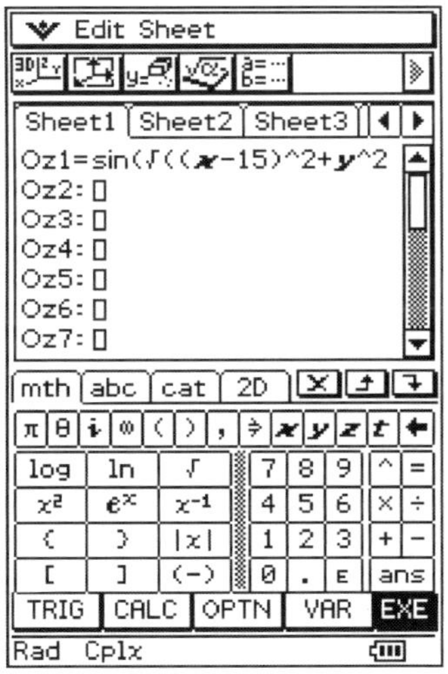

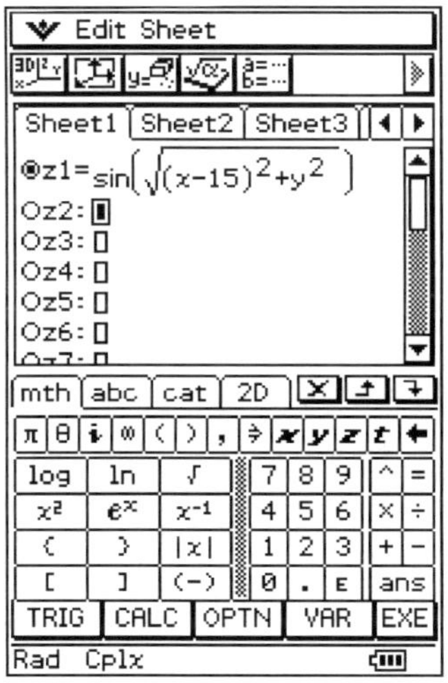

Now hit the graph key – it's in the upper left hand corner. This will automatically close the keyboard to make room for the graph window. After a few seconds of processing, we see the results:

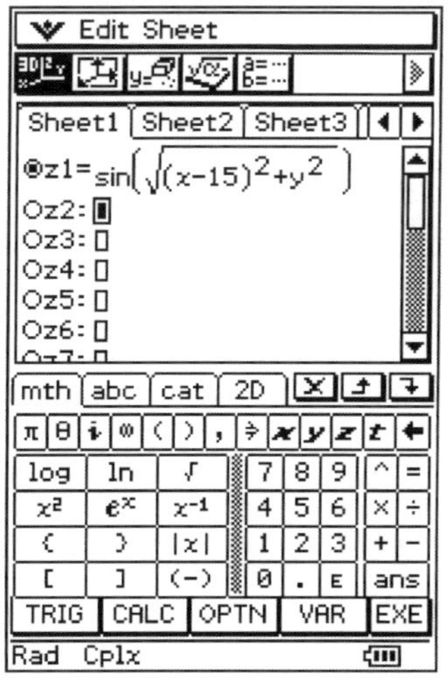

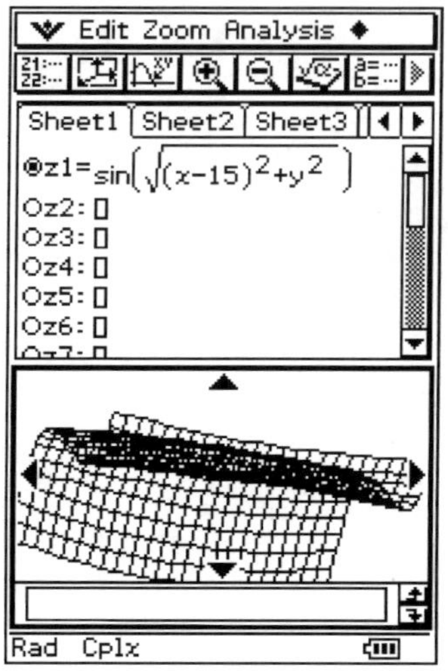

This isn't quite the ripple affect we wanted. Try out the pan arrows in the graph – they are very responsive in the ClassPad. Also try out the zoom +/- buttons at the top. They look like magnifying glasses. Notice that zooming out simply moves the piece we've graphed farther away – it doesn't bring in more x,y space.

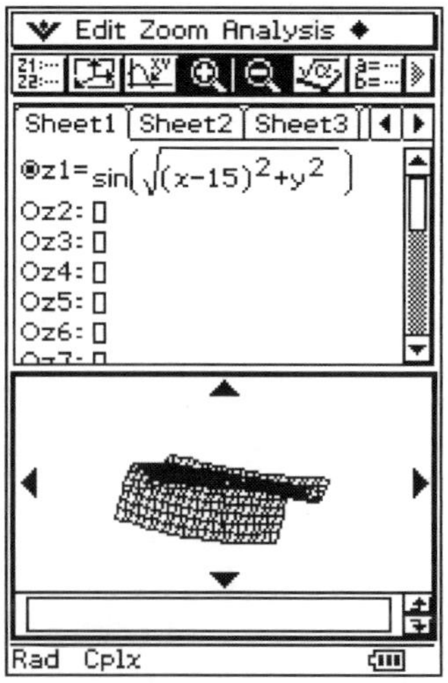

What we need is in the View Window. Press the View Window button – it is second on the left and looks like 3D axes.

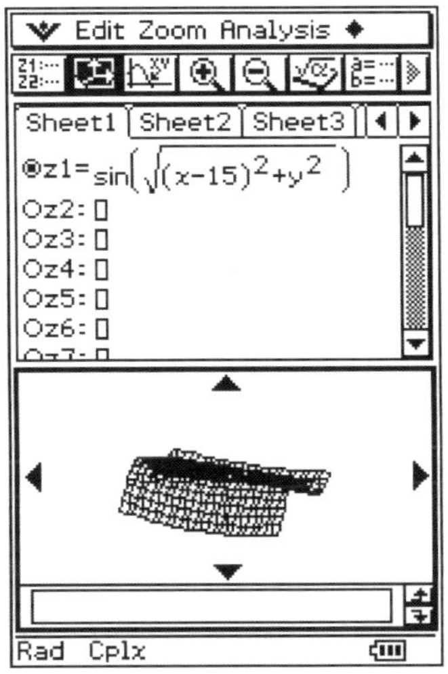

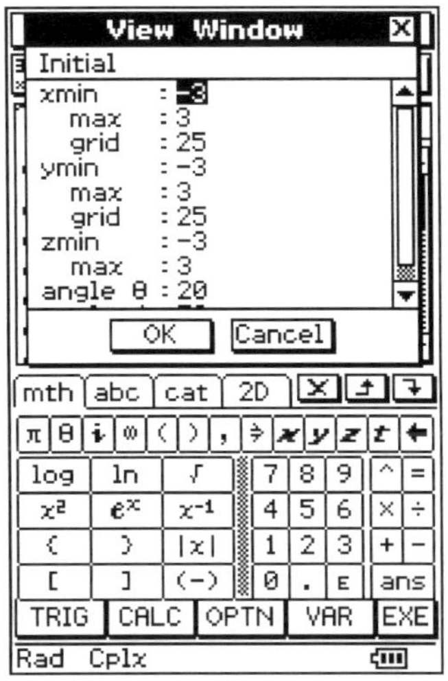

Change the x,y range to (-20,20), the z range to (-10,10).

You can also set the view angle up front instead of using the pan arrows. For this example we found a theta of –150 and a phi of 40 gave a reasonable view. Press OK when you're done.

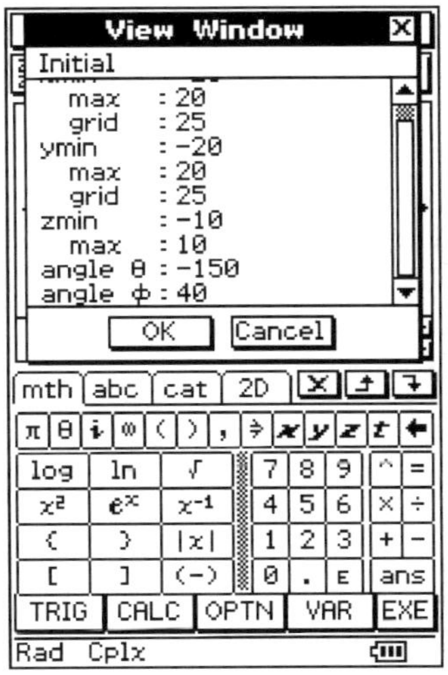

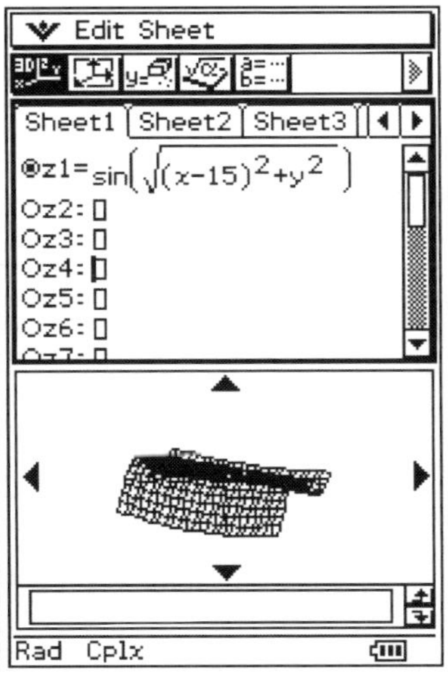

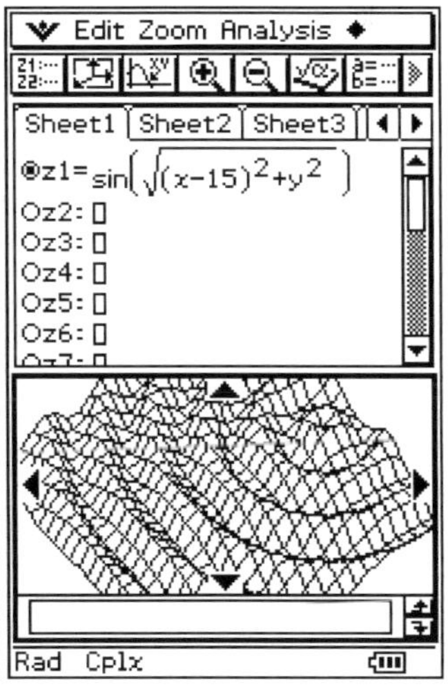

Now the ripples are clearly visible.

Click in the z2 box, and type in the second source, centered on (-15, 0):

sin(sqrt((x+15)^2+y^2

Press EXE when you're done – this will commit the equation and select z2 for plotting. Tapping the graph key will show you the results.

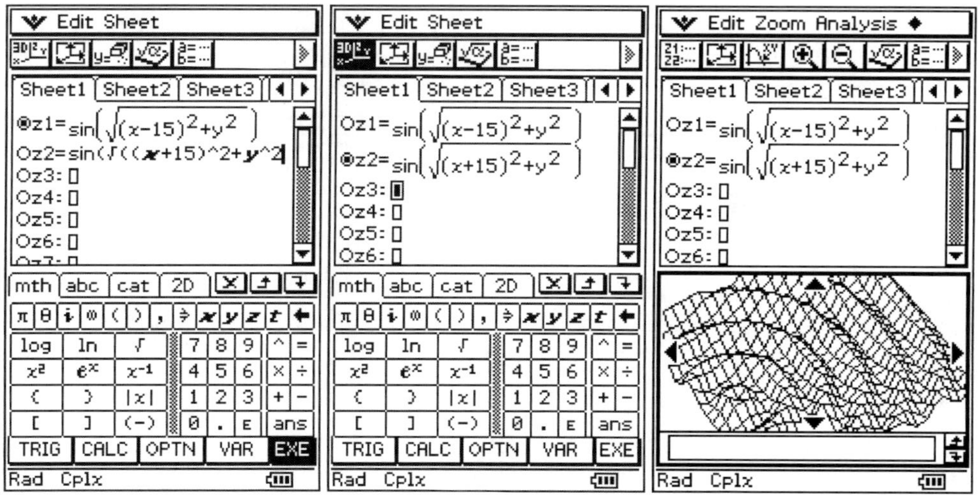

Finally, let's add these two equations in z3. First, drag and drop z1 into z3. Just select the entire z1 equation, releasing the pen when you're done. Then press and hold with the pen on the selection, and drag it into z3. Notice that the cursor appears in z3 as you drag over its box. When you let up on the pen, the selection will pop into place.

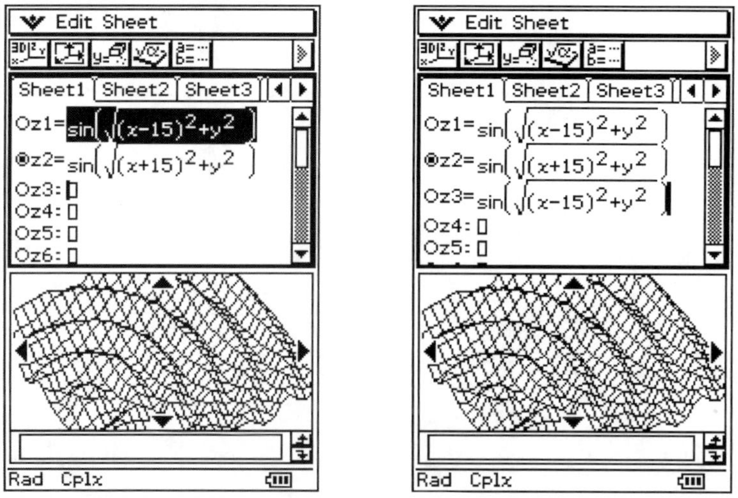

Press the plus key, and we're ready to add z2. Select all of z2, then drag and drop it to the end of z3.

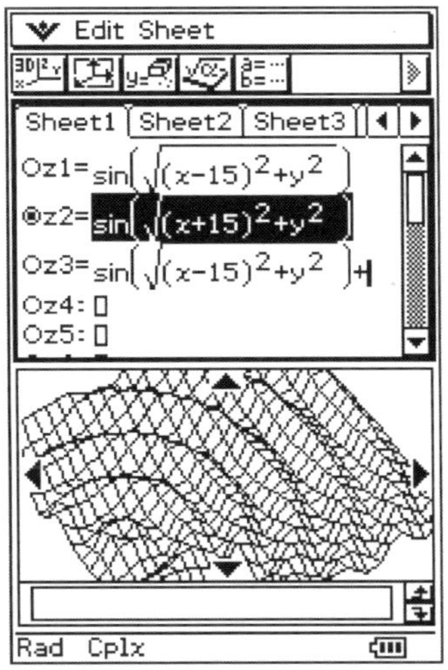

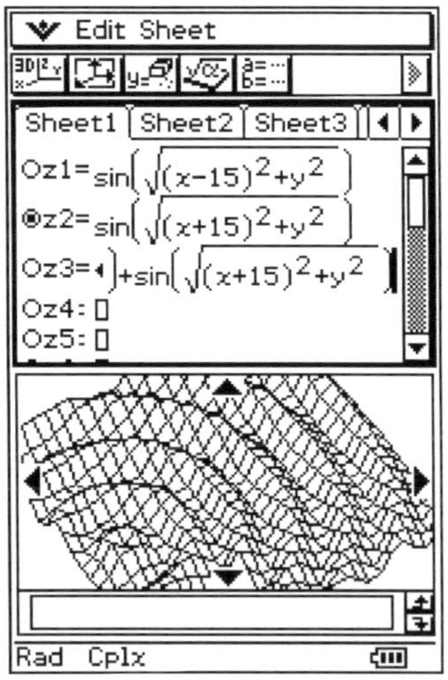

We're ready now to graph the interference pattern.

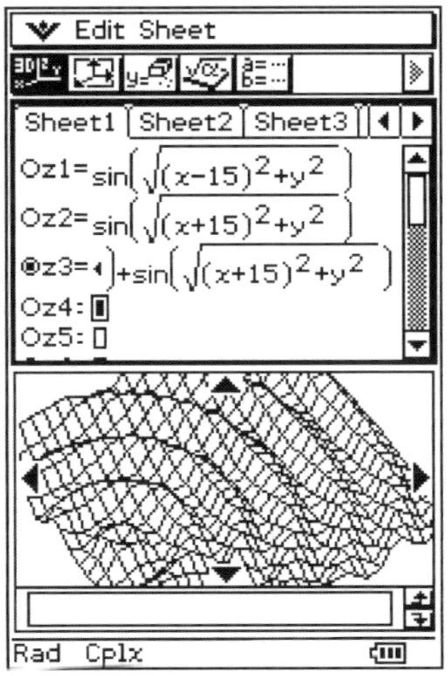

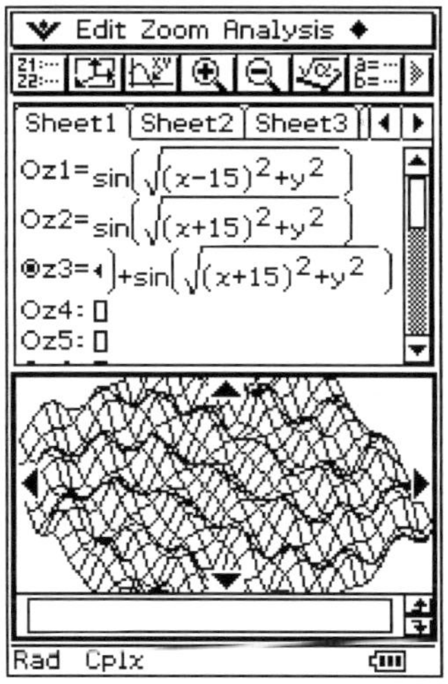

If we wanted to explore the graph in detail, one very useful tool is the point selector. Just click the button that shows x,y in the corner pointing to a graph. A tiny cross will appear over the current point selected. The cross can be hard to pick out at first, try moving it a bit with the arrow pad in the center of the calculator. The x, y and z values of the point pop up at the bottom of the screen.

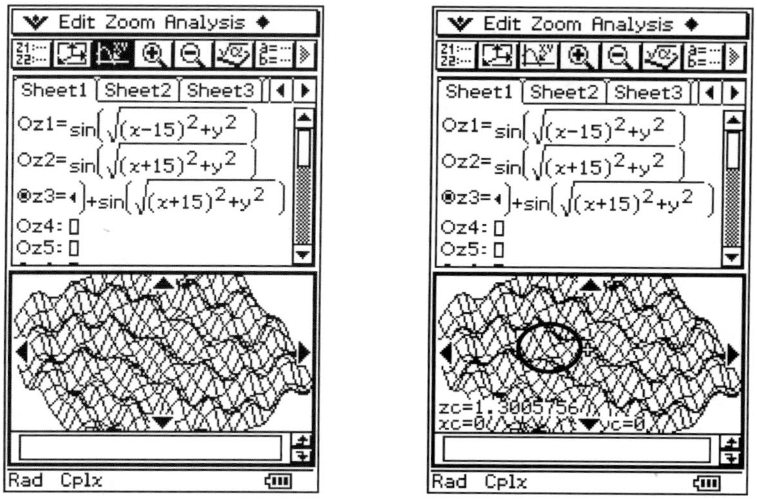

We can take a closer look at the wave interference by changing our view window – for example we can look at the area centered on (0, 15). Just hit that view window key again, and change the ymin, ymax to –5 and 35. Hit OK when you're done. Click somewhere on the graph editor (the top window) and hit the graph key.

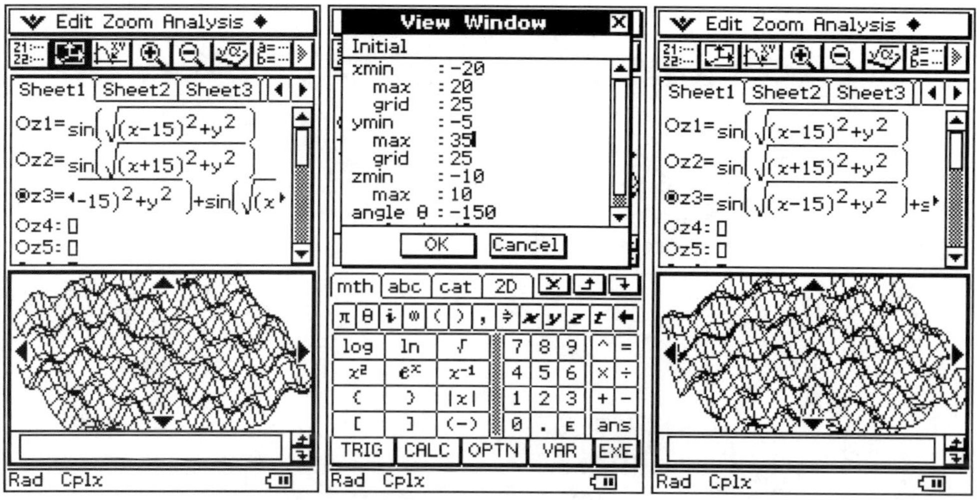

Hit the zoom in three or more times so we can select a point with greater precision. Then press the point select key. Without much trouble we can get the x,y locations for some of the z peaks and valleys.

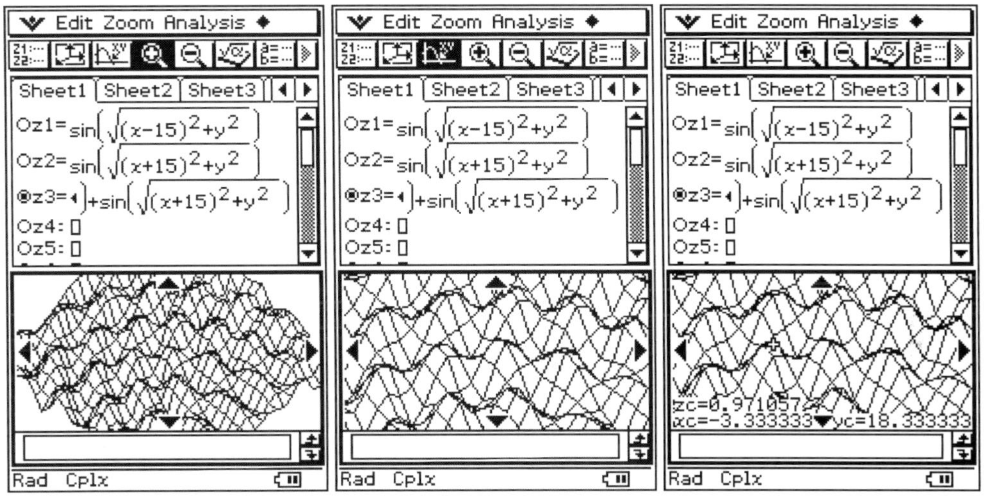

Index

Printed in Dunstable, United Kingdom

91220879R00165